Public Administration and Information Technology

Volume 24

Series editor

Christopher G. Reddick, San Antonio, TX, USA

More information about this series at http://www.springer.com/series/10796

J. Ramon Gil-Garcia • Theresa A. Pardo
Luis F. Luna-Reyes
Editors

Policy Analytics, Modelling, and Informatics

Innovative Tools for Solving Complex Social Problems

Springer

Editors
J. Ramon Gil-Garcia
Center for Technology in Government
& Rockefeller College of Public Affairs
and Policy
University at Albany, State University
of New York
Albany, NY, USA

Theresa A. Pardo
Center for Technology in Government
& Rockefeller College of Public Affairs
and Policy
University at Albany, State University
of New York
Albany, NY, USA

Luis F. Luna-Reyes
Center for Technology in Government
& Rockefeller College of Public Affairs
and Policy
University at Albany, State University
of New York
Albany, NY, USA

ISSN 2512-1812 ISSN 2512-1839 (electronic)
Public Administration and Information Technology
ISBN 978-3-319-61761-9 ISBN 978-3-319-61762-6 (eBook)
DOI 10.1007/978-3-319-61762-6

Library of Congress Control Number: 2017950075

Printed on acid-free paper

This Springer imprint is published by Springer Nature
The registered company is Springer International Publishing AG
The registered company address is: Gewerbestrasse 11, 6330 Cham, Switzerland

To my Amores Nadia, Dante, and Julieta,
with my love, gratitude, and admiration
—J. Ramon Gil-Garcia

To my dear husband Manuel, with love
—Theresa A. Pardo

To Gabriela, my lifetime partner and friend
—Luis F. Luna-Reyes

Foreword

This volume is an important addition to the growing literature on the diverse ways in which the collection, storage, and retrieval of data and their analysis and synthesis, visualization and interpretation, and dissemination are influencing how we think about evidence and its implications for the policy process. There is little doubt that information and communication technologies have transformed how we communicate and interact with each other. These technologies are also altering the ways in which governments and their citizenry interact. For instance, through crowdsourcing, local government is finding novel ways to be responsive to the needs of its citizenry. Through ready access to certain types of data regarding the daily functions of government, citizens find that there is also greater transparency regarding how the government functions. In spite of this progress, it remains unclear whether e-government, the utopian vision of the use of information and communication technologies to enhance public participation and democratic governance, is becoming a reality. What is clear, however, is that we have begun to be able to combine these technologies with tools, techniques, and data to address common, yet complex, policy issues.

With these new tools and techniques, we are currently, perhaps, at the stage where approximately a century ago Ronald Fisher, W. S. Gosset, Jerzy Neyman, Egon and Karl Pearson, and their contemporaries were when they were laying the foundations of the theory and applications of statistics. Over the last century, statistical thinking and methods have transformed how we have approached the systematic study of public policies and their implications. The chapters in this book continue that tradition in the new context of policy analytics, computational modeling, informatics, simulation, visualization, and the use of administrative, blended, and integrated data.

These chapters offer a grand tour of the potential of the novel combination of technology, data, and methods to assist researchers, decision-makers, and practitioners in their quest to address old and new policy concerns. They give us a flavor of the energy and creativity of the growing policy analytics community of researchers collaborating with practitioners to inform and support decision-making and administrative practice. This volume is more than a compilation of novel methods and illustrations of their use. We also learn from these pages that there are a sufficient number of use cases and associated methods that it is possible to develop a set of principles and procedures that will guide the use of policy analytics to address public sector issues. These principles and procedures will also inform how the government and its citizenry interact with each other and how that interaction can occur in a transparent fashion that is equitable and fair while preserving individual privacy and maintaining security.

Big data and associated information systems alter how analysts detect violations and enforce laws and regulations. They also help decision-makers respond systematically and promptly to crises as well as embark upon the longer, slower process of recovery. The combination of geographical information systems with timely data on changing conditions and mobile resources has enhanced the ability of governments to deploy resources to mount effective responses in the face of what, in the past, would have been insurmountable uncertainty attributable not only to ignorance but also to poor communication and an inability to properly assess the received information and react accordingly.

Policy simulations can serve multiple purposes. There are examples here of how they can be used to describe a system, make explicit underlying processes, analyze how that system might function, help generate hypotheses, and explore multiple scenarios under varying assumptions. Perhaps, more useful than the analytical rigor of the simulations is that collaborating on developing them allows modelers and practitioners to mutually agree upon what is important to include in the model and to clarify and make explicit values and underlying assumptions. This early involvement by the potential users of the simulation's output develops a realistic understanding of the simulation's strengths and limitations. Such understanding is essential as tools and techniques become more sophisticated and the potential gap between those who conduct the analyses and those who would use their results also grows. The systematic and collaborative exploration of the complexities regularly encountered in policy contexts builds trust between analysts and practitioners and potentially reduces the risk of misuse and misunderstanding of policy analytics.

This need for collaboration between analysts and practitioners also suggests a change in the traditional role of data scientists, modelers, and analysts who are most comfortable with rational reasoning and the production of rigorous evidence. As indicated by the examples from the diverse policy contexts discussed in this volume, policy discourse is much more complex; the arguments are often based on political feasibility and pragmatism and involve belief, thought, and value systems that do not necessarily adhere to rational norms. Some of the discussions in these chapters illustrate how one might bridge the gap between the producers and consumers of sophisticated analyses. Slightly outside their usual comfort zone, there is a new role for analysts as knowledge and evidence brokers. The technology and the methods provide a pathway for analysts to become brokers who not only help frame policy questions such that they lend themselves to rigorous analysis but also produce knowledge in a form that is usable and used.

This volume illustrates the potentially transformative power of technological innovation for enhancing the ways in which analysts inform and support policy-making processes. At about the same time as the analytical developments of a century ago in the United Kingdom, Woodrow Wilson, in the United States, was writing on public administration and the role of professionals as reformers. A lesson to be learned from that is that while it is important to get things "right," it is also important to recognize the limitations of what we know and to implement analytical innovations with an appropriate level of flexibility to account for these limitations. The authors here seem to be aware of that as they develop novel ways to build the evidentiary basis for effective public policies and their implementation and evaluation.

Anand Desai
Professor, John Glenn College of Public Affairs
Ohio State University
Columbus, OH, USA

Section Head, Evaluation and Assessment Capability
National Science Foundation
Arlington, VA, USA

Acknowledgments

This book is the result of the dedicated effort of many wonderful people who worked together and apart to bring this volume to fruition. We take this opportunity to express our sincere and deep regards and appreciation to all of those who helped and supported us in the conception and completion of this work. Thank you to all the authors for sharing your knowledge through this editorial project and your interest in policy analytics, modeling, and informatics. This volume would not have been possible without the hard work and collegiality of authors; reviewers; the series editor, Christopher Reddick; and the staff at Springer. A special thank you to all the reviewers who not only gave their time and effort but also share their knowledge through very useful and constructive comments that enhanced the book's overall quality and contribution to the field.

From the staff at Springer, we would like to mention the dedication and commitment of Lorraine Klimowich during the entire process. We always received support and useful guidance from her. We want also to thank Ana Catarrivas, whose dedication and diligent effort have been instrumental for the completion of this book. Her role has been central in this editorial effort, and the editors are highly pleased with the final result. We are also grateful to the Center for Technology in Government and the Rockefeller College of Public Affairs and Policy, University at Albany, from which we have received strong institutional support and great encouragement and motivation from colleagues and friends.

Finally, we send love and gratitude to our families. They have tirelessly encouraged us and wholeheartedly supported our academic adventures.

J. Ramon Gil-Garcia
Theresa A. Pardo
Luis F. Luna-Reyes

Contents

Policy Analytics: Definitions, Components, Methods, and Illustrative Examples

J. Ramon Gil-Garcia, Theresa A. Pardo, and Luis F. Luna-Reyes

Abstract There are many ways to define policy analytics. In fact, different terms are used for what could be considered the same phenomenon: the use of data and analytical techniques to make policy decisions. Policy analytics, policy modelling, and policy informatics are just some of the most used terms by scholars and practitioners. There is no clarity, however, in terms of where the boundaries of this concept lie and what main analytical methods it includes. First, regarding the conceptual boundaries, some experts argue that policy analytics includes analysis-related tasks only and has nothing to do with data preparation, management, governance, and stewardship. For other experts, policy analytics encompasses all the activities in the data lifecycle and includes elements outside of the analytic sphere, such as information technologies, stakeholder involvement, and a deep understanding of the context of use and application domains. Second, in terms of analytical methods, some analysts consider data mining, machine learning, and other computer science approaches to be the only valid tools. For others, statistical analysis as well as simulation approaches should also be called analytics. In this chapter, we propose a comprehensive and integrative view in which policy analytics goes beyond data analysis and includes management and preparation of data as well as very diverse techniques such as computer simulation, social network analysis, statistics, geographic information systems, and data mining techniques.

Introduction

Governments of all levels and around the world are facing problems with levels of complexity that require new approaches and the development of innovative solution strategies. New technologies, new access to data, and new analytical tools and techniques are creating opportunities to systematically understand and respond to such complex problems; problems that some call wicked or tangled. Wicked problems

J.R. Gil-Garcia (✉) • T.A. Pardo • L.F. Luna-Reyes
Center for Technology in Government & Rockefeller College of Public Affairs and Policy, University at Albany, State University of New York, Albany, NY, USA
e-mail: jgil-garcia@ctg.albany.edu; tpardo@ctg.albany.edu; lluna-reyes@albany.edu

J.R. Gil-Garcia et al. (eds.), *Policy Analytics, Modelling, and Informatics*, Public Administration and Information Technology 24,
DOI 10.1007/978-3-319-61762-6_1

are hard to define because of dynamics and contradictory requirements (Conklin 2006). Tangled problems, on the other hand, find their main source of complexity in the diversity of stakeholders and in their interdependencies in finding negotiated solutions (Dawes et al. 2009). Many of these analytic techniques and tools have been identified and studied in recent literature. For example, computer simulation, geographic information systems, machine learning techniques, statistics, and social network analysis have been used to define and address social problems and researchers from multiple disciplines have been refining strategies for the use of these tools in policy analysis. The use of technology, data, and analytical tools could have an impact on any stage of the policy process, changing, in some cases, the very nature of the process, the involvement of actors, and the outcomes derived from the implementation of a public policy or government program.

Contemporary societies face complex problems that challenge the sustainability of their social and economic systems. Such problems may require joint efforts from the public and private sectors, as well as from the society at large, in order to find innovative solutions. In addition, there is a recent wave of data access to promote innovation through transparency, participation, and collaboration (Janssen and Helbig 2016). Social networking and new media platforms, advanced simulation websites, and serious gaming tools are used in incremental and "full-scale" ways to involve citizens, obtain their opinions, and engage them in political processes (Koliba et al. 2011). There is an opportunity to combine emergent information technologies, new analytical methods, and open data in order to develop innovative solutions to some of the pressing problems in modern societies. Policymakers have always faced complex problems related to economics, public health, and education, among other policy domains. Governments of all levels and around the world are charged with understanding and responding to the problems whose complexity is related to the embeddedness of those problems in networks of (sometimes circular) causes and effects. This complexity is not new, but the pervasiveness of these problems in modern society creates new imperatives for innovations in policy analytics.

There is a menu of different analytical tools and modelling techniques that can be used for policy analysis, particularly with the intensive use of information technologies in the context of public policy analysis. Puron-Cid et al. (2016) provide a basic list, neither comprehensive nor exhaustive, of the ample technological options for policy analysis: (1) policy-modelling IT tools, (2) Web 2.0 applications, (3) qualitative IT tools, and (4) quantitative IT tools. First, policy-modelling IT tools apply highly sophisticated techniques to simplify abstractions of complex realities, developed for representation, analysis, or simulation. Second, Web 2.0 tools include social media and other collaborative applications. Third, the qualitative IT tools cover various technological instruments that process information content from unstructured and diverse material such as documents, interview transcripts, survey responses, audio, videos, pictures, forms, reports, or media clips. These tools do not generally rely on numbers, whereas the quantitative IT tools involve combinations of the advances in mathematical, statistical, and computational power.

Technology, Data, and Policy Decisions

Applying mathematical models and empirical data to solve complex problems is not new. There are, however, a range of new approaches made possible by advances in technologies, analytical tools and techniques, and in the application of data analytics to complex social problems. These innovative approaches enable a broader understanding of the causes and effects of strategic and policy problems using a breadth of modeling and analytical techniques. For instance, advances in modeling and visualization techniques have made it possible to gain new insights into the importance of engaging policymakers and other stakeholders in using models, data, and other technical tools to analyze problems and policy alternatives. Some examples are business intelligence and decision support systems, data visualization, data management, big data, data mining, and predictive modeling.

Understanding the massive amounts of data emanating from a complex world requires tools that bridge the technical aspects of computer science, conceptual issues related to human cognition, and questions of how users focus their attention and make sense of complex data. Different theories and perspectives try to explain how data are structured into information, how information is made actionable, and how that action in turn creates new social value and new data. Current interest in policy analytics has been promoted, at least partially, by the abundance of data and the development of new technologies that enable the use of both traditional and novel analytical techniques. The increasing variety of modeling techniques, analytical tools, and data resources creates new opportunities for those engaged or interested in public policy and the study of the public sector. The editors' vision includes five key components in this broader conceptualization of policy analytics as a strategic combination of technology, data, and policy decisions:

1. **Analytical Methods and Techniques.** A model, in general, is a conceptual representation of a problem and it helps policymakers and other stakeholders structure the inquiry process. In many cases, data analysis requires the use of multiple analytical and modeling tools and techniques depending on the problem at hand.
2. **Managing Data.** While data availability creates a host of new opportunities, using these data for data analysis presents many real challenges. For example, the data required for modeling most strategic problems need to be integrated from multiple, disparate sources and usually need to be re-coded in a way that is consistent with the new application and context.
3. **Understanding Information Technologies.** The availability of technology-based tools requires new skills and capabilities to use those tools in the creation of applications for decision and strategy making. Tools include both hardware (such as sensors) and software (such as R, SPSS, and Python, among others).
4. **Stakeholder Involvement.** Stakeholder involvement in the construction and analysis of models to explore decisions, strategies, and policies is another pillar of policy analytics. Stakeholder engagement to facilitate data analytics modeling finds its roots in decision science, soft system approaches, and group decision support systems.

5. **Understanding of Context of Use and Application Domains.** Learning from data occurs in context-specific application domains. Each domain has specific requirements and problems that need to be clearly understood in order to choose appropriate methods and processes for data analysis. In this case, policy analytics needs to consider the specificities of the public sector and the policy process.

Benefiting from policy analytics requires individuals, but more likely teams who possess combinations of all five components. Such teams will need a wide range of skills as diverse as building quantitative models to engaging stakeholders in ensuring that models reflect the deep knowledge of context and the nuances of complex problems embedded in unique and dynamics contexts. They must have members who understand the application domain and the context of use of the analytics; they must have members who know the data and how it has been collected and managed, so when data must be brought together, it can happen guided by this knowledge. The cost of bringing together such teams is high, but the cost of engaging in policy analytics without the appropriate and necessary skills may be costly too and even counter-productive.

Some Methods and Techniques for Policy Analytics

A traditional view of policy analysis emphasizes the use of models and empirical research to better understand causes and effects of policy choices, which has been dominated in the last 40 years by an economics orientation. In the last decade, new approaches to policy modelling, like policy informatics, have moved from this original view into a broader understanding of the causes and effects of policy problems and the effectiveness of policy choices and instruments for dealing with them (Puron-Cid et al. 2016). These new lenses promote interdisciplinary approaches to policy analysis and include diverse stakeholders. The possibilities brought forward by open government and open data programs, plus the boom of social networks and media, triggered these developments and the hype surrounding them. Hence, the tools, applications, and opportunities for policy modelling have become of important focus on the public policy agenda. Analytical and IT skills include a variety of evolving competencies, drawn from disciplines such as statistics and computer science, for the management and analysis of both structured and unstructured data (Chen et al. 2012). From the literature and case studies of public projects, we identified the main policy analytics methods and techniques that currently aid and shape public action in different branches and levels of government. Figure 1 lists some of the most relevant methods and techniques; this figure is not a comprehensive overview of all the possible tools and applications, but it includes some of the most used and studied.

A computer simulation is a computer program that reproduces simulated behavior of a real-world problem using a simplified mathematical model of the problem

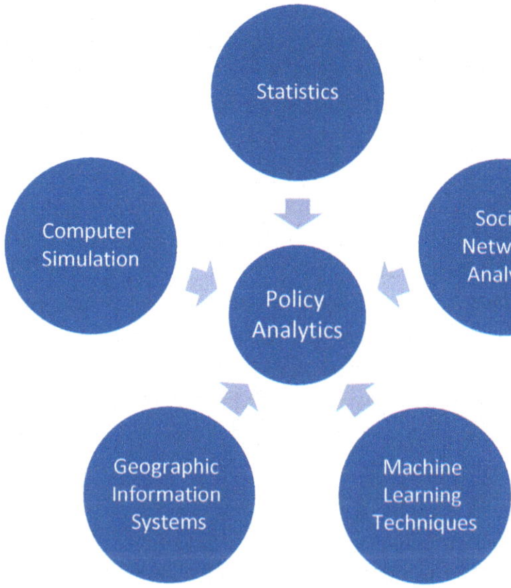

Fig. 1 Key policy analytics methods and techniques

(Richardson and Pugh 1981). The computer simulation allows the policy maker to explore the main impacts of alternative policies in a safe environment, allowing him or her to learn about the problem and policy impacts (Morecroft 2007; Luna-Reyes et al. 2017). Computer simulation allows policy and strategic problems to be represented more completely and comprehensively because it combines hard data with qualitative elements of the system, which are quite often intangible (Forrester 1992). There is a diversity of modeling and simulation techniques, depending on selections of levels of aggregation and time-managing techniques. Some of the most widely adopted simulation techniques are agent-based modeling, discrete event simulation and system dynamics (Sayama 2015; Gilbert and Troitzsch 2002). The use of computer simulation is a valuable tool for researchers and public policy analysts, but it requires specialized knowledge and abilities. One way to gather the necessary skill and knowledge set to understand policy options is through Group Modeling Sessions, where policy makers and simulation experts collaborate in the process of problem modelling and policy simulation (Richardson et al. 2015). In addition, there are clear advantages in the use of computer simulation since it allows for the analysis of different public policy alternatives.

Social networks are formally defined as a set of nodes (or network members) that are tied by one or more types of relations. Social network analysis (SNA) is neither a theory nor a methodology. It takes as its starting point the premise that social life is created primarily and most importantly by relations and the patterns they form. Unlike a theory, social network analysis provides a way of looking at a problem, but it does not predict what we will see, though it does provide guidance on where to look for answers (Marin and Wellman 2011). More specifically, as part of the policy analytics domain, the SNA approach combines a relational measure of the organizations' centrality with a measure based on collective perceptions of the organization's powers in terms of decision-making (Dörry and Decoville 2013). For instance, literature from the digital government field finds local governments tend to use more mechanisms that permit direct citizen involvement, making social networks increasingly relevant in that context (Gibson 2010). This research encourages local governments to use bidirectional communication to better interact with stakeholders and to become more efficient in their response to stakeholders' demands, thus providing greater accountability.

Geographic information system (GIS) technology is an extensively used computer-based tool to deal with spatial issues. GIS technologies integrate a range of geographic information into a single analytical model, in which diverse data are "georeferenced" to cartographic projections (Maliene et al. 2011). GIS technology integrates common database operations, such as query and statistical analysis, with the benefits of unique visualization and geographic analysis offered by maps. These abilities make GIS technology valuable in explaining events, predicting outcomes, and planning strategies (Guan et al. 2011). Recently, the usability of GIS technology in mobile phones, global positioning systems, and online interactive digital maps has opened the door to the wider use of GIS by non-expert users (Maliene et al. 2011). Such widespread acceptance of GIS creates new openness to and opportunities for the use of these technologies in public policy research and practice and across the sectors.

Statistical analysis techniques are adopted for association analysis, data segmentation and clustering, classification and regression analysis, anomaly detection, and predictive modeling in various policy analytics applications (Chen et al. 2012). Grounded in statistical theories and models, multivariate statistical analysis covers analytical techniques such as regression, factor analysis, clustering, and discriminant analysis. These analytical techniques have been used successfully in examinations of and research about public policy, and are being used again under the new label of policy analytics.

Finally, data mining and other computer science approaches are increasingly being used due to the volume and diversity of data now available. As an example, machine learning can be broadly defined as computational methods that use experience to improve performance or to make accurate predictions. Machine learning consists of designing efficient and accurate prediction algorithms. As in other areas of computer science, some critical measures of the quality of these algorithms are their time and space complexity, but in machine learning there is also the notion of sample complexity (Mohri et al. 2012). Statistical machine learning, often based on well-grounded mathematical models and powerful algorithms, includes techniques such as Bayesian networks, hidden Markov models, support vector machine, reinforcement learning, and ensemble models, which have been applied to data, text, and web analytics applications (Chen et al. 2012).

Purpose, Audience and Content of This Book

This book provides one of the first comprehensive approaches to the study of policy analytics, modelling, and informatics. The volume compiles peer-reviewed chapters covering frameworks, studies about the research community, decision-making, democracy and governance, tools, and applications all related to the theme of policy analytics, modelling, and informatics. The chapters encompass theories, methodologies, case studies, and practical recommendations. The target audience of the book is academics and professionals who want to improve their understanding of policy analytics, modelling, and informatics at all levels and branches of government and across a range of political, economic, and cultural contexts.

First, the book helps researchers and students in the digital government field and various disciplines such as public policy, public administration, political science, communication, information science, administrative sciences and management, sociology, computer science, and information technology, among others, who need comprehensive understanding of topics related to policy analytics, policy informatics, or policy modelling. Second, policy analysts, government officials, and public managers will find practical recommendations based on rigorous studies that will contain insights and guidance for the development, management, and evaluation of public policies and government programs, based on innovative uses of information technologies, data, and analytical techniques.

The chapters collectively address the use of data at different stages of the policy process, from agenda setting to design, implementation, and evaluation. The overall results could be translated into actionable lessons and recommendations and are explained in context. More generally, academics and professionals who want to improve their understanding of policy analytics, informatics, and modelling at all levels and branches of government and in very different political, economic, and cultural contexts will find valuable information.

This book is divided into four main sections. The first section includes two chapters dedicated to *Frameworks and Research Community* that study policy modelling, policy analytics, and policy informatics. In the second section, we present four chapters on data-driven *Decision-making.* The third section includes four chapters related to *Democracy and Governance.* The final section includes seven more specific chapters on *Tools and Applications* referring to policy analytics, modelling, and informatics with case studies.

In chapter, entitled **Background on Frameworks for Policy Analytics,** Hamza and Mellouli present different frameworks developed in the literature related to policy-making analysis and policy modelling. The authors explain how these frameworks are generally spread across multiple disciplines such as public policy, political science, computer science, and the social sciences. Policy frameworks offer guidance for theoretical analysis by identifying the elements that impact policy and their relationships. These elements can include, for example, governance structure; policy process; stakeholders; and institutional structures. Hamza and Mellouli argue, however, that there is no standard categorization or classification for these frameworks. And with the growth in policy analytics and policy modelling, there is a rising need to review existing policy frameworks and develop categorization criteria able to classify frameworks concerned with policy-making analysis and policy modelling. This chapter goes through the main frameworks used in understanding the policy making process to generate a general overview of the available frameworks for policy analytics.

In third chapter, Brooks, Janssen, and Papazafeiropoulou explain how policy-making in the digital age is an area which needs knowledge found in communities that traditionally do not connect with each other. The creation of a research community is a challenging endeavor and needs to address both physical and online elements. In communities, groups of people share common interests and are often facilitated by interactions online and face-to-face. The activities should result in a sense of belonging among the community members. The chapter, **eGovPoliNet: Experiences from Building a Policy Informatics Research Community**, outlines the community-building activities of creating a policy informatics community as part of the FP7 eGovPoliNet project. The eGovPoliNet project community organized community building events and provided a platform for sharing experiences and knowledge, which addresses the fragmentation of research communities, as well as the fragmentation among different disciplines, by building a common network where researchers from different disciplines and countries can interact. The aim was to engage different stakeholder groups to work together in exchanging ideas and information.

Opening the section on decision making, Zinner Henriken argue that a central goal of e-government policy is to increase efficiency in public administration, and one way to increase efficiency is via the increased use of automation and rule-based decision making. Chapter "**One Step Forward and Two Steps Back: e-Government Policies in Practice**", explains how the use of data to streamline processes has a prominent role in public policy today. This chapter specifically discusses the role of discretion in data-driven public administration. The empirical setting for this discussion is the Danish public sector, which has been among the first innovators to implement e-government solutions. Denmark has a long tradition of issuing public policies that define goals for front-office e-services and back-office digitization. Rule-based decision-making systems represent an ideal driver for actualizing the visions outlined in these policies. This chapter presents a case study where a rule-based decision-making system was introduced in an agency which handles complex cases requiring in-depth discretion by specialized professional caseworkers. This experience provides a platform for discussing possible challenges when implementing policy goals in an organizational context. This chapter also addresses the concept of "digital nomos", the administrative norms of a digitized public administration.

The next chapter, **Development and Use of Data-centric Information Systems to Support Policymakers: Application to Criminal Justice Systems** from van den Braak and Choenni, maintains that reliable management of information is invaluable for policymakers and advisers to make informed policy decisions. Depending on the information needs of the policymakers and the characteristics of the data required, different approaches may be used to exploit different ways to process data from multiple sources. In this chapter, the authors describe three information systems currently in use in the Dutch criminal justice system. The first system is based on a dataspace approach and uses aggregate data to provide a view of the current state of the Danish criminal justice system. This is particularly useful for evaluating current policy and monitoring the implementation of new policy. The second system uses a data warehouse to integrate individual level data and to look back at older cases. Therefore, it is suitable for evaluating policy. Finally, the third system exploits time series data to forecast the resources needed in the near future, which allows for planning new policy and monitoring its implementation. Then, the authors provide a set of guidelines for developing management information systems in the public sector.

Sixth chapter discusses how an efficacious policy and planning process must be focused on enhancing the ability of decision makers to make sense of an uncertain and complex environment. Porter argues that one tool that could prove useful in this process is system dynamics modeling. **The Value of System Dynamics Modeling in Policy Analytics and Planning** analyses how the use of small system dynamics models (with each module containing ten stocks or less) as a decision support tool has recently been explored in three areas of regional planning: modeling a regional economic and education strategy for Central Coast California; the modeling of U.S.-China relations; and the modeling of violent extremist activity. In each case, an integrated system dynamics model was created or planned that included multiple modules that comprise a strategic system. The models allowed decision-makers to use a "flight control simulator" or "dashboard" to better understand potential, non-linear, behavioral outcomes over time. When used in concert with other methods and tools of evaluation, system dynamics may provide enhanced understanding and key insights into problems previously thought too complex for this level of analysis and may force decision makers to examine a longer time horizon in overcoming policy resistance and establishing system stability.

The next chapter by Misuraca explains how the recent financial crisis has put the economy and society of several European Union Member States under enormous pressure at a time when the demand for social services is growing due to the aging of societies. State budgets were decreased during the crisis and there were considerable cuts to social services. In **Deconstructing Social Policy Innovation Through the Use of Complex Systems Theory: A Methodology for Modelling and Simulation of the Impact of ICT-enabled Innovation in Support of the Redesign of European Social Policies and Protection Systems**, Misuraca argues that this situation calls for the adoption of innovative, long-term social policy strategies and for modernized welfare systems, which foster more efficient, responsive, and appropriate social services. Though many initiatives have been launched and funds allocated, there is limited evidence about the results obtained. This chapter presents the ongoing development of a methodological framework to analyze the economic and social return on investment of initiatives in social policies and services. This framework—called i-FRAME—proposes a novel approach that combines system dynamics and agent-based modelling simulation in a hybrid model to assess ICT-enabled social innovation.

The next section includes four chapters studying themes related to democracy, participation, and governance. First, Loukis analyzes how the public sector, motivated by multiple success stories of crowdsourcing in the private sector and the increasing complexity of social problems and needs, has started moving in this direction as well, which gives rise to the gradual development of "citizen-sourcing". His chapter, **Citizen-sourcing for Public Policy Making: Theoretical Foundations, Methods and Evaluation,** underlines the importance of developing appropriate policy informatics for this purpose. ICT-based citizen-sourcing must rely on theoretically sound, efficient methods that enable efficient retrieval of policy-relevant information, knowledge, and ideas from citizens. These ideas require advanced processing to calculate useful policy analytics, which can then provide substantial support for public policy making. This chapter provides an overview of the research that has been conducted in this area before the author briefly presents four ICT-based methods developed for "active" as well "passive" citizen-sourcing. Finally, leveraging the experience gained from the development and initial pilot applications of these methods, the author proposes some theoretical foundations from prior political and management sciences research to use for the future development of effective ICT-based citizen-sourcing methods to support public policy making, as well as for their evaluation.

The second chapter of this section from Hagen, Harrison, and Dumas contributes to the development of policy informatics by discussing the benefits of analyzing electronic petitions (e-petitions) as a form of citizen-government discourse with deep historic roots that has recently transitioned into a technologically-enabled and novel form of political communication. Their chapter, **Data Analytics for Policy Informatics: The Case of E-petitioning**, begins by presenting a rationale for the analysis of e-petitions as a type of e-participation that can contribute to the development of public policy, provided that it is possible to analyze the large volumes of data produced in petitioning processes. From there, the authors consider two data analytic strategies that offer promising approaches to the analysis of e-petitions and that lend themselves to the future creation of policy informatics tools. The authors discuss the application of topic modeling for the analysis of e-petition textual data to identify emergent topics of substantial concern to the public. They further propose the application of social network analysis to data related to the dynamics of petitioning processes, such as the social connections between petition initiators and signers or tweets that solicit petition signatures in campaigns; both may be useful in revealing patterns of collective action. The paper concludes by reflecting on issues that should be brought to bear on the construction of policy informatics tools that make use of e-petitioning data.

Tenth chapter from Lindquist is entitled **Visualization Practice and Government: Strategic Investments for More Democratic Governance**. This chapter describes how practitioners and advocates see great potential to lever visual tools of all kinds to illuminate complex challenges, share information, create collective insight, and more efficiently show the results of meetings, data-dives, and social media scans, among others. Advocates of using visualization in public sector contexts typically focus on its potential for describing and analyzing complex challenges, tracking social media and other streams of big data, and using infographics and other techniques to communicate information, insights, and messages to elected leaders and various public audiences. However, apart from digitally-driven services, there has been little attention to how visualization can better show how government works and its fabric, shifting contours, and complexity. This chapter reviews different visual practice domains and their underlying craft logics and motivations. It makes a distinction between the application of visual tools in support of the "instrumental" functions of government (associated with competing for, securing, and wielding power) as opposed to "democratic functions" (advising elected leaders, engaging citizens, and furthering accountability). It argues that the greatest inroads for visual tools has been in the more "instrumental areas"—supporting political parties and sitting governments for monitoring and service delivery—as opposed to furthering the larger aspirations of "democratic governance". It concludes that for governments to invest more in visual tools for democratic governance will require political leadership and improved capacity inside government, and that sustained, systematic research is required to monitor the take-up and influence of digital tools in these areas.

Premat closes this section with his chapter, **Can the French Republic be Digital? Lessons from the Last Participatory Experience on the Law-making Process**. The chapter focuses on France's new law on the Digital Republic that was adopted at the end of 2016. It concentrated on the definition of digital rights, net neutrality, and the idea of a public service of data. The law was coproduced by citizens because they could amend the bill proposal online; it was the first time in France that citizens were able to digitally modify a bill. The chapter aims at analyzing the participatory method as a law-making option, asking whether or not this participatory method challenges the traditional way of governing in France. Is it the implementation of citizen control of the law-making process or a single communicative tool? The first part of the chapter describes the relations between the European and the national contexts of this law. The second part of the chapter explains the participatory tool, introduces the actors involved, and analyzes the impact of the method on the law-making process. The third part of the chapter studies the emergence of open government data in France, exploring the competitive context among nations that are going digital. The last part of the chapter confronts how this new open data strategy conflicts with the surveillance laws voted on in 2015 that control Internet tracking for law enforcement.

The final section of the book is dedicated to specific tools and applications for policy analytics, modelling, or informatics. Bolivar starts with a chapter entitled **Relational Benefits and Capacities of Governmental Actors and Stakeholder Participation for Influencing on the Design and Management of Public Sector Services with the Use of Web 2.0 Technologies**. In this chapter, he seeks to identify the perceptions of policy makers in Spanish municipalities about the influence of Web 2.0 technologies on (a) the relational benefits among members of social networks; (b) the capacities of actors to lead these networks and their influence on the design, construction, and management of public sector services; and (c) the capacities of social networks for improving government legitimacy and accountability.

His findings indicate that policy makers do not take advantage of the capabilities of Web 2.0 technologies to enhance the participation of citizens in public sector management and they only use social media tools as new channels for unidirectional communication and representation of the local government, which limits the capacity of these technologies for citizen engagement in public policies.

The next chapter, entitled **Policy Analytics Tool to Identify Gaps in Environmental Governance** by Ekstrom, Lau, and Law, talks about the persistent and emerging social and environmental issues that require new approaches and tools to develop policies that address their inherent complexities. The chapter describes a policy informatics tool, developed iteratively with the feedback of ocean and coastal domain experts. The tool relies on two user inputs: a conceptually modelled ecosystem and a compilation of policy documents that cover a given jurisdiction of interest. With these inputs the user can explore what ecosystem components and linkages are potentially acknowledged in a policy and which are not. The linkage acknowledgement is based on a co-occurrence of key terms. When viewing highly complex ecosystems (or other systems), the tool offers an efficient way to systematically identify potential gaps in policy and where relevant policies do exist that could be leveraged to cover emerging environmental issues. While the tool is simple in its design, the development with the potential user-community (policy makers, government staff, and ecosystem scientists) demonstrates its usefulness even with such a modest form, reflecting the desire for creative, but transparent and accessible, policy analytic tools.

The chapter **On the Spot and on the Map: Real-Time Interactive Model-based Decision Support under Deep Uncertainty** by Pruyt, Islam and Arzt discusses and demonstrates the use of "on the spot" and "on the map" scenario exploration and policy-support in workshop settings. First the authors justify the need for exploratory model-based policy workshops. They then present some methods and techniques needed for these workshops. Special attention is paid to new techniques that the authors believe are crucial for this kind of interactive workshop if time is of the essence, namely (a) techniques to quickly generate small but diverse ensembles of alternative scenarios and (b) techniques to visualize whole-system dynamics on maps by means of geospatial animations. The chapter then describes a workshop related to the 2015–2016 European refugee crises for which this approach and these techniques were developed and used. Finally, it discusses shortcomings of the approach and improvements to deal with them.

Park and Johnston argue policy informatics not only provides new approaches to analyzing policy challenges, but also provides guidance for understanding new forms of organizing in the digital era. Their chapter, **An Event-Driven Lens for Bridging Formal Organizations and Informal Online Participation: How Policy Informatics Enables Just-in-Time Responses to Crises**, investigates how technology accelerates the creation of just-in-time response efforts while also lowering the barriers for joining such efforts to an increasingly diverse set of formal and informal actors who can make meaningful contributions in the context of emergency management. The authors suggest a novel and extended lens, called an "event-driven" lens, for integrating formal and informal responses by reviewing the literature on emergency management, crowdsourcing, open innovation, policy informatics, and digital humanitarianism. They call it an event-driven lens because crises serve as a focusing event that suddenly bring about not only the activation of formal organizations and their latent networks across the levels of government and the sectors, but also the emergence of many informal actors across the globe and from the affected communities to collectively respond to disasters or crises.

The next chapter from Chung and Zeng describes how policy makers face complex challenges such as cybersecurity, infectious disease control, and political rumors, which can quickly elicit large-scale crises in a world connected by social media. Scarce work, however, is available about using social media analytics to support high-impact policy areas that span across nations and population groups. The chapter, **Social Media-based Policy Informatics: Cyber-surveillance for Homeland Security and Public Health Informatics**, presents a framework for social media-based policy informatics and their application to addressing policy issues of concern to governments and the general public. Based on the framework, the authors conducted two case studies. The first study concerns U.S. immigration and border security, which increasingly involves issues of cybersecurity and affects the U.S. economy, national security, and foreign policies. The second study focuses on the West African Ebola disease outbreak, which has caused over 11,305 deaths and is the largest and most complex outbreak in the history of the disease. The chapter concludes with empirical findings obtained from the data analyses and discusses the implications for public policy decision-making.

Ghaffarzadegan et al. present the chapter **Model-Based Policy Analysis to Mitigate Post-Traumatic Stress Disorder**. In this chapter, the authors describe the use of three models for understanding the complexities of post-traumatic stress disorder (PTSD), a major mental illness. The models are (I) a qualitative model describing the social and psychological complexities of PTSD treatment; (II) a system dynamics model of a population of PTSD patients in the military and the Department of Veterans Affairs (VA); and (III) a Monte Carlo simulation model of PTSD prevalence and clinical demand over time among the OEF/OIF population. These models have two characteristics in common. First, they take a systems approach. In all models, the authors set a large boundary and look at the whole system, incorporating both military personnel and veterans. Second, the models are informed by a wide range of qualitative and quantitative data. Model I is rooted in qualitative data, and models II and III are calibrated from several data sources. These models are used to analyze the effect of different policy alternatives, such as greater screening, increased resiliency, and better recruitment procedures, on PTSD prevalence. They also provide an analysis of healthcare costs in the military and the VA for each policy. Overall, the models they developed offer examples of modeling techniques that use a wide range of data sources and inform policy makers in developing programs for mitigating PTSD, a major premise of policy informatics.

The last chapter of the volume is titled **Competition and Collusion Between Government and Businesses in Pollution Management: Analysis Based on Stochastic Differential Game Theory** by Zhang and Jiang. This chapter studies the pollution reduction decision process of local governments under the influence of social media and payments for ecosystem services PES mechanism using stochastic differential game theory. The authors built two differential game scenarios simulating the decision-making process between an upstream area local government and a business. One is a Stackleberg game, which assumes that the business's emission abatement is closely supervised by local government, while the other is a cooperative game that is applied when the business and local government collude to evade emission abatement. Comparing simulation results reveals two main findings. One is that transparency makes local government perform better in pollution abatement. The other is that local government tends to collude with business if there is a lack of transparency and supervision. Based on these findings, policy suggestions are presented to overcome this problem.

Concluding Remarks

As a whole, the book provides evidence about the advantages of using technologies, data, and modelling techniques for understanding complex social phenomena. We argue that this volume will enhance the current understanding of the potential of policy analytics, modelling, and informatics for both academics and practitioners. The book chapters systematically analyze and present how a combination of information technologies, data, and analytical techniques—including computer modelling, statistics, geographic information systems, and social network analysis—could affect the design, implementation and evaluation of public policies and government programs. They support the view that there is no single way to study or approach a social problem; their intrinsic complexity requires diverse tools and perspectives (Gil-Garcia 2014; Luna-Reyes et al. 2017). The chapters collectively make clear that a wide variety of analytical methods, tools, and models exists, each drawing on different perspectives, created for different purposes and drawing on different tools and techniques.

The book, collectively communicates a view that data analytics goes beyond data analysis and includes management and preparation of data as well as a wide range of techniques to engage stakeholders as well as tools to present analytical results as part of a decision making process. Further, the four sections themselves provide an organizing framework for understanding policy analytics. The first section provides an overview of the main policy modelling frameworks and a case study of an effort to build a new multidisciplinary policy informatics community. This section has particular value to readers seeking to understand the basic tools and exploring how best to bring together key actors to create a coherent community of practice. Section two's chapters collectively address the focal activity of policy analytics—decision making. Two chapters introduce two particular kinds of organizational decision making—rule-based decision making systems and evaluation. The authors of these chapters discuss the critical role underlying administrative information systems play in providing the data necessary to drive a range of decision activities. Two additional chapters focus on the use of system dynamics as a tool for complex, dynamic problems such as regional economic policy, violent extremist activity and social policy. Both articles argue that combining system dynamics with other tools can "enhance the ability of decision makers to make sense of an uncertain and complex environment" (Porter 2017).

Section four, focused on democracy and governance, provides four chapters that explore citizen-sourcing, e-petitioning, visualization, and participatory practices. Each chapter explores the use of data and policy analytics as a tool in each of these critical areas of democracy and governance, in the first chapter of the section, the attention is on ensuring that information is available to support citizens, in the second the focus is on using policy analytics to study petitions and, the third, on the value of visualizations in helping stakeholders understand the complexities of a modern government. The final chapter of this section provides a case study of the examination of value of participatory methods for challenging the status quo.

The final section of the books provides the reader with a close look at the use of a number of policy analytics tools and techniques. The chapters collectively provide readers with some food for thought regarding how technology adoption may limit some analytics efforts and how iteration, interactivity, rapid development, and visualization are important aspects of policy analytics efforts. Techniques for focusing communities through the use of an "event-lens" (Park and Johnson 2017), for using social media analytics in addressing concerns about policies that have high impact multi-national consequences, and approaches to conducting new alternatives analysis round out the contribution of the book.

In this chapter, we have proposed a comprehensive and integrative view in which policy analytics goes beyond data analysis and includes management and preparation of data as well as very diverse analytical techniques such as computer simulation, social network analysis, statistics, geographic information systems, data mining techniques, and other computer science approaches. The challenge for all policymakers is matching the right data and the right analytical methods to the decision at hand. The chapters collectively create more clarity around the concept of policy analytics and provide a comprehensive look at the use of policy analytics, in the broadest sense, as critical to understanding and responding to the complex problems facing contemporary society.

References

Chen H, Chiang RH, Storey VC (2012) Business intelligence and analytics: From big data to big impact. MIS Q 36(4):1165–1188

Conklin EJ (2006) Dialogue mapping: building shared understanding of wicked problems. Wiley, Chichester

Dawes SS, Cresswell AM, Pardo TA (2009) From "need to know" to "need to share": tangled problems, information boundaries, and the building of public sector knowledge networks. Public Adm Rev 69(3):392–402

Dörry S, Decoville A (2013) Governance and transportation policy networks in the cross-border metropolitan region of Luxembourg: a social network analysis. Eur Urban Reg Stud. doi:10.1177/0969776413490528

Forrester JW (1992) Policies, decisions and information sources for modeling. Eur J Oper Res 59(1):42–63. doi:10.1016/0377-2217(92)90006-U

Gibson A (2010) Local by social. How local authorities can use social media to achieve more for less. NESTA, London

Gilbert N, Troitzsch KG (2002) Simulation for the social scientist. Open University Press, Buckingham

Gil-Garcia JR (2014) Systemic thinking and system dynamics for the analysis of public policies: fundamentals and recommendations. In: CIDE Mexico Working paper

Guan D, Gao W, Su W, Li H, Hokao K (2011) Modeling and dynamic assessment of urban economy–resource–environment system with a coupled system dynamics–geographic information system model. Ecol Indic 11(5):1333–1344

Janssen M, Helbig N (2016) Innovating and changing the policy-cycle: policy-makers be prepared! Gov Inf Q. doi:10.1016/j.giq.2015.11.009. Epub on 29 January 2016

Koliba C, Zia A, Lee B (2011) Governance informatics: utilizing computer simulation models to manage complex governance networks. Innov J Innov Public Sector 16(1):3–25

Luna-Reyes LF, Gil-Garcia JR, Rich E, Andersen DF (2017) Using system dynamics for the analysis of complex social problems and public policy alternatives: fundamentals and recommendations. In: Chen Y-C, Ahn M (eds) Routledge handbook on information technology in government. Routledge, London, pp 71–92

Maliene V, Grigonis V, Palevicius V, Griffiths S (2011) Geographic information system: old principles with new capabilities. Urban Design Int 16(1):1

Marin A, Wellman B (2011) Social network analysis: an introduction. In: Carrington PJ, Scott J (eds) The Sage handbook of social network analysis. Sage, London, pp 11–25

Mohri M, Rostamizadeh A, Talwalkar A (2012) Foundations of machine learning. MIT press, Cambridge

Morecroft J (2007) Strategic modelling and business dynamics: a feedback approach. Wiley, Chichester

Park CH, Johnson EW (2017) An event-driven lens for bridging formal organizations and informal online participation: how policy informatics enables just-in-time responses to crises. In: Gil-Garcia JR et al (eds) Policy analytics, modelling, and informatics: innovative tools for solving complex social problems. Springer, New York (in this volume)

Porter NW (2017) The value of system dynamics modeling in policy analytics and planning. In: Gil-Garcia JR et al (eds) Policy analytics, modelling, and informatics: innovative tools for solving complex social problems. Springer, New York (in this volume)

Puron-Cid G, Gil-Garcia JR, Luna-Reyes LF (2016) Opportunities and challenges of policy informatics: tackling complex problems through the combination of open data, technology and analytics. Int J Public Admin Digit Age 3(2):66–85

Richardson GP, Pugh AL (1981) Introduction to system dynamics modelling with dynamo. Productivity Press, Cambridge

Richardson GP, Andersen DF, Luna-Reyes LF (2015) Joining minds: system dynamics group model building to create public value. In: Bryson JM, Crosby BC, Bloomberg L (eds) Public value and public administration. Georgetown University Press, Washington, pp 53–67

Sayama H (2015) Introduction to the modeling and analysis of complex systems. Open SUNY Textbooks, Binghamton

Part I
Frameworks and Research Community

Background on Frameworks for Policy Analytics

Karim Hamza and Sehl Mellouli

Abstract Different frameworks related to policy-making analysis and policy modelling have been developed in the literature. These frameworks are generally spread across multidiscipline sciences like public policy; political science, computer science and social sciences. Policy Frameworks address general forms of theoretical analysis, by identifying the elements and their relationships. These elements can include for example: governance structure; policy process; stakeholders; and institutions structure. However, there is no standard categorization or classification for these frameworks. And with the growing development in the policy analytics and policy modelling, there is a rising need to review existing policy frameworks and develop categorization criteria able to classify frameworks concerned by policy-making analysis and policy modelling. This chapter will go through the main frameworks used in understanding the policy making process, in order to make a general overview frameworks for Policy analytics.

Introduction to Policy Frameworks

Policy makers are the persons who take decisions for the well-being of their communities. In order to take good decisions and extremely the better decisions, several frameworks and models have been developed in order to help them in their decision making process. To this end, public policy scholars and policy scientists have developed frameworks, theories and models to better understand policy and policy-making processes (e.g., Sabatier 1991, 2007; McCool 1995; Kraft and Furlong 2007; Smith and Larimer 2009; Birkland 2010).

A Framework refers to concepts to support structured and systematic analysis, design, implementation and assessment/evaluation of a solution. Specifically, a

K. Hamza (✉)
Institute for European Studies (IES), Vrije Universiteit Brussel, Brussel, Belgium
e-mail: khamza@vub.ac.be

S. Mellouli
Université Laval, Ville de Québec, QC, Canada
e-mail: sehl.mellouli@fsa.ulaval.ca

J.R. Gil-Garcia et al. (eds.), *Policy Analytics, Modelling, and Informatics*, Public Administration and Information Technology 24,
DOI 10.1007/978-3-319-61762-6_2

framework identifies elements, identifies the relationships between these elements, and provides a general set of variables that can be used to analyze a proposed solution. On the other hand, models are more precise than frameworks since a model uses specific assumptions about a limited set of variables, identified in the frameworks, to derive precise solutions when combining these variables (Ostrom 2011). Each developed model uses a set of tools. These tools vary from implementation tools like programming languages, to the graphical user interfaces, to the adopted technologies such as multi-agent systems.

This chapter will focus on two abstraction levels: policy-making analysis frameworks, and policy-making simulation tools. At the analysis level, this chapter classifies the frameworks into three main categories: (1) Frameworks focus on Policy Stage; (2) Frameworks focus on Institutions; and (3) Frameworks focus on Human factor and Collations. At the simulation level, this chapter discusses the main simulation tools and their related technologies used for policy-making.

Policy Analytics Frameworks Focusing on Policy Stage

This category of frameworks focus on the process of policy making and on different policy process stages like agenda setting, policy formulation, policy adoption, implementation, evaluation and termination (Lasswell 1956; Brewer 1974; Brewer and deLeon 1983; DeLeon 1999). We present hereafter two main frameworks that are the multiple streams theory of agenda setting by Kingdon in 1984, and Governance Analysis Framework (GAF) by Karim Hamza in 2013.

Multiple Stream Model

The multiple streams model of policy-making has been introduced by John Kingdon and extended by Robert W. Porter. Kingdon argues that for a specific policy to succeed, the following three streams of actions must occur: a problem must be clearly defined, feasible solutions must be offered and political consensus must be obtained (Porter 1995). Moreover, Kingdon and Porter argued that certain policies may fail because (1) problems were narrowly defined only by the government; (2) solutions were one-sided, primarily focusing on the financial aspect of issues and neglecting contextual and implementation problems; and (3) political will existed only at the highest level, whereas the actual capacities of the implementing units were far below the capacities that were needed for successfully achieving reform. The analysis of policies through the lens of the multiple streams model of policymaking is recommended for other policymakers (Kingdon 1984).

Kingdon's perspective claims that policy making can be conceptualised as three largely unrelated "streams": (1) a problem stream, consisting of information about real-world problems and the effects of past governmental interventions; (2) a solution stream/community that is composed of researchers, advocates and other specialists who analyse problems and formulate possible alternatives; and (3) a political stream, consisting of elections, legislative leadership contests and similar data sources. According to Kingdon, major policy reforms are produced if "a window of opportunity" joins these three streams; in other words, in response to a recognised problem, the policy community develops a proposal that is financially and technically feasible and politicians find it advantageous to approve this proposal.

The problems stream denotes which social conditions are perceived by individuals as problems that require resolution by government action. Citizens, the mass media and interest groups often define problems and their potential causes only from their own viewpoint. Additionally, in many cases, it is too difficult to clearly define the problem and its history because many state or non-state actors may lobby for their own views regarding the concerned problem (Kingdon 1984).

The solutions stream consists of policy alternatives that are generated by state actors, state institutions, policy advocates and academics. Policy solutions must be well crafted, include sequence and content development and suggest the timing of reform. The solutions should include (1) the translation of policy directives into implementation programmes, (2) the generation of strategies for the adoption of the policy solution by different actors, (3) provide management strategies to address the solution's opponents and (4) take advantage of supporters of the reform to support the solution (Porter 1995).

The politics stream consists of political events that may or may not be favourable to the policy that is being implemented. Similar to state turbulence or important elections, changes in government ministers and public protests can powerfully influence whether a particular problem will be solved (Kingdon 1984).

The multiple streams model is useful because it argues that numerous real life contextual factors, such as political events, bureaucratic procedures and interest groups, largely shape the future of the political environment. This model forces all interested and involved actors to recognise the realities of policymaking and act pragmatically if they wish to successfully implement these policies. In addition, similarly to the traditional policy stages model, the multiple streams model allows for policies to be dissected and analysed in broad, generalised terms (problems, solutions and politics). This analysis can be highly valuable to policymakers who are faced with social problems that must be fixed in a given political environment.

Governance Analysis Framework (GAF)

Additionally; the Governance Analysis Framework (GAF) was introduced as an analysis framework to asses state stability and ensure that all aspects of governance are covered.The Governance Analysis framework (GAF) includes four main phases. Phase I is the scope and objectives map, which is based on Kingdom and Porter's multiple streams policymaking model, used to examine the problem stream, the solution stream and the political stream. Phase II is the stakeholders map, which

Table 1 Governance analysis framework (GAF)

Phase/component	Phase I Scope map	Phase II Stakeholders map	Phase III Process map	Phase IV Governance map
Objective	Describe the main streams: • Problem Stream • Solution Stream • Political Stream	Describe who are the main actors and their relation with each other inside the governance model: • Citizen Power • Executive Authority • Judicial Authority • Legislative Authority • Non-state Political Power	Identify policy networks collations and their usage of social media: • Identify Control level of main actors on social media • Identify confrontation level on social media	Describe the progress (or not) of the state transformation from transition state toward a state stability
Based on	Multiple Stream Model, (Kingdon 1984; Porter 1995)	Policy Network Analysis [PNA], (Rhodes 1996)	Advocacy Coalition Framework [ACF], (Sabatier and Jenkins-Smith 1999) and Social media	Circle of State Stability

describes the main actors in the policy-making process and how they interact with one another inside the governance model. This phase adapts Rhodes's policy network analysis (PNA) and Sabatier's advocacy coalition framework (ACF). Phase III is the process and tools map, which describes the policy-making process applied within the governance model and it is based on policy network analysis (PNA) school plus the advocacy coalition framework (ACF). Finally, Phase IV is the governance map, which focuses on the general governance structure of the state under study and represents graphically the state's status with respect to turbulence and stability, using the Circle of State Stability model (CSS) (Hamza 2013) shown below in (Table 1).

Phase I—Scope and Objectives Map: The first phase describes the main scope of the governance model under study, which helps break down the model into its main components. The scope of this governance model should reflect the way a society is governed, help create conditions for orderly rule, support collective action and maintain links between the main actors in the political environment (Kooiman 1993, 2003; Hamza 2013).

Additionally, the scope defined must balance certain key values. (1) Openness: working openly and communicating precisely the new policy in the governance model. The language should be accessible and understandable for the general public. (2) Participation: ensuring wide participation throughout the policy process from conception to implementation. Improved participation increases confidence in the end result and the institutions that deliver the new policies. (3) Accountability: the roles in the legislative and executive processes must be clearer. State actors must explain and assume responsibility for their decisions and there is a need for greater responsibility from all those who are involved in developing and implementing policy on all levels. (4) Effectiveness: policies must be effective and timely. (5) Coherence: the policies and actions performed in the governance model must be coherent and easily understood. Coherence requires political leadership and a strong sense of responsibility from the state actors to ensure a consistent approach within a complex system (European Commission 2001; Hamza 2013).

In this phase, Kingdom and Porter's multiple streams policy-making model will be used to examine the scope. The multiple streams model describes policy-making when sets of multiple nonlinear activities (the "streams") are pursued. These streams define the problem, suggest solutions and obtain political consensus, which should occur simultaneously, thus creating an opportunity to facilitate changes (Kingdon 1984; Hamza 2013).

Phase II—Stakeholder Map: Ideally, the governance model should apply to the entire range of stakeholders (actors) involved in the governing process. Thus, continuous two-way communication between the governors and the governed stakeholders can be maintained and the governance model can remain oriented toward the objective (the scope) for which it was designed. In this model, to avoid ignorance of the values inherent in stable governance mechanisms, the distribution of rights and responsibilities among the different stakeholders may require revision to balance the different actors, define the rules and procedures of decision-making and separate the main powers in a governance model, which are represented by the legislative, executive and judiciary powers (Daugbjerg and Fawcett 2010; Hamza 2013).

Therefore, a successful governance model should promote the role of non-state actors in society and public activities and increase the responsibilities of social actors outside the state boundaries (Galeotti and Josselin 2001). Thus, the government's responsibility for the provision of social services must be redefined (Chhotray and Stoker 2009; Hamza 2013).

This phase is inspired by the policy network analysis of (Rhodes 2006). The policy network refers to clusters of different types of actor who are linked in political, social or economic life, have an interest in a given policy action and can help determine a policy's success or failure (Rhodes 1997). This theory seeks to explain policy outcomes by investigating how networks, which facilitate bargaining between stakeholders regarding policy design and detail, are structured in a particular political environment. Additionally, the theory analyses variations in the interest patterns and thus helps map relationships between state and non-state actors and explain how the policy agenda is formed. This model consists of two basic elements: actors and the relations between them (Hamza 2013).

Phase III—Process Map: This phase examines the main process and tools used during the policy-making decisions implemented in the selected governance model to understand the relation between these process and different stakeholders and analyse the impact of technological tools, such as the social media, on this process. The main component of this phase includes the process section and the social media section and focuses on the use of ICT by different stakeholders (actors) of the society (Hamza 2013).

Inspired by Sabatier and Jenkins-Smith's advocacy coalition framework (ACF), the process section addresses problems that involve substantial goal conflicts, important technical disputes and multiple actors (Sabatier and Jenkins-Smith 1999). The ACF framework has found that stakeholder beliefs and behaviour are embedded in informal networks. To enforce coordinated behaviour in the networks, actors form groups. These groups are comprised of individuals from a variety of backgrounds who share a particular belief system and show a degree of coordinated activity over time. This section also pays close attention to collective-action issues (Hamza 2013).

Phase IV—Governance Map: The governance map aims to describe the progress of the governance transformation from the transition state toward a stable state, considering the relation between hierarchical and network governance control based on the Circle of State Stability model described above (Hamza 2013).

The Circle of State Stability model is designed to represent graphically the impact of network governance on state stability in a specific state or country based on the following assumptions. First, there are two main types of governance structure: hierarchy and network governance. These two structures seek to control the state or the country. Market governance is assumed to be part of both hierarchy and network governance because market governance focuses on efficiency, which is an important objective for any type of governance. Second, state instability appears due to four main categories of instability conditions: confrontation, dictatorship, anarchy and failure (Hamza 2013).

Policy Analytics Frameworks Focusing on Institutions

The second category focus on the institution performing the decision making or impacted by the policy making process. This institution can be government; sub-government or other organizations. The most famous framework that focuses on institutions is the Institutional Analysis and Development (IAD) framework developed by Ostrom in 1990.

Institutional Analysis and Development (IAD)

Institutional Analysis and Development (IAD) framework is developed by Elinor Ostrom and other scholars associated with the Workshop in Political Theory and Policy Analysis at Indiana University. This Framework tries to understand the policy process, by outlining a systematic approach for analyzing institutions that govern action and outcomes within collective action arrangements (Ostrom 2011). Institutions are defined within the IAD Framework as a set of prescriptions and constraints that humans use to organize all forms of repetitive and structured interactions (Ostrom 2011). These prescriptions can include rules, norms, and shared strategies (Ostrom 1990). Institutions are further delineated as being formal or informal; the former characterized as rules-in-form and the latter as rules-in-use.

The IAD framework identifies key variables that should be used in evaluating the role of institutions in shaping social interactions and decision-making processes. The analytical focus of the IAD is on an "action arena", where social choices and decisions take place. Three broad categories of variables are identified as influencing the action arena: (1) institutions or rules that govern the action arena, (2) the characteristics of the community or collective unit of interest, and (3) the attributes of the physical environment within which the community acts (Ostrom 1990, 2011). Each of these three categories has been further delineated by IAD scholars into relevant variables and conditions that can influence choices in the action arena (Ostrom 1990, 2011).

The IAD further defines the key features of "action situations" and "actors" that make up the action arena. The action situation has seven key components: (1) the participants in the situation; (2) the participants' positions; (3) the outcomes of participants' decisions; (4) the payoffs or costs and benefits associated with outcomes; (5) the linkages between actions and outcomes; (6) the participants' control in the situation; and (7) information. The variables that are essential to evaluating actors in the action arena are (1) their information processing capabilities; (2) their preferences or values for different actions; (3) their resources; and (4) the processes they use for choosing actions (Ostrom 1999, 2005).

In addition to the types of relevant variables that may help explain collective choices, the IAD has identified multiple levels of institutional analysis: (1) operational level; (2) the collective-choice level; and (3) the constitutional level (Ostrom 1990). The operational level of analysis is where individuals collectively make decisions about day to day activities. The collective-choice level of analysis focuses on decisions about the choice of rules that govern operational activities. The constitutional level of analysis is concerned with the authorized actors for collective-choice decisions and the rules governing those decisions (Ostrom 1990, 2011). Also any one decision-making group or action arena may operate at more than one level of institutional action (Ostrom 1990, 2011).

Policy Analytics Frameworks Focusing on Human Factor and Collations

The third category explores processes from the perspective of the **actions of human agents** and their attainment of goals with respect to devising policies and institutions. For instance, the Advocacy Coalition Framework (ACF) developed by Sabatier and Jenkins-Smith in 1993, which emphasises the role of human agents, considers processes that emerge through conflict or competition among multiple coalitions over long periods of time (Sabatier and Jenkins-Smith 1993). And the policy network analysis (PNA) school that focus on networks, the relationships among policy-making outcomes, the structure of a network and the inclusion or exclusion of certain individuals or groups from the network in question (Fawcett and Daugbjerg 2012).

Advocacy Coalition Framework (ACF)

The Advocacy Coalition Framework (ACF) is a framework of the policy process developed by Sabatier and Jenkins-Smith to address problems that involve substantial goal conflicts, important technical disputes and multiple actors from several layers of government; in other words, this approach seeks to address the entire policy process rather than merely agenda-setting (Sabatier and Weible 2007). This theory was developed as a response to perceived inadequacies in the "stages heuristic" and other traditional approaches to the analysis of the policy process. The ACF theory claims that stakeholder beliefs and behaviour are embedded within informal networks (Sabatier and Weible 2007). To enforce coordinated behaviour within these networks, actors form groups, which consist of sets of people from a variety of positions who share particular belief systems and demonstrate a degree of coordinated activity over time (Sabatier and Jenkins-Smith 1999). This framework also devotes a great deal of attention to collective action issues (Schlager 1999).

Sabatier and Jenkins-Smith note that real-world changes frequently occur in the aftermath of a specific situation and result in the redistribution of political resources and alliances among subsystems. ACF framework is used to enhance the available understanding regarding complex policy processes. The ACF includes the following four main assumptions: (1) the process of policy change and learning occurs over time; (2) the most useful way to consider policy change over time is to focus on policy subsystems, that is, the interactions among actors who seek to influence the policy-process outcome; (3) subsystems must include an intergovernmental dimension; and (4) public policies can be conceptualised in the same manner as belief systems, e.g., as sets of value priorities and causal assumptions about how to realise these priorities (Sabatier and Jenkins-Smith 1993).

As a framework, the ACF's policy subsystem is the primary unit of analysis. A long-term time perspective is needed to obtain an understanding of subsystem affairs; moreover, the expansive set of actors that are involved in policy systems may be aggregated into coalitions and policy designs are interpreted as translations of coalition beliefs. This interpretation will increase interest in understanding and explaining coalition formation, maintenance, stability and structure.

Within the ACF, policy formation and change are functions of competing advocacy coalitions within a policy subsystem. A policy subsystem consists of actors from 'public and private organizations who are actively concerned with a policy problem. The actors within a policy subsystem are grouped into a number of advocacy coalitions that consist of individuals who share particular belief systems, e.g., sets of basic values, causal assumptions and problem perceptions and who exhibit a non-trivial degree of coordinated activity over time.

If coalitions manage to form a structure that different coalitions are likely to adopt (e.g., a loose structure with minimal coordination versus a well-defined structure with high levels of coordination), the stability and continuity of these coalitions will receive the highest levels of attention. One of the shortcomings of ACF is that this theory provides little conception of the strategies that coalitions are likely to pursue in pressing for preferred policies and confounding undesirable policies.

The initial condition for coalition formation is that individuals believe that they will benefit from acting collectively to change policy; once this condition is met, a number of other factors come into play. These factors are characteristics of individuals and the situation that are mutually supportive and promote stable relationships. (Ostrom 1990) suggests the following set of conditions that support the stability and continuity of coalitions: (1) individuals who benefit from the collective goods that are provided by the coalition have clearly been identified; (2) the benefits that individuals receive from these collective goods are related to these individuals' contributions to the provision of the goods in question; (3) the individuals who are most affected by the rules can participate in changing the rules; (4) monitors who actively audit coalition members' behaviours are either accountable to the coalition members or are coalition members themselves; (5) members who violate the coalition rules are likely to be punished or isolated by other members; (6) coalition members have rapid access to low-cost local methods of resolving conflicts among members or between members and officials; and (7) the rights of individuals to form coalitions and to devise policy have not been challenged by external governmental authorities (Ostrom 1990).

These conditions centre on fairly allocating the benefits and costs of collective action and on monitoring and enforcing agreed-upon behaviours. For a coalition to maintain itself over a period of time, it must not only be able to capture the benefits that it produces but also allocate these benefits and their production costs in a fair manner. In addition, the behaviours of group members must be monitored and actions that violate the agreed-upon standards must be punished. Monitoring is critical because individuals continually face incentives to defect by pursuing their own self-interests at the expense of the larger group. Finally, according to Moe, once a coalition gains control of public authority, its intent is to design and implement public agencies and policies that will effectively achieve its policy goals (Ostrom 1990; Sabatier and Weible 2007).

Additionally, Sabatier and Weible identify six categories of coalition resources: (1) formal legal authority to make policy decisions; (2) public opinion, (3) information, (4) the mobilisation of troops, (5) financial resources and (6) skilful leadership. These resources can be hierarchically arranged with respect to their usefulness to coalitions for the purpose of generating policy change (Sabatier and Weible 2007).

Finally, the ACF views policy change over time as a function of three sets of factors. The first of these three sets of factors is the interaction of competing advocacy coalitions within a policy subsystem or community; typically, the advocacy coalition subsystem consists of actors from many state and non-state institutions that share a set of basic beliefs and that seek to manipulate the rules of various governmental institutions to achieve their goals over the course of time. The second set of factors comprises external changes to the subsystem that are caused by other policy subsystems or socioeconomic conditions; these changes may affect advocacy coalitions and may influence the decisions that emerge from the advocacy coalition subsystem. The third set of factors involves the importance of a stable legal system and constitutional rules to regulate the relationships between different actors in the aforementioned coalitions (Sabatier and Weible 2007).

Policy Network Analysis (PNA)

The policy network analysis (PNA) school focus on networks; however, the PNA school are more concerned with micro-level examinations about the relationships among policy-making outcomes, the structure of a network and the inclusion or exclusion of certain individuals or groups from the network in question (Fawcett and Daugbjerg 2012).

The term 'network' is frequently used to describe clusters of different types of actors who are linked together in political, social or economic capacities. Networks may be loosely structured but remain capable of spreading information or engaging in collective action. The literature on networks is often vague or abstract. However, the growing interest in network governance structures reflects the increasing shift of modern societies and economies towards mutuality and against hierarchies (Peterson and O'Toole 2001).

The policy network analysis tries to examine how national policies can emerge from sector networks that link authorities across different levels of government and join public and private actors. Rhodes observes the difficulties that governments experience in attempting to steer disaggregated structures of interdependent organisations and perceives the emergence of network interactions as a common response to these challenges in advanced and industrialised societies. He mentions the term 'policy network', which refers to interest intermediation between public and private actors, a topic that has received considerable interest in the political science literature in recent years (Rhodes 2006; Jordan and Schubert 1992). In addition, Peterson and Bomberg define a policy network as 'a cluster of actors, each of which has an interest in a given policy action and the capacity to help determine policy success or failure' (Peterson and Bomberg 1999). Recently, analysts of governance have sought to explain policy outcomes by investigating how networks, which facilitate bargaining between stakeholders over policy design and detail, are structured in a particular sector. The policy networks theory attempts to define how networks are structured in any policy sector and thereby help explain and predict policy outcomes (Scharpf 1999).

According to Adam and Kriesi, a policy network consists of two basic elements: actors and the relationships among these actors (Adam and Kriesi 2007).

Importantly, actors of regulation are institutionalised actors (Peterson 2003); thus, institutionalised actors constitute the unit of analysis in most governance studies. Institutionalised actors are formally organised and have resources that are distributed within the organisation according to hierarchies or majorities (Scharpf 1989). Policy network analysis begins from three basic assumptions. First, modern governance is frequently non-hierarchical. Few policy solutions are simply imposed by public authorities. Governance involves mutuality and interdependence between public and non-public actors and among different types of public actors. Second, the policy process must classify relationships between various groups and the government; these relationships are dependent on the relevant policy areas that are examined (Rhodes 1996). Third, governments ultimately remain responsible for governance. Policy networks can help to narrow options and shift agendas by pursuing strategies that generate new political and economic forces (Thatcher 1998).

The Rhodes model of policy networks has most likely been employed more often than any other model for the purpose of examining EU governance (Peterson and Bomberg 1999). This model assumes that three key variables determine what type of policy network exists in a specific sector: (1) the relative stability of a network's membership, which refers to whether the same actors tend to dominate decision-making over time or whether network membership is fluid and dependent on the specific policy issue that is under discussion; (2) the network's relative insularity, which refers to whether the network is a cabal that excludes outsiders or a structure that is highly permeable by a variety of actors with different objectives; and (3) the strength of resource dependencies, which refers to whether network members depend heavily on each other for valued resources, such as money, expertise and legitimacy, or whether most actors in a network are self-sufficient and are therefore relatively independent of one another.

This framework help in describing and analysing variations in the patterns of interests and thereby facilitates the mapping of relationships between state and non-state actors and the determination of how the policy agenda is shaped (Rhodes 1996). However, this approach has shortcomings in that it does not explain the changes inside networks; moreover, it does not sufficiently account for the roles of structure and strategic interaction within networks. Policy network analysis is increasingly used to make sense of policymaking environments. A frequent primary aim of this analysis is to determine which interests dominate bargaining within networks (Coleman and Perl 1999).

This school of analysis is primarily concerned with how networks affect power and how these power relationships can privilege certain interests more than others during the making and delivery of public policy (Rhodes 1996). The approach disaggregates the analysis of networks to the sectorial or sub-sectorial level. Accordingly, the state is perceived as fragmented; therefore, the interest of the state varies among different policy sectors (Rhodes 1996) and it is not unusual for various state authorities to hold conflicting views on policy. State capacity also varies significantly across the various agencies of the state (Atkinson and Coleman 1989).

Simulation Tools

In the previous section, different policy frameworks have been presented. These frameworks provide guidelines for the development of a policy. However, one of the main issues that policy-makers are facing is the possibility to simulate the impacts of a policy before its adoption. Doing so, policy makers can ensure, at some extent, that they are taking good decisions. To this end, several policy simulation tools have been proposed in the literature. In this section, we present some of these tools from the following perspectives: Name of the tool, the authors, the objective of the tool, and the related technologies used in the tool.

Literature Review Strategy

Different policy-simulation tools have been propsed in the literature. In order to identify these tools, we conducted a preliminary search on the ABI/inform database. We recognize that this is not an exhaustive literature review of tools used for policy simulation. Our research was on two steps: the first step is on the ABI/inform database then a second step validation with an expert to see if there are other tools that can be added to this literature review. From the first step, we obtained nine tools and then, in the second step, four other tools have been added by an expert in the domain.

For the ABI/inform database, we searched for publications between 2000 and 2016 with the key word: "policy simulation tool". We looked at peer-reviewed articles. The search returned 1487 papers. Then, using ABI/inform options, we refined our search to look only at simulation tools. The refined results gave 373 papers. Finally, using also ABI/inform tools, we refined the categories of these 373 papers to look only on papers related to policy, public policy, government, and public administration. With this final refinement, we obtained 14 English written papers. We went through these 14 papers to only study those related to policy simulation tools. From the 14 papers, only 9 papers were retained. We found that one paper is out of the scope of policy simulation; it was dealing with genetically modified corns. A second paper was about policy simulation but it was totally theoretic with no results or implementations. A third paper wasn't about simulation but more about poverty in Kosovo. A fourth paper was about the importance of simulation tools and how simulation tools can improve policy development but also with no implementation or results. Finally, a fifth paper wasn't electronically available.

The meeting with the expert led us to add four other tools that weren't found in the ABI/Inform search that we conducted. These four tools are: OCOPOMO, Virsim, Urbansim, and Skin.

When looking at the different tools, we found that these tools are applied in different domains such as energy, transportation, or urbanism. In addition, we found that there are three main technologies used in these tools: multi-agent systems, mathematical models, or developed software that combines different tools and

Table 2 Tools, technologies, and application domains

Paper title	Technology	Application domain
A multiregional model of China and its application	Mathematical equations	Economy
Mobilizing for change: simulating political movements in armed conflicts	MAS	Conflicts
Are cash transfers a realistic policy tool for poverty reduction in Sub-Saharan Africa? Evidence from Congo-Brazzaville and Côte d'Ivoire	Ex ante simulation	Economy
"Using a Simulation-Based Learning Environment for Teaching and Learning About Complexity in Public Policy Decision Making"	Learning environment	Policy
OCOPOMO Model (Scherer et al. 2013)	Multi-agent systems	Energy
The Simulating Knowledge Dynamics in Innovation Networks (SKIN) (Ahrweiler et al. 2004)	Agent-based model	Behaviour
"The DH Accident and Emergency Department model: a national generic model used locally"	Discrete event simulation	Health
Improving maintenance decision making in the Finnish air force through simulation	Discrete event simulation	Aviation
System Dynamics Approach as a Risk Management Tool in Analyzing Pension Expenditure: The Case of Malaysian Employees Public Pension	Systems dynamics	Economy
"Gaming and Simulation for Railway Innovation: A Case Study of the Dutch Railway System"	A multi-actor environment	Innovation
"Transportation Modeling as a Didactic Tool: Human Settlement and Transport"	VISUM software	Transport
VirSim (Fasth et al. 2010)	Multi-agent systems	Health
UrbanSim (Waddell 2002)	Python	Urbanism

technologies. The following table summarizes the different tools, their related technologies, and their related application domains (Table 2).

Multi-Agent Based Tools

We present hereafter four policy simulation tools using the multi-agent yechnology. The first tool, called Security Community (Altaweel et al. 2012), is developed in the context of conflicting zones and it was applied in conflicts in Central Asia. It mainly supports policy makers when they decide to move populations in conflicting zones. It simulates the impacts of these movements to explore unknown behavioral qualities relevant to mobilization.

Thes second tool is OCOPOMO (Scherer et al. 2013). It was applied in the context of the energy domain and it was developed in order to simulate the behaviour of key stakeholders and the process of decision making. It provides a basis for testing the effectiveness of various government policies under different conditions such as abnormal climatic phenomena. The simulation takes into account the inter-relations between stakeholders, the economic conditions of the energy domain, and the social dynamics of the relations between the stakeholders. OCOPOMO is developed based on the Declarative Rule-based Agent Modelling system (DRAMS) (Lotzman et al. 2011).

The third tool is the Simulating Knowledge Dynamics in Innovation Networks (SKIN) (Ahrweiler et al. 2004). It is used to simulate the behaviour of agents, who act and interact in a large-scale complex and changing social environment. It was applied for example in industry where agents have to buy and sell goods. SKIN has the advantage to consider the knowledge dynamics in and between the agents.

The fourth tool is VirSim (Fasth et al. 2010). It has the objective to simulate the spread of a pandemic influenza. The simulation is based on real population data in Sweden. It provides policy-makers with tools to evaluate the effect of different measures connected to school closure and vaccination. These tools use advanced reasoning methods considering the population characteristics and its exposure to infection. The main goal is to help policy-makers find the most optimal policies to, for example, the starting time and the duration of school closure as well as the pace and the coverage of population with vaccination. VirSim is developed using the AnyLogic tool (a muti-agent based simulation tool), and different Graphical User Interfaces tools.

Mathematical Based Tools

We present hereafter five policy simulation tools that are based on mathematical models. The first paper is applied in the contxt of China. It presents a tool to simulate the interactions among the provincial economies and the dynamic relationship between the centre and local governments in China (Gu and Chen 2005). It aims to study the impact of fiscal recentralisation in the 1990s that the Government of China has decided. The tool consists of over 1200 equations, and has used a database of annual data from 1985 to 1998 for variables at both national and provincial levels.

The second tool addresses the issue of reducing waiting time at emergencis in UK (Fletcher et al. 2007). The tool was developed as an analytical support to the Departement of Health in UK. The department had the objective in 2002 that by December 2004, 98% of patients arriving at Accident and Emergency (A&E) departments in England should be completed, that is, admitted, discharged or transferred, within 4 h. It was important that targets such as this were seen to be delivered as evidence that NHS performance was being modernized and improved to meet patients' expectations. The tool is based on a discrete event simulation to see the impacts, on waiting times, of any decision that can be taken at any stage from admission, discharge, or transfer.

The third tool is a simulation model to improve the maintenance decision-making in the Finnish Air Force through simulation (Mattila et al. 2008). It helps to see the impacts of maintenance operations on existing resources. It allows then to optimize maintenance operations according to the context of the operations (normal or conflicting conditions). The proposed tool is based on event-driven simulation that describes the flight operations and maintenance of fighter during normal and conflict conditions.

The fourth tool studies the potential role of cash transfers as instruments for poverty reduction and human development in (Hodges et al. 2013) Sub-Saharan Africa. The tool was applied on data collected from two different countries with two different contexts Congo-Brazzaville (middle-income oil producer) and Côte d'Ivoire (low-income country) (Hodges et al. 2013). The simulation is based on the evaluation of different criteria such as efficiency, impacts, cost, cost-effectiveness and affordability of different cash transfer options. Simulation results show that cash transfers would have more impact on monetary poverty reduction than on human development.

The fifth tool simulates pension expenditure for the Malaysian employees (Sapiri et al. 2014). The tool simulates the impact of changes and policy decisions based on demographic and salary risks. The tool is developed using risk management and System Dynamics methodology.

Software Tools

We present in what follows four simulation tools that were developed in the form of a complete software solution. The first software solution is a simulation-based learning environment to teach public managers and decision-makers about dynamic systems and decision making in such complex systems (Minyoung et al. 2012). The objective is to prepare future and current public decision makers for a rapidly changing, complex world. The tool was applied in the case of natural disasters preparedness.

The second software tool builds a framework related to innovation of complex systems in a multi-actor environment (van den Hoogen and Meijer 2015). This software is used to assess innovation processes in the context of future development of railroads. Future innovation means that railways companies need to figure out new ideas to upgrade their infrastructures and improve their performances. The software tool will evaluate these innovations since they need to be invented, explored, tested, and implemented in an incumbent system.

The third software tool is a learning environment (Ohnmacht et al. 2015) to simulate the complexity that actual urban planning committees might face when building transportation plans. The results of this learning environment showed that participants revealed that they unearthed interrelations between settlement, transportation, and society.

The fourth software tool is UrbanSim. It is a decision support system for urban transportation investments (Waddell 2002). The software is now available as an open source software that can be downloaded and deployed by governments. The software deals with different issues from building a new light rail system, to changes in land use policies. UrbanSim integrates a simulater to evaluate the impact of the decisions at the long-term economic, social, and environmental levels.

This preliminary literature review on policy simulation tools, we haven't found any reference to any policy analytic framework. This shows, that there still exists a gap between the framework level and the implementation level. Research is still required in the way to bridge the gap between policy framework and policy tools assuming that tools nave to support frameworks.

Conclusion and Discussion

Most of the frameworks concerned by policy making analysis and policy modelling, are spread across multidiscipline sciences like public policy; political science, computer science and social sciences. Frameworks address general forms of theoretical analysis, by identifying the elements and general relationships among different components and provide a general set of variables that can be used to analyse specific arguments. These elements can include: governance structure; policy process; stakeholders; and institutions structure. It also uses different political models and technological tools, to analyse or explain or predict specific political behaviour.

These frameworks have in common three main elements with different degree of depth in analysis. These elements are: (1) People or the actors, who do what like collation and influence; (2) the process either it is related to agenda setting or one of the stages of the policy cycle; (3) political environment or what are the conditions surrounding the policy making process like political; social or economic conditions. Second is the depth of the analysis either it is Macro-Level like Nation or whole government or whole society; or the Micro-Level like specific sector of society or organization; or sub-government.

The primary frameworks classification is divided into: **First**; Frameworks focusing on Policy Stage. This category focus on the process of policy making and Different policy process stages like agenda setting, policy formulation, policy adoption, implementation, evaluation and termination (Lasswell 1956; Brewer 1974; Brewer and deLeon 1983; DeLeon 1999). The main frameworks can be in this category is the multiple streams theory of agenda setting by Kingdon in 1984 and Governance Analysis Framework (GAF) by Karim Hamza in 2013. **Second**; Frameworks focusing on Institutions. This category focus on the institution performing the decision making or impacted by the policy making process. This institution can be government; sub-government or other organizations. The most famous framework that focuses on institutions is the Institutional Analysis and Development (IAD) framework developed by Ostrom in 1990. **Third**; Frameworks focusing on Human factors and Collations. This category explores processes from the perspective of the actions of human agents and their attainment of goals with respect to devising policies and institutions. The most common framework in this category is the Advocacy Coalition Framework (ACF) developed by Sabatier and Jenkins-Smith in 1993 (Sabatier and Jenkins-Smith 1993), and the policy network analysis (PNA) school (Fawcett and Daugbjerg 2012).

At a second stage, the chapter presented a set of policy simulation tools. The tools were grouped into three families related to the adopted technologies in the tools: multi-agent based tools, mathematic based tools, and package software tools. The presented tools have been applied in different application domains.

This chapter presented different policy-analytics frameworks. It also presented different policy simulation tools and their related technologies. Even if models are used to implement frameworks, the different presented models are not related to frameworks. Hence, this chapter identifies the following future research directions. The first direction is related to the development of new tools that can support existing policy-making frameworks. Doing so, the tools will be considered as a real support to policy-makers since it will support their ways of operating. The second direction is related to the technologies adopted in these models. In fact, nowadays, policy-makers are mobile persons who are always traveling. Hence, new mobile technologies need to be integrated within the models in order to support and reflect this reality. Finally, new methodologies need to be developed to support the development of these models while coping with policy-makers needs so that the tools can provide the required functionalities to policy-makers in order to coordinate their activities.

References

Adam S, Kriesi H (2007) The network approach. In: Sabatier P (ed) Theories of the policy process. Westview Press, Boulder

Ahrweiler P, Pyka A, Gilbert N (2004) Simulating knowledge dynamics in innovation networks (SKIN) (No. 267). Volkswirtschaftliche Diskussionsreihe/Institut für Volkswirtschaftslehre der Universität Augsburg

Altaweel M, Sallach D, Macal C (2012) Mobilizing for change: simulating political movements in armed conflicts. Soc Sci Comput Rev. doi:10.1177/0894439312451106

Atkinson M, Coleman W (1989) Strong states and weak states: sectoral policy networks in advanced capitalist economies. Br J Polit Sci 14(1):46–67

Birkland TA (2010) An introduction to the policy process, 3rd edn. M.E. Sharpe, Armonk

Brewer G (1974) The policy sciences emerge: to nurture and structure a discipline. Policy Sci 5(3):239–244

Brewer G, deLeon P (1983) The foundations of policy analysis. Brooks/Cole, Monterey

Chhotray V, Stoker G (2009) Governance theory and practice: a cross disciplinary approach. Palgrave Macmillan, Houndmills

Coleman W, Perl A (1999) Internationalized policy environments and policy network analysis. Polit Stud 47(4):691–709

Daugbjerg C, Fawcett P (2010) Governance theory and the question of power: lesson drawing from the Governance Network and Policy Network Analysis Schools. In: The Australian political studies association conference. School of Social and Political Sciences, University of Melbourne, Melbourne, pp 27–29

DeLeon P (1999) The stages approach to the policy process: what has it done? Where is it going? In: Sabatier PA (ed) Theories of the policy process. Westview Press, Boulder, pp 19–34

European Commission (2001) European governance: a white paper. Retrieved April 1, 2012, from http://www.europa.eu.int/comm/governance/index_en.htm
Fawcett P, Daugbjerg C (2012) Explaining governance outcomes: epistemology, network governance and policy network analysis. Polit Stud Rev 10(2):195–208
Fletcher A, Halsall D, Huxham S, Worthingt D (2007) The DH accident and emergency department model: a national generic model used locally. J Oper Res Soc 58(12):1554–1562
Galeotti M, Josselin D (2001) Non-state actors in world politics. Palgrave, Basingstoke/New York
Gu Q, Chen K (2005) A multiregional model of China and its application. Econ Model 22(6):1020–1063
Hamza, K. (2013). The impact of social media and network governance on state stability in times of turbulence: Egypt after 2011 revolution. PhD thesis, Institute for European Studies, Vrije Universiteit, Brussels
Hodges A, Notten G, O'Brien C, Tiberti L (2013) Are cash transfers a realistic policy tool for poverty reduction in Sub-Saharan Africa? Evidence from Congo-Brazzaville and Côte d'Ivoire. Glob Soc Policy 13(2):168–192
van den Hoogen J, Meijer S (2015) Gaming and simulation for railway innovation a case study of the Dutch railway system. Simul Gaming 46(5):489–511
Jordan G, Schubert K (1992) A preliminary ordering of policy network labelling. Eur J Polit Res 21(1–2):7–27
Kingdon J (1984) Agendas, alternatives, and public policies. Little, Brown and Company, Boston
Kooiman J (1993) Modern governance: new government-society interactions. Sage, London
Kooiman J (2003) Governing as governance. Sage, London
Kraft M, Furlong S (2007) Public policy: politics, analysis, and alternatives, 2nd edn. CQ Press, Washington
Lasswell H (1956) The decision process. University of Maryland Press, College Park
Mattila V, Virtanen K, Raivio T (2008) Improving maintenance decision making in the Finnish air force through simulation. Interfaces 38(3):187–201
McCool D (1995) Public policy theories, models, and concepts: an anthology. Prentice Hall, Upper Saddle River
Minyoung K, Roderick HM, Deborah LA, Anderson FD (2012) Using a simulation-based learning environment for teaching and learning about complexity in public policy decision making. J Public Affairs Educ 22(1):49–66
Moors A, Ellsworth PC, Scherer KR, Frijda NH (2013) Appraisal theories of emotion: state of the art and future development. Emotion Rev 5(2):119–124
Ohnmacht T, von Arx W, Schick N, Wegelin P (2015) Transportation modeling as a didactic tool human settlement and transport. Simul Gaming 46(5):563–590
Ostrom E (1990) Governing the commons: the evolution of institutions for collective action. Cambridge University Press, New York
Ostrom V (1999) Cryptoimperialism, predatory states, and self-governance. In Polycentric governance and development: readings from the Workshop in political theory and policy analysis. University of Michigan Press, pp 166–185
Ostrom E (2011) Background on the institutional analysis and development framework. Policy Stud J 39(1):7
Peterson J (2003) Policy networks. IHS Political Science Series. The Department of Political Science, Institute for Advanced Studies, Vienna
Peterson J, Bomberg E (1999) Decision-making in the European Union. Palgrave, Basingstoke/New York
Peterson J, O'Toole L (2001) Federal governance in the United States and the European Union: a policy network perspective. In: Nicolaïdis K, Howse R (eds) The federal vision: legitimacy and levels of governance in the United States and the European Union. Oxford University Press, Oxford/New York
Porter R (1995) Knowledge utilization and the process of policy formation: toward a framework for Africa. USAID, Washington

Rhodes RA (1996) The new governance: governing without government. Polit Stud 44(4):652–667
Rhodes R (1997) Understanding governance: policy networks, governance, reflexivity and accountability. Open University Press, Buckingham
Rhodes RA (2006) Policy network analysis. In: Moran M, Rein M, Goodin R (eds) The oxford handbook of public policy. Oxford University Press, New York, pp 425–447
Rhodes RA, Bache I, George S (1996) Policy network and policy making in the Europen Union: a critical appraisal. In: Hooghe L (ed) Cohesion policy and European integration: building multi-level governance. Oxford University Press, Oxford, pp 367–387
Sabatier PA (1991) Toward better theories of the policy process. Polit Sci Pol 24(2):147–156
Sabatier P (2007) Theories of the policy process. Westview Press, Boulder
Sabatier P, Jenkins-Smith H (1993) Policy change and learning: an advocacy coalition approach. Westview Press, Boulder
Sabatier P, Jenkins-Smith H (1999) The advocacy coalition framework: an assessment. In: Sabatier P (ed) Theories of the Policy process. Westview Press, Boulder, pp 117–166
Sabatier PA, Weible CM (2007) The advocacy coalition: innovations and clarifications. In: Sabatier P (ed) Theories of the policy process, 2nd edn. Westview, Boulder, pp 189–220
Sapiri H, Kamil A, Tahar R (2014) System dynamics approach as a risk management tool in analyzing pension expenditure: the case of Malaysian employees public pension plan. Singap Econ Rev 59(5)
Scharpf F (1989) Decision rules, decision styles and policy choices. J Theor Pol 1(2):149–176
Scharpf F (1999) Governing in Europe: effective and democratic? Oxford University Press, Oxford/New York
Schlager E (1999) A comparison of frameworks, theories, and models of policy processes. In: Sabatier PA (ed) Theories of the policies process. Westview Press, Boulder, pp 233–260
Smith K, Larimer C (2009) The new policy theory primer. Westview Press, Boulder
Thatcher M (1998) The development of policy network analyses: from modest origins to overarching frameworks. J Theor Polit 10(4):389–416

Karim Hamza is currently an external expert for the European Union Parliament, at the Parliamentary Research Services for Impact Assessment and European Added Value, affiliated as Senior Associate Researcher at Institute for European Studies in Brussels and a fellow researcher at Center for Technology in Government in New York. His research and publications focus on Smart Government Technology; and Sustainable Development. He authored different book chapters published by top international publishers like Springer and Palgrave. Additionally, he is affiliated as program committee with different international conferences specialized in Government Technology in Public Administration, such as the International Conference on Theory and Practice of E-Governance (ICEGOV). He is engaged in different projects with United Nation & European Union; most recently, he took part in the European Union project "EGOVPOLINET", which focused on standardizing ICT Technology standards and methodology for policy analysis and governance for public sector in European Union countries.

Sehl Mellouli is a full professor at the department of Information Systems of Laval University. His main research interests are related to smart cities, smart government, data analytics, and social media analysis. Professor Mellouli is one of the founeder of the Smart Cities Smart Government Research and Practice Consortium. He has over 70 publications in peer reviewed journals and conferences. Professor Melloui hold a Ph.D in computer science from Laval Univesity.

eGovPoliNet: Experiences from Building a Policy Informatics Research Community

Laurence Brooks, Marijn Janssen, and Anastasia Papazafeiropoulou

Abstract Policy-making in the digital age is an area which needs knowledge that can be found in communities that traditionally do not connect with each other. The creation of a research community is a challenging endeavour and needs to address both physical and online elements. In communities, groups of people share some common interests and are often facilitated by interacting with each other through the Internet and face-to-face meeting (Stewart, Behav Inform Technol 29(6):555–556, 2010). The activities should results in a sense of feeling of belonging to the communities. In this paper we outline the community-building activities of creating a policy informatics community which were part of the FP7 eGovPoliNet project. The eGovPoliNet project community organized community building event and provided a platform for sharing experiences and knowledge, which addresses the fragmentation of research communities, as well as the fragmentation among different disciplines, by building a common network where researchers from different disciplines and countries can interact. The aim was to engage different stakeholder groups to work together in exchanging ideas and information. The focus was on e-Government, information systems, complex systems, public administration and policy research and social simulation research communities, although persons from other research communities were also involved.

The eGovPoliNet community building process consisted of three phases namely: Initiating (period 1), Growing (period 2), and Sustaining (period 3) In the initiating phase (period 1), the European and international multidisciplinary research landscape was outlined by identifying the key players in terms of ICT for Governance

L. Brooks (✉)
Centre for Computing and Social Responsibility, De Montfort University, The Gateway, Leicester LE1 9BH, UK
e-mail: laurence.brooks@dmu.ac.uk

M. Janssen
Faculty of Technology, Policy, and Management, Delft University of Technology, P.O. Box 5015, 2600 GA Delft, The Netherlands
e-mail: M.F.W.H.A.Janssen@tudelft.nl

A. Papazafeiropoulou
Department of Computer Science, Brunel University London, Uxbridge, Middlesex UB8 3PH, UK
e-mail: Anastasia.Papazafeiropoulou@brunel.ac.uk

J.R. Gil-Garcia et al. (eds.), *Policy Analytics, Modelling, and Informatics*, Public Administration and Information Technology 24,
DOI 10.1007/978-3-319-61762-6_3

and Policy Modelling R&D and by determining the targeted communities. The growing phase (period 2) focused on developing the community by organising events by the project members and involving key players. New members were attracted by organising community building activities at various conferences, organising tracks, workshops, panels and PhD Colloquia. The sustaining phase (period 3) focused on maintaining the community through collaboration type events and PhD colloquia which resulted in further growth of the community. Finally, plans for future sustainability of the community were formulated. Community building is a time-intensive process, as people should gain a sense of belonging to the community. Different type of activities are appealing to different people. Following-up activities is essential, which involves people creating content and organizing events.

Abbreviations and Acronyms

BU	Brunel University London
CERTH	Centre for Research and Technology Hellas
COMPASS	The University of Auckland
CTG/SUNY	Center for Technology in Government/State University of New York
DG INFSO	Directorate General Information Society and Media
dg.o	Digital Government Society Conference
DGS	Digital Government Society
EC	European Commission
ECCS	European Conference on Complex System
ECEG	European Conference on e-Government
ECMS	European Conference on Modelling and Simulation
ESSA	European Social Simulation Association
ICDGS	International Conference on e-Democracy, e-Government and e-Society
ICEBEG	International Conference on e-Business and e-Government
ICEE	International Conference on e-Business and e-Government
ICEGOV	International Conference on Theory and Practice of Electronic Governance
ICT	Information and Communication technology
ICT4GOV	Information and Communication Technology for Governance
IFIP	International Federation for Information Processing
INNOVA	INNOVA SPA
IPR	Intellectual Property Right
IRSPM	International Research Society for Public Management
IS	Information Systems
ISA	International Sociological Association
IST	Information Society Technology
IT	Information Technology

MRSU	MOSKOWSKIJ GOSUDARSTVENNIJ OBLASTNOJ UNIVERSITET
MS	Milestone
PIDS	Project Information and Dissemination Service
PPP	Public Private Partnership
PPT	Power Point Presentation
PUC-PR	ASSOCIACAO PARANAENSE DE CULTURA APC
R&D	Research and Development
RC33	Research Committee on Logic & Methodology of ISA
RG	Rijksuniversiteit Groningen
SEO	Search Engine Optimization
TBC	To Be Communicated
TUD	Technische Universiteit Delft
TUK	Technical University Kosice
UCD	University College Dublin, National University of Ireland, Dublin
UKBRUN	Brunel University London
UKL	Universitaet Koblenz-Landau
ULAVAL	Universite LAVAL
UNU-IIST	UNU International Institute Software Technology UNUIIST
UTS	University of Technology Sydney
VOLTERRA	Volterra Partners LLP
VUB	Vrije Universiteit Brussel
WCSS	World Congress on Social Simulation
WP	Work Package

Introduction

The fields of governance and of policy modelling are fragmented as different disciplines meet at their own conferences in their specialist fields. Unfortunately, the distinct approaches to investigate the area have led to a wide set of definitions and understandings. Realizing that most policy problems are multidisciplinary, it is critical to stimulate scientists from different disciplines to align to define policy problems and develop effective research programs, therefore the FP7 eGovPoliNet aims to set up an international community in ICT solutions for governance and policy modelling. To achieve this, eGovPoliNet will build on experiences accumulated by leading actors bringing together innovative knowledge of the field (Majstorovic and Wimmer 2014; Janssen et al. 2015). This chapter focuses on the community building aspects in which the e-government, information systems, complex systems, social simulation and public administration & policy research domains were targeted.

Online communities can be defined as "computer-mediated spaces where there is a potential for an integration of content and communication with an emphasis on member-generated content" (Hagel and Armstrong 1997). Communities refer in general to a group of people who share some common interests, interacting with each other through the Internet and are facilitated by face-to-face meeting.

Communities must preserve intimacy among members and a sense of membership continuity to make the community sustainable (Hagel and Armstrong 1997). Communities consist of generated content but also of hooks such as calendar events and membership directories, which encourage increased community interaction (Jones and Rafaeli 2000). Therefore creating community building activities was an essential part of eGovPoliNet project.

Within the EU funded eGovPoliNet project, one element was to address the fragmentation of research community, as well as the fragmentation inherent in different disciplines, by building a common network where practitioners and researchers from different disciplines and countries could interact. This effort involved setting up the necessary communication structures for ensuring multi-disciplinary research, and development. The aim was to engage all stakeholder groups to work together, through two-way interaction between various scientific communities. The focus was on research rather than practitioners.

eGovPoliNet sought to establish closer working practices between the target groups by starting the discussion of future projects. In period 1 the focus was on recruiting the initial members, whereas the main activities for periods 2 and 3 were related to the organisation of face-to-face and virtual meetings and extending and integrating scientific communities. While period 2 focussed on expanding the community, period 3 focussed on continuity and sustainability of the community.

Achievements

The community building strategy is shown schematically in Fig. 1 and consists of three main phases. The first phase (period 1) outlined the European and international multidisciplinary research landscape by identifying the key players in terms of ICT for Governance and Policy Modelling R&D and by determining the targeted communities.

The second phase (period 2), which took approximately 18 months, focussed on growing the community by organising events by the project members and involving key players. New members were attracted by organising community building activities at various conferences, organising tracks, workshops, panels and PhD colloquia. The final phase (period 3) focussed on sustaining the community through collaboration events and PhD colloquia, leading to further growth of the community.

In the third phase community building activities at various conferences were organised resulting in collaborations among members from different communities. Three PhD colloquia were organised at three different conferences to stimulate interdisciplinary research in this field. Also workshops and panels were organised, bringing together people from different academic communities and practitioners.

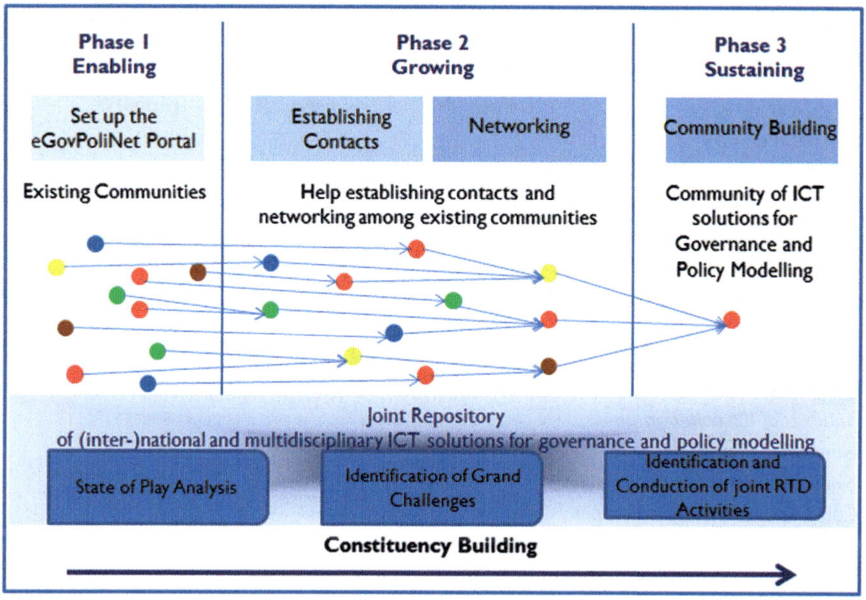

Fig. 1 Overall community building strategy for the eGovPoliNet project (based on eGovPoliNet (2001))

Over the duration of the project, the strategy of community building included online activities as well as face-to-face meetings:

- workshops and panels to engage researchers coming from different disciplines;
- joint papers, comparative cases and best practices;
- monthly virtual meetings between the eGovPoliNet partners to develop content, track events and coordinate activities.

Our premise was that content was needed to attract people and let them contribute to content development. Success depended on incorporating existing practices and exploring new practices. Progress was measured by collecting data at the end of each period and calculating the metrics for determining the status of the community, see Table 1.

In period 1 the initial members were recruited and several workshops and panels were organised. In periods 2 and 3 a large number of community building activities were conducted. The network grew considerably during the last period with a continuous stream of new members subscribing to the LinkedIn community. The increase in new members can be attributed to reaching *a critical mass*; once a sufficient size was reached it became more attractive for new members to join. In the second and third period, there was more collaboration among members than expected which resulted in a slight increase in the network density, despite the growth in members. In the second period the network closeness slightly increased, as there were many new members that did not know each other. In particular in the

Table 1 Overview of measures and values at the end of period 1, period 2 and period 3 (end of the project)

	Initiating (end of period 1)	Expanding (end of period 2)	Sustaining (end of period 3)
LinkedIn: number of members	267	1290	2740
Portal: number of members	0	53	163
Portal: number of unique visitors	0	219	612
Analysis of the social network			
Network size ('knowing')	160	485	513
Network size ('collaborating')	42	91	187
Network density	0.019	0.021	0.024
Network closeness (average geographic distance)	2.94	3.06	2.93
Analysis of the collaboration			
Number of joint papers	6	28	141
Number of workshops and panels	8 (2 panels)	12 (4 panels)	15 (4 panels)
Collaboration leading to a paper	4	28	59
Number of PhD colloquia organised	0	4	3
Number of PhD proposals at colloquia	0	33	13

LinkedIn community there were many members that did not participate actively and only followed discussions passively.

In period 3 the focus was on collaboration and ensuring the activities became focused on sustainability. Furthermore, non-eGovPoliNet partners were involved in the organisation of events to ensure that the activities would sustain after the project ended. As we were creating and shaping this new field, the need for having a solid knowledge base and a curriculum to translate the developments in existing education programmes arose. A book was edited laying the knowledge foundation for this **field** and a curriculum was developed which provides a reference for implementation educational programmes in this area. All these activities aimed to make a sustainable community that would continue after the project ended.

The policy-making 2.0 LinkedIn community has become a channel to announce activities and to share new ideas. Familiarity with members, perceived similarity with others, and trust in other members was demonstrated by Zhao et al. (2012) to be important in communities. In the community building process the familiarity among members has been built and members who were previously in different communities started collaborating. Not only continuing online, but also to keep organising physical meetings is important to keep the community running, which still continues.

The remainder of this chapter focuses on the development of the community and the events that are already planned for after the project ending. The structure is to first give an overview of the community building strategy in the next section. In the following section, the communities for period 1 and 2 are first discussed followed by the events and community developed and the metrics for measuring the growth. In the final section, the community building events that were planned for taking place after the project ended are presented.

Community Building Strategy

eGovPoliNet was funded by the European Commission (EC) 7th Framework Programme and was aimed at setting up an international community in ICT solutions for Governance and Policy Modelling. The consortium partners were drawn from various countries both within and outside of the European Union (EU), working together to share ideas, experiences and practices in the field.

eGovPoliNet Objectives

eGovPoliNet had five key objectives: (1) To establish a global multi-disciplinary digital participation, governance and policy modelling research community. (2) To integrate the currently fragmented research in digital public participation, governance and policy modelling. (3) To stimulate joint research and practice in the eGovPoliNet agreed research areas. (4) To disseminate eGovPoliNet findings among public governance and policy modelling stakeholders. (5) To provide a barometer for effectiveness for public governance and policy modelling in Europe and worldwide by establishing a corpus of knowledge and lessons-learnt resources to evidence what kind of projects have delivered what kind of results and have thereby been considered effective for digital public governance and policy modelling.

To achieve these objectives, eGovPoliNet built on experiences gained by leading actors bringing together the innovative knowledge in the field. The intended activities were to:

- establish a dynamic network of researchers;
- encourage international community building of relevant stakeholders working in relevant areas;
- encourage multidisciplinary community building;
- expand the social networking and Web 2.0, as well as exploit mass cooperation platforms for networking stakeholders;
- identify new tools and technologies, concepts and approaches, good and bad practices which help address complex societal issues and provide findings at the eGovPoliNet portal;
- make efficient the collection of feedback from public sector organisations on the contents provided by the eGovPoliNet portal.

The aim of the eGovPoliNet was to grow towards an interdisciplinary community. Therefore, criteria were developed to evaluate the development of the network (i.e. demonstrate that the community is growing and collaborating, see Janssen et al. 2012). The added value of connecting different actors, from different backgrounds and operating in different communities lies in the idea that they can learn from each other in terms of background, methods, projects, and practices. This section provides a brief overview of a strategy for expanding the network.

Community Building Objectives

Within eGovPoliNet, the aim for growing the network was: *Seeking collaboration between different actors that are from different backgrounds and operate in different communities.* The specific objectives were:

- Expand the network to include more disciplines and to get a better representation of under-represented disciplines;
- Encourage collaboration between researchers of multiple disciplines;
- Expand the network to include more practitioners/policy makers and to get a better view of the networks they provide access to;
- Encourage collaboration between researchers and practitioners;
- Encourage international (comparative) research (many countries were represented; this provided a great opportunity);
- Encouraging the joint organisation of workshops, panels, special issues etc.

These specific objectives were used to formulate the detailed strategy for community building.

Strategy for Community Building

Community building is ill-researched and there is a limited number of strategies available. Brown (2001) successfully applied three phases for community building in distance learning classes. Each of the phases should result in a greater degree of engagement.

1. *Making friends*: connecting on-line with others with whom the students felt comfortable communicating.
2. *Conferment*: making participants' part of a long, thoughtful, threaded discussion on a subject of importance after which participants felt both personal satisfaction and kinship.
3. *Camaraderie*: only achieved after long-term or intense association with others involving personal communication

Researchers and practitioners need to work together in order to tackle policy challenges by integrating different perspectives, developing comparative studies, and sharing their experiences. Zhang et al. (2011) identified a number of challenges.

1. a lack of shared interests and sense of urgency to collaborate;
2. issues with forming and maintaining personal relationships (Zhang et al. 2011; Kraut et al. 1986);
3. disciplines have different traditions, norms, values, whereas interdisciplinary research has relative fewer established outlets for publication

The more varied the potential members of the community are the more difficult it might be to create a coherent community. Of vital importance is that the potential members have something in common like shared interests, experiences, goals, values or vision (Brown 2001). Successful communities "are well-balanced systems that oscillate between exploring new practices and exploiting existing ones "(Probst and Borzillo 2008). There are 3 dimensions that are important for communities (Zhao et al. 2012):

1. The *structural dimension* can be reflected by the extent and quality of relationships and familiarity. Familiarity is "the extent to which members of a community know each other based on interaction" (Lu et al. 2010). Familiarity with other community members is viewed as a condition for developing the community.
2. *The relational dimension.* This dimension looks at personal relationships between individuals which develop through repeated interactions between members. This contributes to building trust among participants. In the community building activities, the fostering of personal relationships is key to growing the community.
3. The *cognitive dimension* relates to perceived similarity among members. Similarity is defined as "the extent to which that community members perceive sharing common characteristics such as shared goal and vision one perceives with other members" (Lu et al. 2010). Similarity is important, but members should also be sufficiently different to foster variety and to add value to the eGovPoliNet community by bringing in their expertise and knowledge.

By having a focal point on policy-making problems as experienced by practitioners, a clear and shared objective is created in which different disciplines should contribute to the same practical challenge. The forming and maintaining of personal relationships is accomplished by having online and face-to-face community building. By having a 3 year strategy consisting of various phases the difference in values should become accepted.

eGovPoliNet therefore exploits online and face-to-face meetings to connect and establish the community. Physical meetings will mostly serve to strengthen the community through social relations. These meetings were organised in conjunction with important conferences and other events relevant to the community and served as points of reference, where results and information gathered in the recent period were discussed, structured and amended, and plans for the subsequent period were confirmed from the work plan or revised accordingly. Regular virtual discussions (online and by phone) were used to support the achievement of eGovPoliNet's objectives to strengthen the community.

A key part of the strategy was that partners seeks collaboration with other parties. For each partner, it was expected that they recruited additional members to the (online) network and that each partner organised a workshop (at different conferences and events) with people from other communities. In the first project period, partners invited people from other communities to a workshop or event. In the second and third periods, workshops were organised in communities to which people from other communities were also invited.

Strategy of Events for Community Building

eGovPoliNet aimed at letting the community grow. Relevant players from various communities were targeted. The community building activities were always targeted at a minimum of two communities. The event ensured that people from at least two different communities are involved (see Fig. 2), to ensure that these communities started to get to know each other and joint activities were stimulated. Figure 2 summarises the events organisation protocol. Each event should result in a *measurable output* of the event and report this in the template. The mechanism used to track this was a 'community building template' (see Appendix: Community Building Activities). Ideally the template should be filled in before *and* after each activity. However, in practice the template was often only filled after the activity took place. The advantage of filling in the template before the activities took place was that it can be used to explain, share and discuss the plans. After the activity the template should have been filled in to evaluate the actual impact (this must be very specific such as the list of participants, outcomes like joint papers, cases etc.). The community building reports delivered by partners were used for the social network analysis and collecting other metrics.

The reports contain the participants list, sometimes pictures of the events and titles of the papers/abstracts/PhD proposals. This provided us with insight into who attended the events and what the direct effects of the events were. There might be indirect effects (for example writing joint project proposals) which are harder to measure and are only known afterwards (such as, when a project is accepted).

The basic idea of realising this strategy is that each partner organised community building activities. Activities target always at least two communities to bring them together. For these activities *persons* (name, email address, affiliations) were identified from the communities that should be involved. If papers, abstracts or PhD proposals were part of the output, then these were uploaded in the portal whenever possible (i.e. sometimes copyright issues prevent this). The ambition is that at least the title, author(s) and abstract are uploaded to enable community members to know each other.

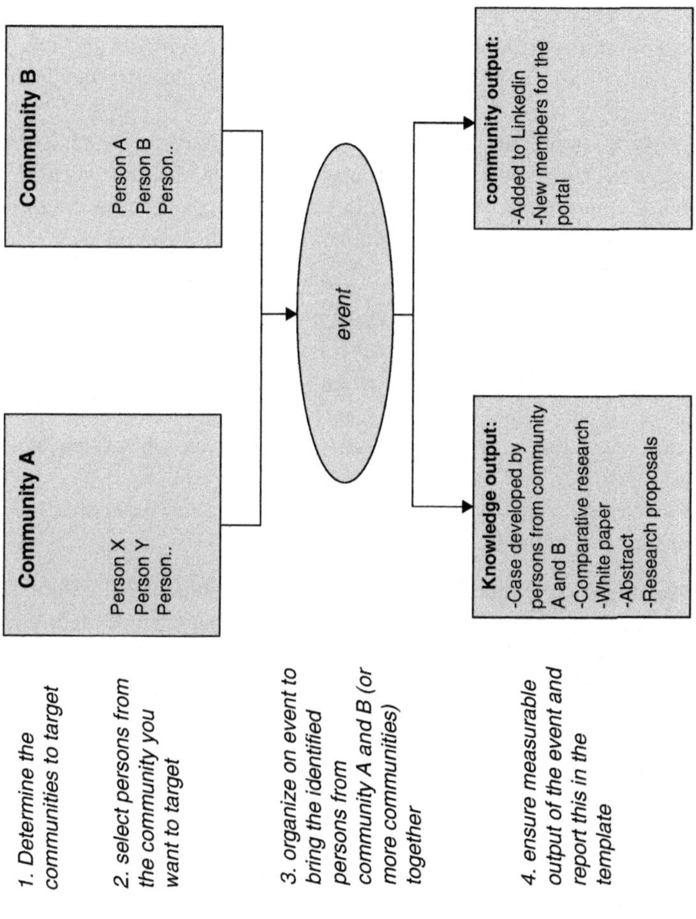

Fig. 2 Event organisation protocol

Online Community Building Strategy

The online community consists of two types of community building focus points. One activity was open to everybody and was used to create awareness of the network, show some of the activities and stimulate discussions. For some persons this would be what they desire, whereas others wanted to collaborate with each other in-depth. Therefore, the second online community building was focussed on in-depth knowledge exchange, the sharing of findings and detailed activities. The results of the community building activities should be that members are confident in contributing, feeling valued and feel part of the community and that they learn from each other.

The LinkedIn eGovPoliNet community group 'Policy Making 2.0'[1] aimed at attracting a large user base of people who are interested in bridging scientific communities. Online community building requires the setting of some conditions to make it work. We used the following guidelines (based on Brown 2001).

1. Strategies to create an environment that fosters openness, respect and trust.
2. Create interest, support, sincerity, and understanding of the existing disciplines.
3. Share relevant experiences as well as information helpful to others.
4. Formulate responses positively, even when provocative ideas and opinions are presented.
5. Provide timely feedback, provide support and stimulate discussion by asking questions.
6. Continue threaded discussions that were going and to keep them alive.
7. Communicate with individuals directly if necessary.

In the beginning, the community was kept small to enable the eGovPoliNet partners to create content and prepare. In this way the community can be made attractive before inviting people and having a large user base for which limited content can be offered. In future years the goal was to boost the online efforts and all partners were asked to follow a plan and contribute in four different ways.

1. To post a comment concerning the eGovPoliNet related research one is working on. This could be an example, development, reference to relevant report or an open discussion on a specific topic.
2. To recruit somebody from an external research community to post something. Community building requires the involvement of other organisations than those who are part of the consortium. The member should recruit somebody from another community and ask them to post something in that week
3. This is similar to 2, with the exception that this is targeting the practitioners' community. Somebody from practice should be recruited to post something.
4. Comment on a posting (contribute to discussion on this topic and make it lively).

[1] See https://www.linkedin.com/groups/4165795 (last accessed 11/05/2016).

These actions should ensure that the community shows activities and is attractive. Once there are activities of non-eGovPoliNet partners the community should become self-sustainable.

The *portal* is aimed at stimulating sharing among eGovPoliNet members who are actively working on integrating communities by working on best practice and research crossing communities. Traditionally, people tend to do things in their own disciplines. Coalitions having participants from various disciplines might breed new ideas, have more problem solving capacity and view the problems from different disciplines. The portal is first filled with more information before a large number of members was invited. A certain critical mass of knowledge is necessary before these can be developed.

To stimulate this collaboration and in-depth knowledge sharing, virtual meetings took place each month. In these meetings two partners were asked to give a short presentation of their contribution as a case, paper or other community building activities. The virtual meeting space (Clickmeeting) offers a collaborative, interactive, and mobile learning environment. It helps to create virtual classrooms, offices and meeting spaces that offer the opportunity to talk (voice) and see each other (video), present slides, chat and work together on a shared whiteboard. These facilities should stimulate collaborations among eGovPoliNet members. The meetings were recorded, minutes were made and the minutes, slides and recordings were stored in the shared workspace.

The basic idea was that eGovPoliNet partners would contribute in cooperation with someone from another community (practitioner, scientific). In this way the activity itself already contributed to the community building activities. All activities were summarized using the template presented in Appendix: Community Building Activities Template. This enabled partners to understand the events and to determine its impact. Some of the results of these activities were stored and made available in the portal. This provided the content of the portal to make it attractive for others to join. The basic idea was that others who used the content would also start contributing to the portal and the activities would become self-sustainable (after period 3).

Face to Face Community Building Strategy

Apart from the online community building there were face-to-face meetings to share ideas, to gain understanding and appreciation of other disciplines. Therefore monthly online meetings were organised in which partners discussed their activities face-to-face. Physical meetings were used to build the community through social relations. These meetings were organised in conjunction with important conferences and other events relevant to the community and served as points of reference, where results and information gathered in the recent periods were discussed, structured and amended, and plans for the subsequent period were confirmed from the work plan or to be revised accordingly.

An important task of the face-to-face community building meetings was the organisation of PhD Colloquia. PhD research provides the basis for any scientific field. Stimulating research in this field, providing feedback, and ensuring the various disciplines are considered in the research to provide a foundation for the eGovPoliNet field.

Resulting outputs were the results of both online and offline community building strategies. Output was created by members of different communities who used the output to work together. The type of output typically contained comparative work which compared practices or compared efforts within communities. This was aimed at analysing differences and similarities among communities.

Other output was joint work in which persons from different communities collaborated with each other. This had different forms, such as a description and analysis of a policy-making practice, writing a white paper, writing a scientific paper to be published at a conference or journal or a special issue containing input from different disciplines.

For each output contribution, the following three requirements should be satisfied.

1. The work should have been conducted within eGovPoliNet;
2. The work should contribute to the objective of eGovPoliNet community building;
3. The work should result in community building (outcome).

The latter requirements should be described by each community building activity. How it contributes to the community building. Finally, having tracks, special issues and writing of proposals between members of the formerly fragmented communities demonstrated the collaboration between various communities and should ensure long term sustainability.

Community Development

An overview of the community development over the years is presented. Periods 1 and 2 are summarised, whereas the events for period 3 are presented in detail.

Targeted Communities

To mitigate the risk of targeting a too broad range of communities which are less relevant, the focus has been on targeting five communities that provide the core field for ICT-enabled Policy-making. A summary of the main communities targeted is given in Table 2.

Table 2 Main communities targeted

Main communities	Contributing insights to the domain
E-government (EGOV)	E-government is the interdisciplinary field that tackles ICT and public administration aspects in a broad sense (this includes integrated service delivery, web 2.0, etc.). E-government is considered to be interdisciplinary by nature and is open for eGovPoliNet type of work which needs elements from public administration, policy-making, simulation, and complex systems. Within this field the IFIP WG8.5 working group on Public administration & ICT, international community on theory and practice of governance (ICEGOV) and digital government society (DGS) were targeted
Information systems (IS)	Information systems bridges business and computer science and studies both the technical system as social system. The Association for Information Systems (AIS) serves society through the advancement of knowledge and the promotion of excellence in the practice and study of information systems. This field is targeted by focussing on the European Conference on Information Systems and UKAIS conference
Complex systems (CS)	The study of systems built of individual agents that are capable of adapting as they interact with each other and with an environment, and especially the attempt to understand how the individuals affect the system-level responses (Auyang 1998). In recent years, CS has attracted much interest in management and organisational related literature. Complex systems view organisation as an entity that emerges over time into a coherent form, and adapts and organises itself without any singular entity deliberately managing or controlling it
Public administration & policy research	Political science studies the political system and political behaviour of state, government, and politics. It aims to analyse and understand, revealing the relationships underlying political events and conditions. Public administration houses the implementation of government policy and an academic discipline that studies this implementation and that prepares civil servants for this work. Public administration is "centrally concerned with the organisation of government policies and programs as well as the behaviour of officials (usually non-elected) formally responsible for their conduct". The focus is on International Research Society for Public Management (IRSPM) and Association of Public Administration (APA)
Social simulation	Modelling, simulation and visualisation provides the instruments and tools for being able to gain an understanding of the phenomena and being able to visualise what is going on. The focus of these communities is often not on policy-making, but on advancing the modelling constructs and visualisations. The focus is on The Society for Modelling and Simulation Europe (SCS)

The 'Community' in Period 1 (start)

A qualitative and quantitative survey was conducted during the start of the project. The survey consists of two parts: first, for each respondent it inventories disciplines, core communities, known communities, collaboration communities, research topics, methods used and expectations of the project. Secondly, it inventories relationships with members of the international network, serving as the initial measure for the social network analysis of the survey that will be repeated multiple times in the course of the project.

Social network analyses were carried out using *NodeXL* (Smith et al. 2010), an MS Excel based open source based tool which has been used for conducting similar analysis (Welser et al. 2009), and has integrated visualisation options and can be learned within a short timeframe (Hansen et al. 2011). Figure 3 shows the network from the start as analysed using NodeXL. The nodes represent the people who are part of the eGovPoliNet network and their relationship with each other. This graph shows that most of the people who took part in the project then already knew each other, or at least several other people in the network. There were also exceptions, i.e. people who only knew a few other people.

Figure 4, shows the network after period 1. The data is based on the participants of the events organised. In the first period key people in the targeted communities were identified to connect to and events were organised to facilitate this. Figure 4 shows that several communities have been connected to the core of eGovPoliNet by focussing on key stakeholders (linking pins). For example, the red nodes are the information systems community, which shows that four eGovPoliNet partner representatives are connected to this community (ie, the four lines originating from the centre of the red cluster) and six key people from this community are involved (ie. the six red dots in the cluster).

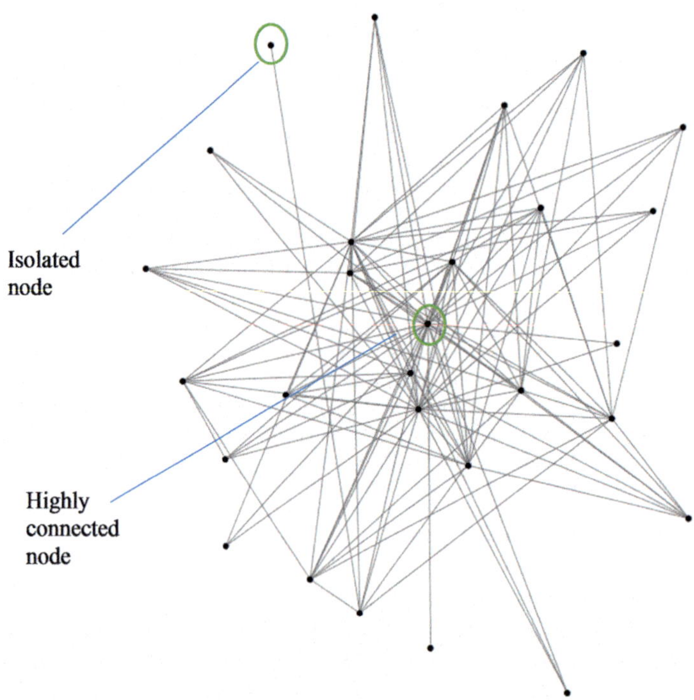

Fig. 3 Social network analysis of the eGovPoliNet members at the start of the project

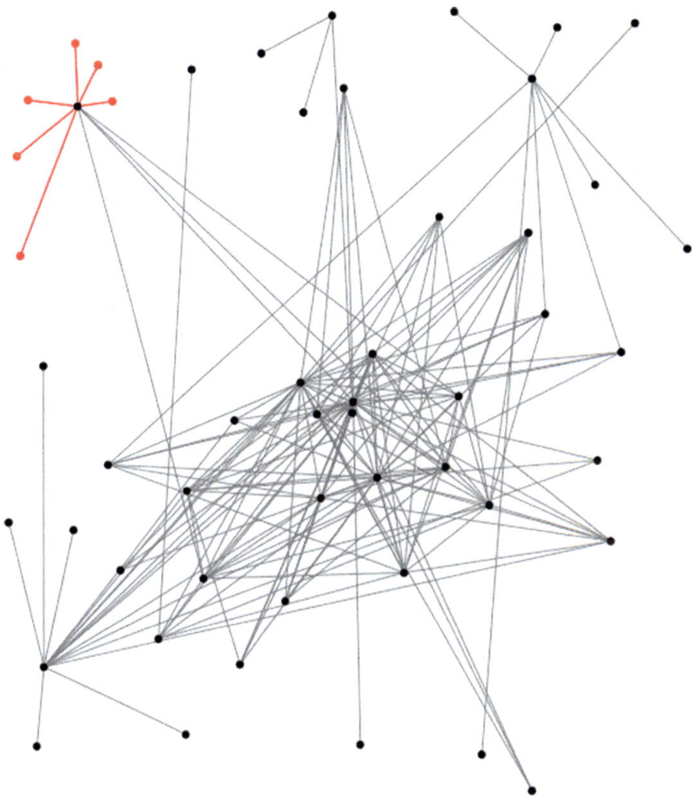

Fig. 4 Social network analysis of the eGovPoliNet members after period 1

Overall, there are seven groups/communities pre-defined in the analysis, (1) eGovPoliNet (the partner representatives), (2) eGovernment, (3) Information Systems, (4) Complex Systems, (5) Public Administration & Policy research, (6) Social Administration (only those who are connected are included in the analysis, as not all persons within these communities are known and can be added) and (7) Practitioners (those who participated in events). The connection to the e-government community is strong, whereas the connection to complex systems community is the weakest.

The 'Community' in Period 2

In period 2 a variety of events were organised. In Fig. 5, as in the figure from period 1 the red nodes are the information systems community, which shows the growth from this community into the eGovPoliNet community. It shows an increase in ties to the starting communities (on the right hand side in the figure). The graph also shows that only a few members from this community are connected to other communities (i.e. the one at the top is linked to the complex systems community).

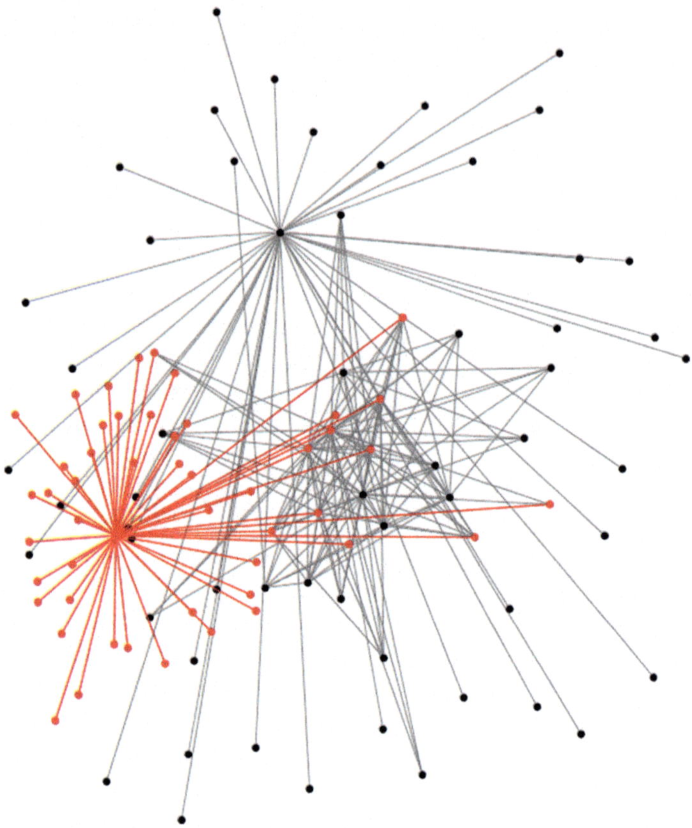

Fig. 5 Social network analysis of the eGovPoliNet members after period 2

The *network density* is the proportion of direct ties in a network relative to the total number of possible ties (Emirbayer and Goodwin (1994) cited in Zhang et al. 2011). Although we expected that the network density would have decreased due to the growth of the network, it did not. There was more collaboration among members than expected, which results in a slight increase in the network density.

Collaborative ties between actors refer to things such as writing papers together, writing grant proposals together, collaborating in a project. Just knowing each other is enough for having ties, but not sufficient for collaborative ties. Whereas the network closeness is calculated by the distances between pairs of actors (Hanneman and Ridddle 2005). The network closeness has slightly increased, as there are many new members that do not know each other. In the LinkedIn community there are many members that do not participate actively and only listen. Community building activities and collaboration in period 3 should result in a decrease in the distance among members, so a more coherent community will be created.

Community Building in Period 3

In addition to recruiting persons at the individual level, a large number of community building activities has been organised. Three types of community building activities can be distinguished:

1. Community building events: aimed at letting the community grow;
2. Community building events for collaboration: Aimed at stimulating collaboration within the community;
3. Community building events for PhD student: PhD Colloquia aimed at involving PhD Students in this field;

Community Building Events

This type of community building events are aimed at recruiting members for the community and keeping existing members active. At all events the participants have been asked to fill in a presence list including their name and email address. These lists are used to invite participants to the LinkedIn community and become active. In total 485 attendees participated in the events (see Table 3). From these persons, 29 became new community members for the LinkedIn group. Indirectly there might be more members, but we are not able to trace this. Also a large number of participants were already a member as most events are a continuation of the events organised in the previous period (and as such could not be added again).

Please note that one person can be involved in multiple events. For example the panel and plenary discussions at ICEGOV has likely overlapping audience as the total number of conference attendees was about 350 people. We tried to take this into account, but had to make guesses as no detailed attendance list was available. In period 3 the number of new participants added to the LinkedIn community due to these activities is less than in period 2, as already many attendees have already become members previously.

Community Building Events for Collaboration

The community building events for collaboration were aimed at stimulating the writing of papers by members of different communities and presentation of papers from one community to another community. In total, 123 papers and 5 journal papers have been developed by persons from different communities (mostly non-partners). Apart from the collaboration, these events also resulted directly in attracting 36 new community members, mainly in the field of policy research, for the LinkedIn group. Indirectly there might be more members, but we are not able to trace this. Table 4 gives an overview.

Table 3 Community building events organised in period 3 and assigned to targeted communities

Event	Impact	E-government	Information systems	Complex systems	Public administration & policy research	Social simulation	Total number of participations	Participants added to The community (direct)
Panel at ICEGOV	LinkedIn members, minutes	X			X		100	15
Plenary discussion at ICEGOV	LinkedIn members, minutes	X			X		183 (5 panellists)	Unknown
tGov workshop presentation	LinkedIn members, minutes	X	X	X	X		51	3
Curriculum development policy informatics	curriculum	X	X	X	X	X	32	3
Samos summit for policy-making	presentation	X		X	X		22	(All were members)
ECIS panel—information systems in the public sector	LinkedIn members, minutes	X				X	12 (5 panellists)	3
panel at Dg.o—understanding and improving the uptake and utilization of open data	LinkedIn members, Collaboration	X			X		16 (2 organisers)	2

SKIN workshop—joining complexity science and social simulation for policy (SKIN 3)	LinkedIn members, Collaboration						44	1
NASPAA panel—#OpenData #BigData: data, big and small, in the public affairs curriculum	LinkedIn members, Collaboration				X		35 (4 panellists)	2
Total		7	2	3	7	2	485	29

Table 4 Types of community building activities, number of collaborative engagements of community members along targeted communities and impact achieved

Event	Types of community building activity	E-government	Information systems	Complex systems	Public administration & policy research	Social simulation	Total number of participations	Participants added to the community (direct)
Track at ICEGOV	Collaboration, papers, abstract to portal	X			X		6 papers 14 authors	2
Track at dg.o	Paper, proceedings	X	X				10 papers 30 authors	None[a]
SKIN workshop—joining complexity science and social simulation for policy (SKIN 3)	Papers proceedings	X		X			16 Papers 5 posters 23 authors	17
Track at the 17th international conference “Internet and modern society” and its part “e-Governance in Information society”	Collaboration, new LinkedIn, abstract to portal		X			X	54 papers 63 authors	9

Papers at IFIP EGOV/ePart conference	Papers proceedings	X		X		X	7 papers 22 authors	None[a]
t-gov workshop—Co-Creating Public Services of the Future:	EU project meetings	X	X	X		X	12 papers 31 authors	1
Total		5	3	3		4	113 papers 175 authors 5 posters	29

[a]Those engaged were already members

Community Building Events by Having Special Issues of Journals

The previous activities showed that there are many conference publications. In the short life time of this project, we managed to have two special issues with 9 peer-reviewed publications, as listed in Table 5.

Policy Informatics Curriculum

As this new field and knowledge base are being created and shaped, the need for an academic curriculum has arisen. There are no standard curricula and developing a curriculum demands input from various disciplines. A workshop was held to explore integration of data-intensive analytical skills in public affairs education. This workshop should provide the basis for the uptake of new developments in existing programmes.

The workshop "Policy Informatics in the PA Curriculum: A workshop to explore integration of data-intensive analytical skills in public affairs education" was held on 09 May 2014 at the Center for Technology in Government (CTG), State University of New York (SUNY), University at Albany. The event was supported by a grant to CTG from the National Science Foundation (NSF) and by the eGovPoliNet Consortium. The workshop had the following goals:

1. To understand the analytical needs of policy makers and program managers.
2. To share approaches to educating public administration and policy analysis students in the types, uses, and limitations of policy informatics.
3. To explore new methods for policy informatics education.
4. To consider curriculum recommendations for public affairs schools.

Public administration and public policy curricula need to confront these trends and develop ways to train professional analysts and managers to understand and address them. This workshop showed the needs and opportunities in the emerging data-intensive science and decision-making environment and explored ways to integrate them into public affairs education.[2]

Springer Book "Policy-Practice and Digital Science"

To take advantage of these developments in the digital world, approaches are changing and new methods are needed, which are able to deal with societal and computational complexity. This requires the use of knowledge originating from various disciplines including public administration, policy analyses, information systems, complex systems and computer science. All these knowledge areas are needed for policy-making in the digital age and were integrated in the book 'Policy-Practice

[2] Further information about the workshop can be found at http://www.ctg.albany.edu/news/events?eventID=72 (last access: 11/05/2016).

Table 5 Number of peer-reviewed publications in two special issues

Event	E-government	Information systems	Complex systems	Public administration & policy research	Social simulation	Total number of participations	Participants added to the community (direct)
International Journal of E-Government Research (IJEGR), Special Issue on Policy-making: a next challenge in e-government research	X	X	X	X	X	5 papers 21 authors	None
Journal of Policy Analysis and Management (PAM) Special Issue on policy informatics			X	X		4 papers 18 authors	7
Total		2	3	2	1	9 papers 39 authors	7

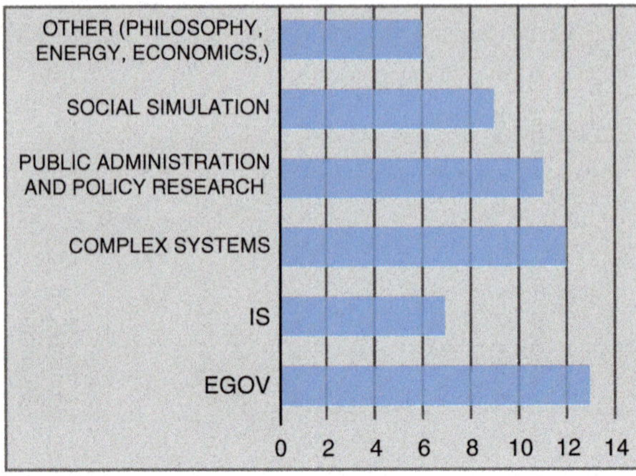

Fig. 6 Overview of the disciplinary background of the chapter authors

and Digital Science" by Janssen et al. (2015). The aim of this book was to provide a foundation for this new interdisciplinary field, in which various traditional disciplines are blended together with the curriculum development. The book provides a foundation for this growing field.

In total 54 different authors were involved in the creation of this book. Some chapters have a single author, but most of the chapters have multiple authors. The authors represent a wide range of disciplines as shown in Fig. 6. The focus has been on targeting five communities that provide the core field for ICT-enabled governance and policy making. A sixth category was added for authors not belonging to any of these communities, such as philosophy, and economics. Figure 6 shows that the contribution of authors are evenly distributed among the communities. A large part of the authors can be classified as belonging to the e-government/e-participation community, which is by nature interdisciplinary.

PhD Colloquia

As in period 2, the PhD colloquia were organised at conferences in the e-government community. These types of conferences are interdisciplinary by nature and the organisers were open for interdisciplinary research. Furthermore, these conferences attract persons coming from various communities, and are of interest for persons from various communities.

The Role of LinkedIn and the Portal in the Community

Community building was supported by a LinkedIn group and by developing a portal. Figure 7 presents an overview of the discussions started and commented on in the LinkedIn community, starting from November 2011. A steady initiation of

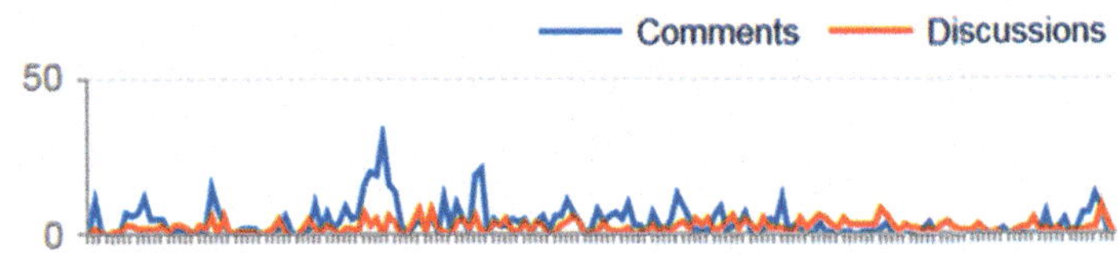

Fig. 7 Overview of discussions and comments in the LinkedIn community over time

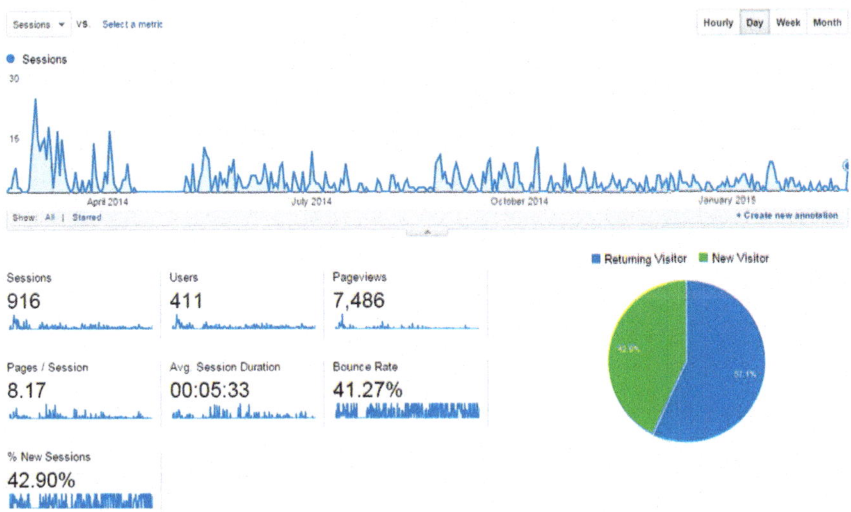

Fig. 8 Overview of the traffic of the portal

discussions is shown, whereas the responses (comments) to the initiated discussions vary a lot. Partly this can be attributed to the topic, as a discussion can include the announcement of an event, the sharing of new work, etc. The sharing of ideas and discussion about ideas is posted less frequently. This also shows that a large number of members are ‘listeners’, they follow the discussions but do not actively contribute.

In the LinkedIn community a ‘network effect’ or ‘network externality’ seems to have occurred. Network effects or *network externalities* refer to the dependence of the value of a good or service on the number of other people who use it (Katz and Shapiro 1985). A positive network externality happens as being part of a community becomes more valuable as more users joined the eGovPoliNet community. The network effect can explain the ongoing growth of the LinkedIn community, although our efforts were not focused on letting it grow anymore in the final period. The large number of members will ensure that there is a sufficient number of participants to maintain interactions and participation.

The google analytics for the project portal for 29 May 2014 to 25 September 2014 are shown in Fig. 8. It shows the activities and the users. The analytics in the figure at the top shows the new persons entering the community. At the beginning there is a large influx which slows, and then the events resulted in a more steady

Table 6 Overview of the LinkedIn and Portal communities

	Initiating (end of period 1)	Expanding (end of period 2)	Sustaining (end of period 3)
LinkedIn: number of members	267	1290	2740
Portal: number of members	0	53	163
Portal: number of unique visitors	0	219	612

Table 7 Collaboration at the end of period 3

	After period 1	After period 2	After period 3
Number of joint papers	6	28	141
Number of workshops and panels	8 (2 panels)	12 (4 panels)	15 (4 panels)
Collaboration leading to a paper	4	28	59
Number of PhD colloquia organised	0	4	3
Number of PhD proposals at colloquia	0	33	13

inflow of new members. Table 6, below, shows an overview of the incredible growth of the community in terms of LinkedIn members and portal members and visitors. This tables shows that the community has considerably developed over time.

Analysing the Community at the End of Period 3

The collaboration is analysed based on the metrics determined in period 1. The number of joint papers is calculated by counting the 113 conference papers, 9 journal papers, and 19 book chapters which resulted in 141 joint papers. The previous tables show that 15 events were organised from which 4 are panels. Some events took place at the same outlet (for example there were 2 panels and a track at ICEGOV). The observed collaborations resulting in a paper were estimated at 59. As there are 141 joint papers the actual collaboration should be higher (Table 7).

Based on the events and collaborations a social network analysis (SNA) was conducted. As with the figures for period 1 (Fig. 4) and period 2 (Fig. 5) the red nodes are the information systems community. Fig. 9 shows the community at the end of period 3 and although not represented directly on the figure, it also reflects that more and more members from this community are connected to other communities. Indeed there might be even more connections which were not administrated and fall outside our scope of analysis (e.g. events organised by others, events in which attendee lists were not completed and conference/journal papers not indexed).

Figure 10 shows the social network and the members of the eGovPoliNet community. It shows that the network has considerably expanded beyond the original eGovPoliNet members which are depicted using the red colour, whereas the

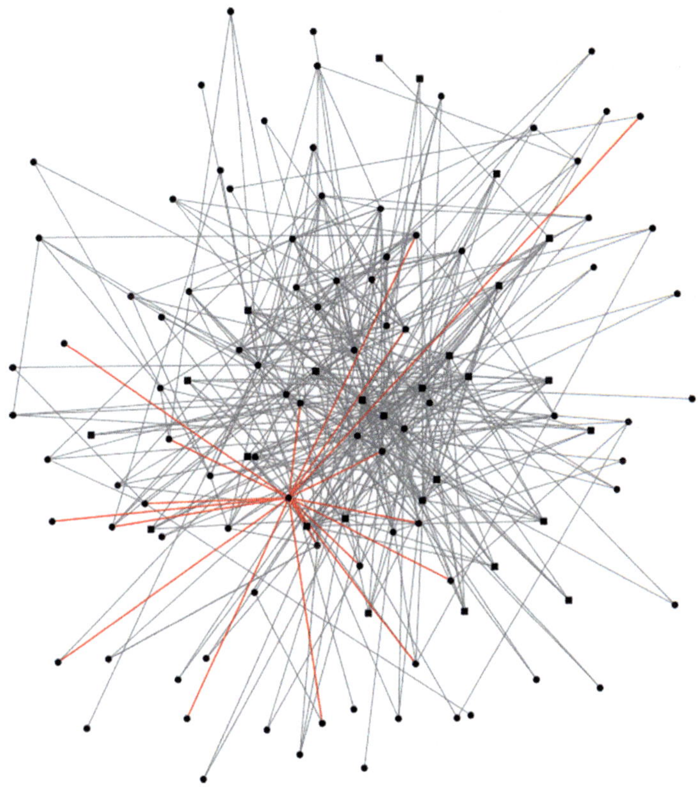

Fig. 9 Social network analysis of the eGovPoliNet members after period 3

information systems persons are blue. The circle on the outside shows the persons who are 'listeners', who do not actively engage in content-generation, but consume the context and incidentally contribute to a discussion on LinkedIn. There are more persons who are only 'listeners' than visualised in this figure. We did not opt for including them and limited the analysis to 477 persons from which 385 persons can be classified as active.

There are various other *limitations* in the SNA analysis. First, we did not include the visitors to the portal. Second, sometimes persons cannot easily be allocated to a certain community. Some persons fit within two or even more communities which makes it difficult to determine how communities are collected. Third, collaboration can involve papers having multiple authors. Only key authors might be open and collaborate with other members, whereas some authors might only provide their expertise. Nevertheless all authors are included in the analyses. Finally, we had two events in period 3 in which the attendees list were not collected and we had to guess the number of attendees.

Table 8 shows the development of the community. The *network size* is calculated by counting the number of different persons who attended the events over the years.

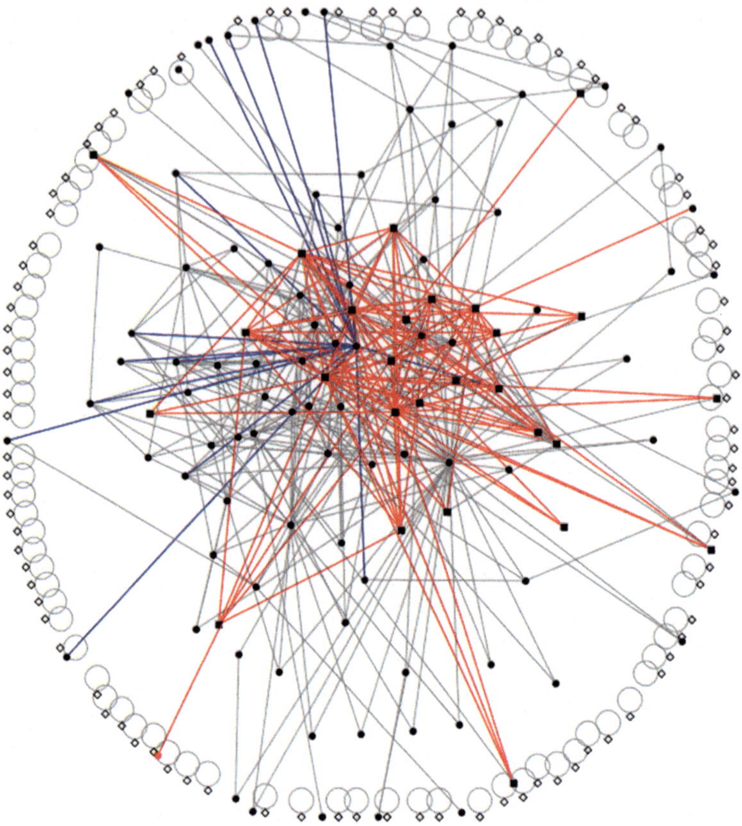

Fig. 10 Social network and the eGovPoliNet members after period 3

Table 8 Social network at the end of each period

	Start of the project	Initiating (end of period 1)	Expanding (end of period 2)	Sustaining (end of period 3)
Network size ('knowing');	0	160	485	513
Network size ('collaborating');	0	42	91	187
Network density;	0	0.019	0.021	0.024
Network closeness (average geographic distance);	0	2.94	3.06	2.93

The network size for knowing is 513 which is calculated by adding up 485 persons attending events and 28 person attending the PhD colloquia. Papers have 1 or more authors, whereas PhD proposals have only one author (PhD supervisors are not included in the network size).

The *network density* is the proportion of direct ties in a network relative to the total number of possible ties (Emirbayer and Goodwin (1994) cited in Zhang et al. 2011). Table 8 shows the changes in network density over the periods. The network grew in period 3, but at the same time there are many collaborations among members which resulted in a slight increase in the network density. In the LinkedIn community there are many members that do not participate actively and only listen. If the 'listeners' were left out this number would be much higher.

By collaborative ties between actors we mean things such as writing papers together, writing grant proposals together, collaborating in a project. Just knowing each other is enough for having ties, but not sufficient for collaborative ties. Whereas the *network closeness* is calculated by the distances between pairs of actors (Hanneman and Ridddle 2005). Table 8 shows that the network closeness has decreased, as the links between core members are closer, even for those who are new and entering the network.

Community Building Activities Post-eGovPoliNet

In the first period, a number of community building activities have taken place which were focussed on analysing and understanding the community. In the second period, the community building activities focussed on expanding the community. In the third period, the focus was on sustainability; this was done by ensuring that persons from outside the eGovPoliNet project were involved in the organisation of events. In period three, the focus was on continuing key events and enlarging the impact of these events. Several of the events organised have become 'accepted' by these conferences and considered as 'belonging' as part of these conferences. The conference organised provided invitation for running the track for another year without having to ask. This has resulted in a large number of events that are continued after the project ended.

Table 9 indicates the planned community events for collaboration. These are more than in period 2 to ensure that community members collaborate. Furthermore, the same outlets as in period 2 are targeted as this ensures a recurring presence. The idea is that people will get to know the events and will start considering this as a periodically occurring event. Apart from eGovPoliNet partners, other people will be involved in the organisation of these events to ensure sustainability after the project ends.

Although the number of events varies per community, this does not mean that the impact in any one community might be less. For example, the eGovernment Policy/Policy informatics minitrack at AMCIS might have a huge impact as AMCIS is visited by 700–1000 information systems experts.

Three PhD colloquia were organised by eGovPoliNet partners at primarily e-government type of conferences. Students could submit a PhD proposal to be presented at the colloquia and from these appropriate proposals were selected and accepted for presentation and discussion. The idea was to attract PhD students from all communities to those colloquia to ensure that PhD students from various disciplines meet each other in a multidisciplinary setting. Table 10 indicates the plans for future PhD colloquia.

Table 9 Indication of planned community events to sustainably support collaboration of the Policy Community

Event	Expected impact	E-government (EGOV)	Information systems (IS)	Complex systems	Public administration & policy research	Social simulation
Track at ICEGOV	Collaboration, abstract to portal	X			X	
Track/ at Dg.o	Collaboration	X			X	
ESSA—social simulation	Collaboration			X		X
eGovernment Policy/Policy informatics minitrack at AMCIS	Collaboration	X	X		X	
Joining Complexity Science and Social Simulation for Policy (SKIN 3)	Papers in proceedings			X		X
Policy Modelling and Policy Informatics Track at IFIP EGOV/ePart	Papers in proceedings, Platform for networking	X		X	X	
tGov workshop	EU project meetings	X	X	X	X	X
Total		5	2	4	5	3

Table 10 Indication of planned PhD colloquia of the sustained Policy Community

Event	Expected impact	E-government (EGOV)	Information systems (IS)	Complex systems	Public administration & policy research	Social simulation
PhD colloquium ICEGOV	Collaboration, abstract to portal	X	X	X	X	X
PhD colloquium at dg.o	Collaboration, abstract to portal	X	X	X	X	X
PhD colloquium at IFIP EGOV/ePart	Collaboration, abstract to portal	X	X	X	X	X
Total		3	3	3	3	3

Conclusion

This chapter has detailed the creation and growth of an academic community, through the stages (phases) of the process of the lifecycle of the eGovPoliNet project and the development of a ‘policy informatics’ community. Creating the community was done in three phases: Initiating (period 1), Growing (period 2), and Sustaining (period 3). The experiences show that developing a community is time-intensive and could only be successful by organizing many online and physical activities. Key to the success is the organization of physical meetings in which people from different disciplines come together and ensuring sustainably by retaining key persons which function as linking pins between communities. Developing a community has to be done carefully, and in the multi-phase way, to initially attract people to join the community and then sufficient interaction/relationship development to enable people to stay. While this is tricky to measure, various metrics have been employed, including the social network analysis of the Policy Modelling 2.0 LinkedIn group; this shows that there is some development across existing communities. Creating communities is not easy, and in some there is much that pulls apart, as much as pushes together, but in this context, the eGovPoliNet can justifiably claim to have made a good start on the development of a Policy Informatics community, whether and how it survives, only time will tell.

Acknowledgments The eGovPoliNet Coordination Action (Grant Agreement N°288136) was co-funded by the European Commission, Information Society and Media Directorate General, under the Seventh Framework Programme (FP7), Cooperation Theme three, “Information and Communication Technologies”. The authors wish to acknowledge the Commission for their support, the efforts of the partners and the contributions of all those involved in eGovPoliNet.

Appendix: Community Building Activities Template

Field	Details (Your Data Here!)
Description (to be filled in before the event)	
\Id	This is a unique identifier of the activity.
Title	
Topic	Description (Who, Why, What, When, Where, How)
Purpose	The purpose of the event related to the objective of eGovPoliNet community building. For example the purpose is 1) participation and/or 2) integrate the currently fragmented research by involving both policy-researchers as well as complex systems researchers)

(continued)

Field	Details (Your Data Here!)
Communities involved:	e.g. complex system researchers and policy-makers from government
Type	Knowing or collaboration (in time this should shift to the latter)
Location and date	What is the location and date (e.g. at IFIP EGOV Conference in Koblenz September 2013), including URL (if applicable)
Set-up event:	Draft agenda (related to the purpose to be achieved, including name of presenters, name of presentation and other detailed information)
Who	Who is the organiser who are the collaborators
Actual impact	
Communities involved:	e.g. complex system researchers and policy-makers from government, including list of names
Feedback:	e.g. minutes, who is going to collaborate with whom
Outcomes	Quantifiable outcomes related to KPI after the event took place (eg. Event resulted in XX linked in members, 2 case studies, …)
Dissemination (only if it was also a dissemination activity that goes beyond the persons mentioned before)	
Field	Details (YOUR DATA HERE!)
Short description of work performed	(1–2 lines. It should include some info such as number of copies produced, languages covered etc.)
Reason why the material was created (Objective)	(1–2 lines)
Relevant WP(s)	List here the specific WPs for which this material was produced. If the material was produced to disseminate the whole project's results you should write "PROJECT"
Partners that created the material	The partner (organisations' name) that created this material
Other partners involved	
Type of audience the material is designed for	Preferably a list of participants names, their function and affiliation
Number of audience reached	(see above) The total number distributed over groups like policy-makers, researchers, elected politicians, public managers etc.
What impact is to be reached according to the project objectives	
More info	
Attachment	Provide the material in electronic form to enable easy assimilation

References

Auyang SY (1998) Foundations of complex-system theories in economics, evolutionary biology, and statistical physics. Cambridge University Press, New York

Brown RE (2001) The process of community-building in distance learning classes. J Asynch Learn Netw 5(2):18–35

Hagel J, Armstrong AG (1997) Net gain: expanding markets through virtual communities. Harvard Business School Press, Boston

Hanneman RA, Ridddle M (2005) Introduction to social network methods. http://faculty.ucr.edu/~hanneman/nettext/Introduction_to_Social_Network_Methods.pdf

Hansen DL, Shneiderman B, Smith MA (2011) Analyzing social media networks with NodeXL. Insights from a connected world. Elsevier, Amsterdam

Janssen M, Klievink B, Deljoo A (2012) Community and constituency building report year 1: the initial eGovPoliNet landscape, stakeholders and method for constituency building evaluation

Janssen M, Wimmer MA, Deljoo A (2015) Policy practice and digital science: integrating complex systems, social simulation and public administration in policy research, vol 10. Springer, Cham

Jones Q, Rafaeli S (2000) Time to split, virtually: 'discourse architecture' and 'community building' create vibrant virtual publics. Electron Mark 10(4):214–233

Katz M, Shapiro C (1985) Network externalities, competition and compatibility. Am Econ Rev 75(3):424–440

Kraut R, Galegher J, Egido C (1986) Relationships and tasks in scientific research collaborations. Paper presented at the ACM conference on computer supported cooperative work, New York, NY

Lu Y, Zhao L, Wang B (2010) From virtual community members to C2C e-commerce buyers: trust in virtual communities and its effect on consumers purchase intention. Electron Commer Res Appl 9(4):346–360

Majstorovic D, Wimmer MA (2014) A collaborative approach to study policy modelling research and practice from different disciplines. Paper presented at the joint proceedings of ongoing research, posters, workshop and projects of IFIP EGOV and ePart 2014, Dublin

Probst G, Borzillo S (2008) Why communities of practice succeed and why they fail. Eur Manag J 26(5):335–347

Smith M, Milic-Frayling N, Shneiderman B, Mendes Rodrigues E, Leskovec J, Dunne C (2010) NodeXL: a free and open network overview, discovery and exploration add-in for Excel 2007/2010/2013/2016. http://nodexlcodeplexcom/ from the Social Media Research Foundation, http://wwwsmrfoundationorg

Stewart T (2010) Online communities. Behav Inform Technol 29(6):555–556

Welser H, Gleave E, Smith M, Barash V, Meckes J (2009) Whither the experts? Social affordances and the cultivation of experts in community Q&A systems. In: International symposium on social intelligence and networking (SIN '09), 2009. IEEE Computer Society Press

Zhang J, Luna-Reyes LF, Nakashima M, Gil-Garcia JR, Sayogo DS, Mellouli S (2011) Building and sustaining a transnational and interdisciplinary research group: lessons learned from a North American experience. In: 44th Hawaii international conference on system sciences (HICSS), Hawaii, 2011. IEEE

Zhao L, Lu Y, Wang B, Chau PYK, Zhang L (2012) Cultivating the sense of belonging and motivating user participation in virtual communities: a social capital perspective. Int J Inf Manag 32(6):574–588. doi:10.1016/j.ijinfomgt.2012.02.006

Laurence Brooks is a full Professor of Technology and Social Responsibility in the Centre for Computing and Social Responsibility, Faculty of Technology, De Montfort University, UK. His research interests include: the role that social theory plays in our understanding of information systems in areas such as health information systems and e-government, the social and digital economy, information systems in a developing countries context, as well as ethics of emerging technologies. He is a Past-President of the UK Academy for Information Systems (UKAIS) and a member of the AIS and President of their SIG Philosophy of IS, plus a member of both IFIP 8.2 and IFIP 9.4. He has presented at numerous highly rated peer-reviewed conferences and published in the Information Systems Journal, Information Systems Frontiers, the Journal of Business Research and the European Journal of Information Systems. He can be reached at: laurence.brooks@dmu.ac.uk.

Marijn Janssen is a full Professor in ICT & Governance and chair of the Information and Communication Technology section of the Technology, Policy and Management Faculty of Delft University of Technology. His research interests are in the field of orchestration, (shared) services, open and big data and infrastructures. He is Co-Editor-in-Chief of Government Information Quarterly, conference chair of IFIP EGOV series and is chairing mini-tracks at the DG.o, ICEGOV, HICCS and AMCIS conferences. He was ranked as one of the leading e-government researchers in surveys in 2009, 2014 and 2016, and has published over 380 refereed publications. More information: www.tbm.tudelft.nl/marijnj.

Anastasia Papazafeiropoulou is a Senior Lecturer in the Computer Science department at Brunel University, UK. She has been involved in a number of European and UK funded research projects with emphasis on electronic commerce, small and medium size enterprises (SMEs) and their involvement in IT open innovation and entrepreneurship. She has 14 years of post-doc research experience in the study of diffusion and adoption of electronic commerce, e-government, broadband Internet, Enterprise Recourse Planning Systems (ERPs), IP-telephony, and mobile TV. She is currently involved in research related to the study of mindfulness practices as they can be used to overcome technostress. She can be reached at: Anastasia.Papazafeiropoulou@brunel.ac.uk.

Part II
Decision Making

One Step Forward and Two Steps Back: E-government Policies in Practice

Helle Zinner Henriksen

Abstract A central goal of e-government policy is to increase efficiency in public administration, and one way to increase efficiency is via the increased use of automation and rule-based decision making. The use of data to streamline processes also has a prominent role in public policy today. This chapter specifically discusses the role of discretion in data-driven public administration. The empirical setting for this discussion is the Danish public sector which has been among the first movers to implement e-government solutions. Denmark has a long tradition of issuing public policies defining goals of front-office e-services as well as back-office digitization. Rule-based decision-making systems represent an ideal driver for actualizing the visions outlined in these policies. This chapter presents an experience where a rule-based decision-making system was introduced in an agency which handles complex cases requiring in-depth discretion by specialized professional caseworkers. This experience provides a platform for discussing possible challenges when implementing policy goals in an organizational context. This chapter also addresses the concept of "digital nomos", the administrative norms of a digitized public administration.

Introduction

Policy informatics where data is utilized in the policy making process (Tsoukias et al. 2013) is multi-faceted by nature (Johnston 2015) because most policy problems are trans-disciplinary, involving social, economic, and political dimensions to name a few (Barrett et al. 2011). Barrett and his colleagues (Barrett et al. 2011) suggest that policy informatics has to accommodate multiple stake-holders, incorporate multiple sources of data, model large networked systems, and retain a high level of flexibility to cope with a large number of possible interventions. The domains of policy informatics and policy modelling focus on participation from various stakeholders, with the underlying assumption that their involvement

H.Z. Henriksen (✉)
Department of Digitalization, Copenhagen Business School, Frederiksberg, Denmark
e-mail: hzh.digi@cbs.dk

J.R. Gil-Garcia et al. (eds.), *Policy Analytics, Modelling, and Informatics*, Public Administration and Information Technology 24,
DOI 10.1007/978-3-319-61762-6_4

provides a positive contribution to the process (Johnston 2015). The focus of policy informatics and policy modelling is often on creative processes for addressing difficult problems by bringing multiple sources into play (e.g., via crowdsourcing Johnston 2015; Barrett et al. 2011), rather than the more mundane processes administrators use to implement policies in their day-to-day routines. Whereas policy informatics and policy modelling mainly focus on mobilizing interest and involvement from central stakeholders (Johnston 2015; Barrett et al. 2011), policy analytics accommodates the entire policy cycle, ultimately leading to the implementation of the policies (Tsoukias et al. 2013). Policy implementation is when the visions of the policies are carried out in the real world setting. The policies are interpreted and the intentions of the policies are translated into the specific context (Janssen et al. 2015). From a policy analytics perspective, the policy implementation stage is the point in the process where mapping of resistances and side effects of the policy takes place (Tsoukias et al. 2013). This chapter focuses on the implementation of the policies and how it was received by employees in an organizational context. And further on the institutional pressures from external stakeholders caused by the policy implementation.

Since 2001, when the first Danish national digitization strategy was published, there has been a continuous effort in getting more service out of a limited public budget. The driver for the digitization agenda has been increased efficiency and effectiveness (Henriksen 2016). The Danish public organizations therefore witness pressures to streamline their processes using technologies to support the automation of processes by applying tools, like evidence-based management and data-driven processes. These are all a continuation of the more traditional office automation.

Automation is not new within the domain of public administration and e-government research. Research on artificial intelligence in administrative discretion (Barth and Arnold 1999), system-level bureaucracy (Bovens and Zouridis 2002), decision factories and information refinery (Jorna and Wagenaar 2007), computerized bureaucracy (Angell and Samonas 2009), and algorithmic supported processes (Smith et al. 2010) all involve various degrees of IT-supported and automated processes. Their findings have all been part of the academic discussion for some time now. More recently, new concepts have also entered the vocabulary, all referring to the automation of processes in public administration. Here, citizens are expected to play a more active role in the case-handling process. Concepts include self-service government which assumes that people can govern themselves by using data stored in governmental databases (Paulin 2013) and Smart Cities which suggests that data becomes the core driver behind government-supported citizen services (Gil-Garcia et al. 2015). Those two examples involve citizens in the case-handling processes. Policy informatics takes the involvement of citizens even further and includes citizens along with other stakeholders in the policy-making process (Johnston 2015; Janssen et al. 2015). The assumption is that citizens are increasingly more technology savvy and therefore they can be expected to play a more proactive role in the development of policies (Johnston 2015).

This shift, which is driven by an increased involvement of a broad range of stakeholders, calls for an understanding of the tools and techniques which are discussed in other chapters of this book. Likewise, it also calls for an understanding of what in this chapter is referred to as digital nomos, the administrative norms of a digitized public administration. It is of relevance because public administration is exposed to fundamental changes due to the increased involvement in the policy processes from stakeholders who may not necessarily have a thorough understanding of norms and procedures within the administration. The term nomos originates from Greek and has a poly-semantic meaning, relating to laws, customs, and traditional social norms. These values are all cornerstones in the public administration of modern constitutional states (Cover 1983). Public administration is characterized by a long tradition of nomos most well-known from the work of Max Weber. Noted philosophers have argued that great legal civilizations are marked by a richness of nomos (Cover 1983). However, the richness of digital nomos is still to be explored and understood.

Critics may argue that a discussion regarding automation in organizations is passé. As far back as almost three decades ago Shoshana Zuboff (1988) explored the potential impact of automating processes, thereby making manual work processes obsolete. One of her conclusions was that increased automation in the workplace would lead to either de-skilling or re-skilling. She based her empirical analysis on technology rather than information technology. With the massive increase in the ability to collect, store, and compute data from various sources, it is now worthwhile to revisit the significance of workplace technologies. Workplace technologies driven by data and automatic processes are of particular relevance to the public sector which is characterized by a large number of information-intensive processes and tasks.

The main reason for putting digital nomos on the agenda is precisely the emergence of technologies which are capable of supporting automated processes in the daily work-flow of public administration and accommodating input from citizens and other stakeholders from various structured (self-service systems) and unstructured (e.g., social media) sources. It is no longer a discussion of futuristic scenarios of what may happen when policies requiring automated decision-making hit the back-office of public administration; it is happening now.

In a discussion of digital nomos set in a fully digitized environment, the role of discretion is uncertain. The introduction of automated decision making may totally eliminate the discretion (Jorna and Wagenaar 2007) leading to that case handling practices completely disappear or vanish as stated by Bovens and Zouridis (2002). However, this chapter builds on the assumption that there is still a need for skilled caseworkers in an automated organization.

This chapter aims to:

- Describe how policy goals related to e-government translate into specific IT-solutions and how the policy goals are handled in practice.
- Analyze the consequences of policy implementation in a specific organizational context where the two conflicting goals of automation and case handling meet.
- Discuss whether e-government policies can support the development of a "digital nomos".

An empirical example of a Danish government agency with approximately 50,000 complex cases annually is used to illustrate some implications of the implementation of a rule-based decision-making system in a sector characterized by a high level of individual discretion in the case-handling process. This agency was one of the first movers in the Danish public sector experimenting with a sophisticated rule-based back-office system. The system failed! In this context, the failed system is used as a subject for discussing policy implementation. The objective is not to generalize but rather to have a platform for analyzing some implications of introducing automated case handling in an environment which traditionally has been populated by highly skilled employees.

The remainder of this chapter is organized as follows. The next section presents: the traditional role of discretion in the public sector and how it broadly links to policy informatics and policy implementation; research relating to discretion in the context of street-level bureaucracy and the shift towards data-driven public administration; and the concept of shared service centers which is anticipated to play a central role in the implementation of an effective, digital public administration. Then, the empirical context of case handling in a government agency is described and analyzed. Later is a discussion of the trends identified in the agency experience. Finally, the chapter is wrapped up with some concluding remarks and practical recommendations.

The Role of Discretion in Public Administrative Practices

Policy informatics has commonly been applied to multi-faceted problems. Examples include environment and sustainability (Zhang et al. 2016), epidemic outbreaks (Barrett et al. 2011), and areas where engagement and expertise from diverse stakeholders are of vital importance (Johnston 2015). Policy informatics is the first stage in a process leading to implementation of policies (Tsoukias et al. 2013). The novel approach is that policy development can be driven by specialists and non-experts in collaboration (Susskind and Susskind 2015) and can involve innovative ways of interacting via social media, crowdsourcing etc. (Barrett et al. 2011). However, in the policy implementation stage more traditional values such as accountability, reliability and robustness still play a central role (Janssen et al. 2015). These are core values in public administration which have to be taken into account regardless of how the policy initiatives were developed. A persistent problem of public administration is the burden of costly processes. E-government has been perceived as an ideal solution to streamline processes and achieve efficiency. In fact, most nations have issued digitization policies to meet the challenges of costly processes. In the Danish context the digitization policies have included a range of stakeholders in the development process of the policies (Henriksen 2016).

One central cost driver in public administration is case handling. Case handling is characterized by cost-intensive manual processes, often handled by skilled and highly paid professionals (Bovens and Zouridis 2002). The automation of administrative practices for example case handling has been discussed within the domain of legal informatics, specifically focusing on how systems comply with the law (e.g., Ingolfo et al. 2014) and the challenges of translating norms into algorithms supporting machine learning and artificial intelligence (Bench-Capon 2015). The domains of legal informatics and machine learning relate mainly to the detailed technical challenges involved in an automated public administration and involve significant elements of computer science.

Another much-discussed aspect of automated administrative practices is "street-level bureaucracy", a term originally introduced by Lipsky (1979). He argued that it is the front-line workers who implement policies and focused on how public employees administer gate-keeping through administrative procedures and exercise the power of discretion. Several studies followed on the topic of discretion discussing the influence of IT on administrative practices (e.g., Buffat 2015; Reddick 2005; Bovens and Zouridis 2002).

In general though, more e-government research has focused on the "e" part, with reference to the technical solutions supporting digitalization, than on the "government" part, referring to the administrative component of e-government (Taylor and Lips 2008). E-government research has generally focused on the specific capabilities and impacts of technologies in the public sector (Heeks and Bailur 2007). The impact of technology has received much attention both in relation to citizens' use of—and sometimes failure to use—digital services, whereas the implications for public employees have received less attention.

Bovens and Zouridis (2002) gathered empirical evidence to illustrate the shift from street-level bureaucracy (case handling driven by traditional manual processes), to screen-level bureaucracy (case handling assisted by decision support systems), and finally to system-level bureaucracy (automated case handling practices) at the turn of the millennium. Today, fifteen years later there is a significant shift in the capabilities and range of IT systems and the potential use of data and IT systems in public administration have changed radically. Bovens and Zouridis (2002) concluded that it would be system developers who would be in control of the design of the discretionary outcomes of day-to-day case handling. IT would be decisive in the execution and control of processes, leaving out the human interference and discretion in case handling. The consequence for highly skilled caseworkers would be far-reaching: they would be out of a job. They would simply vanish from the organization.

One pivotal concept in the discussion of the role of IT in relation to caseworkers is discretion (Tummers and Bekkers 2014; Reddick et al. 2011; Bovens and Zouridis 2002; Jorna and Wagenaar 2007). The next section presents the core features of discretion from a traditional legal position.

The Concept of Discretion

Public administration relies on public employees being in a position of interpreting and deciding between alternatives to deliver the best possible outcome in a given case (Cover 1983). This practice is labelled discretion. There is a distinction between juridical discretion and administrative discretion. Juridical discretion is exercised in the court-room and is not addressed further in this chapter. Administrative discretion on the other hand is embedded in administrative practices and relates to the exercise of professional judgement (West, 2008). The traditional objective of discretion is to make decisions in accordance with standards set by a particular authority (Dworkin 1978). Dworkin further explains that "a man has discretion if his duty is defined by standards that reasonable men can interpret in different ways" (ibid, p. 69). In legal terminology, going outside the boundaries of these standards is referred to as "abuse of discretion". The abuse of discretion relates to the failure to exercise reasonable judgement (West, 2008). Discretion and the avoidance of abuse of discretion are at the core of traditional administrative nomos where interpretive commitments play a central role (Cover 1983). Dworkin (Dworkin 1978) uses the metaphor of a doughnut to define discretion. According to him, the hole of the doughnut is discretion, i.e., the interpretative freedom of caseworkers. The doughnut itself is the limitations of the interpretative freedom. It is the surrounding belt of regulation, practices, and norms: the administrative nomos. In an environment with increased use of data and simulation of case handling processes, the rules and practices of the IT-system would act as the surrounding belt, possibly leaving little need for the capabilities of skilled public employees practicing discretion.

The discretion is exercised on a daily basis by public employees who exercise their power in deciding on for example clients' access to resources and benefits, choice of treatment, and level of service (Lipsky 1979). Employee discretion is exercised in cases where there is no established rule which encapsulates the complexity of the situation (Dworkin 1978), or where decision needs to be taken instantaneously (e.g., police reacting to irresponsible driving or in oral exams where grading is granted immediately after the performance of the student). However, in the near future, the traffic example may be taken over by sensors and Internet of Things (IoT) technologies, and oral exams may also soon be an extinct practice with the help of artificial intelligence. Another example where the employees may not be needed is in nursing care where robots can take over demanding tasks (Brynjolfsson and McAfee 2014). The traditional discretionary power of public employees is mainly exercised in direct interaction with citizens. The empirical examples provided by Lipsky (1979) are taken from professions such as prison officers and teachers. IoT, robots, and artificial intelligence may soon play a central role in what used to be important manual tasks exercised by human labor (Brynjolfsson and McAfee 2014) and also for more administrative tasks performed by professionals (Susskind and Susskind 2015). In this context focus is mainly on those professionals who perform administrative tasks involving some level of complexity in the case handling process and which require some degree of discretion. They are not necessarily in direct contact with citizens but they most often handle cases of vital significance to citizens. The empirical example presented in the next section involves professionals. They are employees who are highly specialized. They have often had long academic training and have developed specialized expertise in handling a specific domain of legislation. They have developed administrative nomos.

The ideal scenario of discretion is that rules are interpreted and that the outcome of the exercise is the best possible outcome, all circumstances considered (Dworkin, 1978). The work of Lipsky (1979) questions this neutrality of employees and provides examples of "gate-keeping" where employees exercise their power to safeguard their own position rather than providing the best possible solution to clients.

IT and Discretion

Before looking into the specific aspects of IT and discretion, a brief discussion regarding the general perception of IT in public sector is provided. In their review of views on technology in e-government literature, Heeks and Bailur (2007) placed causes of impacts associated with new technology on a continuum from technological determinism, to contingent or socio/technical, and finally to social determinism. They classified the types of determinism in relation to three types of impacts caused by technology: pessimistic, neutral, and optimistic. Their analysis suggests that an optimistic stance in relation to technological determinism was more prevalent than any of the other categories in the 3 × 3 matrix. The positive attitude reflected in the reviewed articles may be attributed to that fact that the articles focused more on technological capabilities than the administrative part of e-government (Taylor and Lips 2008) and that e-government researchers have more enthusiasm about technology in general. Literature on automated decision-making in public sector suggests a more diverse picture.

Research shows that an automated administration has many of the traditional benefits of IT, including streamlining of processes and less need for manual control (Jorna and Wagenaar 2007), less errors (Paulin 2013), and better support of operations and administrative procedures (Cordella and Tempini 2015). Benefits specific to public administration include improved accountability and impartial and objective algorithmic processes (Smith et al. 2010), prevention of corruption (Cordella and Tempini 2015), increased trust (Paulin 2013), and fairness (Reddick et al. 2011). Buffat (2015) argues that e-government applications influence discretion both negatively by eliminating frontline policy discretion and positively by providing possibilities for direct (on-line) interaction for both citizens and employees. However, terms like infocracy (Zuurmond and Snellen 1997), decision factories (Jorna and Wagenaar 2007), and ICT-exacerbated accountability dysfunctions (Smith et al. 2010) give negative connotations to automated and rule-based decision making. The combination of IT and public administration has been criticized in a number of aspects, but critique has largely focused on the internal processes in public administration rather than the changed interaction with citizens. One exception is Bovens and Zouridis (2002), who argue that advanced IT systems alienate citizens from the public sector. Another exception is Tummers and Rocco (2015), who find that frontline caseworkers, when there are shortcomings in the IT systems, focus on serving clients rather than depending on the system or making compromises on case handling.

The term infocracy was introduced relatively early in the history of IT in the public sector by Zuurmond and Snellen (1997). It refers to structures with high levels of control in bureaucracies influencing the democratic processes negatively. A fundamental problem is that automation formalizes decision-making processes and constrains human discretion (Smith et al. 2010), and leads to operational discretion rather than individually tailored decisions (Jorna and Wagenaar 2007). The level of automation and formalized processes is however a choice which politicians need to make according to Jorna and Wagenaar (2007). Jorna and Wargenaar point to a fundamental dilemma caused by the introduction of what they label decision factories: the public sector has to choose between moving towards operational discretion or towards tailored decisions. From a policy informatics stance, it is also a choice between ideology and technology. Policies can be driven by the ideological stances of elected politicians or policy analytics where data points to an ideal solution (Tsoukias et al. 2013).

One turn towards the operational discretion where technology is at the driver's seat is an increased use of shared service centers (SSCs).

Shared Service Centers

Shared service centers (SSCs) can be utilized to share services between organizations. SSCs have been applied both in private and public sectors and research suggests that there are similarities (Janssen et al. 2007). SSCs represent potentials for resource sharing and cutting costs (ibid.), better cross-data processing, and outsourcing processes to private sector (Greve 2015). The success of SSCs relates to more precise decisions which are less prone to local differences in outcome of judgements (Greve 2015). Fundamentally, SSCs appear to be an ideal way of streamlining processes with the help of information technology.

However, SSCs in the public sector require changes in public policies in order to best utilize this management idea (Ulbrich 2010). SSCs require simple legal rules which can be interpreted by machines, reducing the room for discretion significantly. Often, the number of rules and exceptions to rules mushroom over time. Reforms of rules and regulations are therefore necessary to harvest the potential of SSCs. Another aspect relating to centralized digital solutions such as the SSC relates to local influence. Local managers may have to give up their influence over data and processes that are transferred to an external unit (Greve 2015). That in turn may lead to managers losing sight of the overall goal of their unit. The aspects of power struggles (Greve 2015) and the necessary changes in public policies (Ulbrich 2010) represent central aspects in the discussion of digital nomos because they address potential sources of resistance to change caused by IT.

The above brief introduction of discretion, IT and discretion, and SSCs provides the foundation for discussing the real-life example presented in the next section.

Policy Implementation: Digital Rule-Based Case-Handling Testing

Over the past decade, the public sector in Denmark has harvested a number of successes from digital solutions and it has managed well the delicate balance between voluntary adoption and successful enforcement of digital solutions (Andersen et al. 2007). Even though the mandatory shifts have not gone unnoticed among front-end workers (Berger 2014), digitization in the public sector is still viewed as successful if measured in terms of positioning in international e-government rankings.[1] The driver for digitization has been the aforementioned long chain of digitization policies (Henriksen 2016). These policies have first and foremost focused on establishing robust digital infrastructure for front-office digital processes. The digital infrastructure includes a secure digital signature and a digital post box for citizens, thereby establishing a digital window for citizens to interact with the public sector (Henriksen 2015). It has been a central objective to include a broad range of stake-holders in the development of the policies (Henriksen 2016). The increased use of e-services in society has influenced the administrative nomos because citizens are now more actively involved in filing in information needed in case-handling processes. Citizens are in charge of performing parts of the case-handling process, such as filling in personal data and selecting general options from drop-down menus in on-line systems (resembling the self-service e-government suggested by Paulin (2013)).

In this study, an empirical example from a Danish agency is used to illustrate the challenges of policy implementation involving a rule-based decision system where citizens are actively involved in entering data. The idea of actively involving citizens aligns well with the overall e-government policy (Henriksen 2015, 2016).

Research Method

The initial interest for this study came from media headlines and intensive discussions about the termination of a public IT project on rule-based and automated case handling worth approximately DKK 300 million. The media headlines led to a

[1] Denmark was number 16 in the latest UN global e-government ranking from 2014. Although the nation has dropped significantly from a 4th place in 2012 to a 16th place in 2014, it is still among the top 20 globally.

systematic search for publicly available sources describing the content and the context of the project. A reading of the publicly available sources helped form an understanding of the technical constraints, the outcomes of project evaluations performed by consultancy companies, and the escalating costs over time. Since the project was publicly funded, it also received political attention from elected politicians, who held the minister in charge accountable for the massive loss from the terminated project. A further search in the Danish Parliament's open archives led to a more detailed understanding of the specific costs which caused budget overruns. The escalating costs were mainly caused by the challenging tasks of system-integration and the need to buy IT-consultancy expertise. However, the media (including IT-professional media) and the documentation from the Parliament did not mention anything about the organizational aspects of the system introduction. To learn more about rule-based case handling within the organization, a senior shop steward in the agency was asked how the organization's employees had viewed the pilot testing of the IT-supported procedures. The interview took place during spring 2016, 2 years after the termination of the project. The interview was organized broadly around the key concepts presented by Bovens and Zouridis (2002) in relation to modes of case handling ranging from traditional street-level bureaucracy to screen-level bureaucracy to system-level bureaucracy. The interview—along with secondary data including public evaluation reports, press clips, and documents from the open archive of the Danish Parliament constitute the empirical material used in the case presentation and analysis below.

Scope and Content of Cases in the Agency

The agency is an impartial authority which makes decisions on individual compensation claims from citizens. The agency processes a total of about 50,000 claims each year, of which about 40,000 are new cases. Of the new cases, there is an almost even split between two types of compensation (21,000 cases vs. 19,000 cases) but the outcomes of the two types are significantly different; about 80% of the first type result in compensation to a citizen, whereas that happens with only 20% of the second type. The remaining 10,000 annual cases are reassessment of older cases, where compensation was earlier awarded for a limited time-period and the situation is now being reassessed to determine if the basis for compensation is still valid. It is vital for citizens who are dependent on compensation that their cases are handled with care. At the same time, every compensation awarded to a citizen represents a monetary cost for society.

Similar to other public agencies, there is only a small margin for error and there is an extensive system of controls related to the case handing outcomes. The agency is similar to any other organization in the Danish context that issues legally binding decisions governed by the Danish Public Administration Act. The Public Administration Act grants any applicant the opportunity to appeal against a given decision which is not in favor of an application. The agency has an appeals board.

For any case under appeal, it is the responsibility of the agency to provide all relevant documentation to the appeals board.

The agency currently employs approximately 200 lawyers who perform case handling. Due to the high complexity involved in determining if a citizen is eligible for compensation, case handling involves substantial discretion from highly specialized and skilled professionals. The job requires a high level of consistency in case handling and a thorough knowledge of the constantly changing regulation in the domain. Apart from the 200 lawyers, there are about 50 non-academic employees in charge of preparing the cases; this involves collecting and registering data for the cases, record-keeping, and archiving.

The Rule-Based Decision Making Project

The agency was among the early first-movers in Denmark to experiment with a rule-based case handling system. The system was launched as a digital spearhead project in 2008. The project aligned well with the overall national digitization policy document named "Shared Infrastructure and One Point of Access", published in 2007 (Henriksen 2016). It received initial funding from central government in 2009, but also over the following years due to substantial budget-overruns. Even though the project was similar to many large-scale IT-projects suffering from budget-overruns and delays (Flyvbjerg 2014), the business proposition for the project appeared to be convincing to the minister-in-charge for the first few years. It was only after five years the plug was pulled and the project was terminated.

The project description emphasized that the system could lead to significant reductions in labor costs. The IT system was an advanced and ambitious project, aimed at automating the case handling of about two-third of the 40,000 new cases which the agency received annually. The project description outlined the features of the system and listed the specific tasks which could be automated and thus reduce human interaction. It specifically mentioned that it aimed at supporting rule-based case handling. It said that the idea was to have computer supported processes which would lead to faster and more homogeneous outcomes of the case handling process. The rule-based case handling would base decisions on objective facts which were independent of human interaction and discretion. The system was anticipated to provide fast case handling thus streamlining processes benefitting both employees and citizens. Hence, the outcome of the system would be a reduction in the need for administrative resources in complex administrative procedures. In brief, the features outlined in the project description reflect those features mentioned in literature on rule-based decision systems (e.g., Angell and Samonas 2009; Smith et al. 2010).

A year after the launch of the project, an application for further funding was submitted to the Danish Parliament. The application mentioned that:

> The system will generate automatic decisions in standard-cases, it will lead to a more structured statement of facts, and lead to reduction in cost of training of employees in the use of the system...The increase in productivity is harvested in two ways: automatic collecting of

> information needed for case-handling and automatic preparation of the decisions. The automation of the processes is driven by combining data of the actual case for example diagnosis, sector of industry, and occupation. It is estimated that it is possible to automate 80 percent of the tasks in standard-cases which are 20 percent of all cases handled by the agency.

The project was terminated after the pilot-tests, resulting in significant financial loss (approximate DKK 300 million) and an organization with a painful experience in policy implementation of rule-based case handling. In the aftermath of the failure, the minister was asked to give an account of the project to the Parliament. Apart from the explanation of the enormous bill to taxpayers, the minister stated that:

> …the possibilities for automatization of case-handling in x-agency have been overestimated. (minister in charge of the agency)

It is against this backdrop that shop steward of the agency explained the employees' views on data-driven decision making and objective case handling.

> …it all ended with a massive failure, a scandal and the agency ended with a loss of almost DKK 300 million…. we expect that the agency has learnt something from this. A machine can still not make discretion. (shop steward in x-agency)

Reflections After the System-Failure

To recover from the enormous bill, the agency had to cut its staff. This led the government attorney to start an investigation into the agency's case-handling procedures. The attorney concluded that more data was necessary to provide a robust foundation for decision-making. The conclusion led to a political demand for more individual discretion in case handling, which resulted in the agency again recruiting more lawyers to handle the cases. It was obviously necessary for the minister in charge of the agency to demonstrate political will and determination in the slipstream of the failed system which left a large bill for taxpayers.

> We came from a system where we had a model [for discretion] which was less detailed and very concise and then we streamlined with an IT-system, but it collapsed and also due to [a reprimand from] the attorney to the government did we go back to an enormous discretion procedure. We have never exercised more discretion than we do today. (shop steward in x-agency)

The quote suggests that the failed rule-based decision-making system ignited institutional pressures, both political (from the minister-in-charge) and procedural (from the government attorney). The institutional pressures reinforced the need for individual discretion from highly qualified professionals thereby maintaining a focus on the traditional administrative nomos rather than a shift towards a digital nomos where the rule-based system would have a more prominent role.

It was originally anticipated that the agency could harvest benefits from the system because citizens could play a more active role in the application procedure and case preparation.

> … we could get the machine to do a lot, if only the citizen filled in forms correctly, then the machine could do the rest. (shop steward in x-agency)

This view was completely in line with societal trends. By the time the system started, citizens were already well versed in doing their travel arrangements and banking online. With one of the most technology and internet savvy populations in the world, Denmark should have been in a good position to place some minimum data requirements on its citizens. However, one problem with this assumption was that citizens who are technology literates are not necessarily administration literates (Belanger and Carter 2009). Citizens may be capable of surfing the web but not of reading and understanding forms which mirror legal requirements. Also, while citizens may be technology savvy in repetitive tasks such as doing online banking and e-shopping, these competences may not necessarily be transferrable directly to a situation where the citizen has to fill in forms of vital significance for their economic situation. Furthermore, filling the on-line forms from the agency typically happens once in a lifetime and in an unfamiliar system where even minor mistakes can have far-reaching consequences. One of the first observations in the citizen-system interaction was that filling forms represented a major obstacle.

> Citizens do not act the same way as a qualified lawyer. Firstly, the citizen does not fill the form as a lawyer does it. The citizen adds text and might not understand something and writes something in a parenthesis and all those things. Secondly, it was observed that only a fraction of the cases were suitable for this [the rule-based system] – 2% of the cases are based on objective things. The rest is based on individual assessment. (shop steward in x-agency)

A fundamental requirement for correct outcomes is correct data input. It is important to have information quality, representational quality, and a concise and consistent format which is machine interpretable and understandable (Dawes and Helbig 2015). Apart from the challenges of getting citizens to enter correct data, the quote also suggests that the system was of limited use in the majority of processes, including case handling. As a consequence, only a fraction of processes were suitable for the rule-based system and discretion was still a central component in the daily work leaving very little room for delegating tasks to the system.

Cover (1983) argues that the human factor plays a central role in the discretionary landscape, both in relation to creating and maintaining practices. Maintaining practices in the highly specialized environment of case handling requires knowledge sharing. IT-supported knowledge sharing or knowledge management has played a central role in the IS debate for more than two decades (Nonaka and Takeuchi 1995).

> …we also have an information system where employees can search for information, and that is fine… we have people updating the system, but honestly I do not know how much it is used. What you really learn from is the time-consuming one-to-one training, courses and seminars and meetings where cases are discussed…that is something which works, searching in a database sounds smart but it is questionable if it works. (shop steward in x-agency)

The agency does have a knowledge sharing system. But similar to other knowledge intensive environments, the system is utilized very little. Internalization

(Nonaka and Takeuchi 1995), where explicit knowledge is converted into tacit knowledge in a process of learning by doing, is the preferred mode of knowledge sharing and transfer in the agency. The knowledge sharing system was not a part of the rule-based system but it is illustrative of how the agency works with its systems, and it illustrates how the agency creates and maintains its practices. The system has very low priority in the organization because feeding information into the system takes resources away from actual case handling and the system does not capture the richness of the practices

> Interviewer: "It sounds like miles apart from evidence-based or data driven administration."
>
> Informant: "Yes! We have had the same [ERMS - Electronic Record Management System] system for 25 years, it is adjusted and pretty advanced and we are happy with it. For the case-handling procedure in relation to the production of decisions - I do not believe that technology can take us much further…you may get some plug-ins which can lead to marginal savings. But not much perhaps 5%…but it relates to the non-academic staff." (shop steward in x-agency)

Directly asked about the agency's position on a continuum from street-level, to screen-level, to system-level bureaucracy (cf. Bovens and Zouridis 2002), the informant suggested that the agency is positioned between street-level and screen-level bureaucracy. Practices are highly dependent on traditional decision support systems like the 25-year old ERMS. The agency has examples of screen-level bureaucracy where the role of IT is leading and where there is partial human interaction in the case-handling processes. These tasks are mainly undertaken by the non-academic staff, which according to the informant could be an endangered group of personnel in the future. However, the core activities are best classified as street-level bureaucracy (Lipsky 1979; Bovens and Zouridis 2002). The role of IT in street-level bureaucracy is supportive and there is full human interaction with individual cases.

The informant expressed lukewarm enthusiasm towards rule-based case handling after the organization's recovery from its fatal first-mover experience. The experience has led to skepticism in the organization towards automated case handling. The intention was to implement a rule-based case handling system where citizens could enter some data, other data should come from shared government repositories of data, the system would calculate the compensation, and employees would only control the final decision This vision did not come true. Instead, the employees became aware of the strengths and in particular the weaknesses of rule-based case handling systems.

> We are very reserved towards [increased digitization] and the case-workers are not nervous about it in reality. We have experienced that it doesn't work. It may be that it will work in 50 or 100 years, but it is not possible today. (shop steward in x-agency)

The reference to 50 or 100 years reflects little trust in the capabilities of IT. Gatekeeping (Lipsky 1979) in general and resistance to change caused by IT (Markus 1983) in specific are well-known side effects in organizations. Managing resistance to change and associated implementation tactics play a central role when introducing IT in private (Markus 1983) and public organizations (Fountain 2001). However, primary and secondary data indicate that in this example institutional factors played a more central role. The legislation (Public Administration Act) and rules imposed by the government attorney challenged the legitimation of the project. Legitimation plays a central role for policy makers (Tsoukias et al. 2013) and it was the minister in charge who decided to terminate the project.

However, the failure of this project has not stopped politicians in their pursuit to increase IT-driven efficiency in case handling. The first step out of the shadow of the failed system has been to transfer tasks from the agency to a SSC. The specific SSC has successfully taken over tasks based on objective criteria. In Denmark the shared service center model is implemented in many cases where there is "objective case handling" (Greve 2015). Examples of objective case handling include application for benefits like retirement pension and child benefit, which citizens are entitled to if they fulfill objective criteria such as a certain age or parenthood. The objectivity refers to decisions that can be made based on criteria and values which are machine readable. The SSCs should generate efficiency gains in terms of both costs and turnaround time of case handling.

The delegation of tasks to the SSC reduces the requirement of individual discretion exercised by highly specialized professionals in the agency. This challenges the entire situation which led to the intervention by the government attorney. The shop steward did not hesitate in stating that the only possible way out of that dilemma would be to change regulation thus reducing the legal requirements to level of discretion.

> … and now we [the agency] get involved in a [Shared Service Centre] system and a legal reform… it is obvious that new winds are blowing politically, because they have seen how expensive it has become. (shop steward in x-agency)

Discussion

This study suggests that policy implementation can be challenging at the organizational level. The first mover experience has backfired and left the organization with a degree of skepticism toward fully-automated case handling processes which was the aim of the project. It also shows that the experiment with a rule-based system ultimately increased the need for individual discretion which is similar to what was observed by Jorna and Wagenaar (2007) and further required employment of additional highly skilled professionals. What appeared to be a technically sound idea in the first case turned out to be problematic when it was applied in a specific context. It is beyond the scope of this chapter to discuss if the system had too many limitations and did not live up to the detailed regulation administered by the agency or if the failure should be interpreted through the lens of organizational power struggles and change-management challenges. At this point it is only observed, that the increased focus on individual discretion was initiated by a citizen writing a letter to a newspaper editor, airing his dissatisfaction with the level of scrutiny in case handling of the agency. Though a singular complaint from a citizen should be interpreted with care, it demonstrates that media attention had massive impact on administrative practices in the agency. In the Danish society, where legal rights play a central role and where public administration is governed to prevent abuse of discretion, there is a significant delegation of power to the government attorney. The government attorney takes action if he or she notices a potential violation of citizen rights. In the specific case the government attorney demanded more detailed case handling, thus challenging the ambitions of the digitization policies.

The case echoes conclusions from Jane Fountain's seminal book, *Building the Virtual State* (Fountain 2001). Fountain suggests that two types of forces influence technology-driven initiatives. One force relates to objective technologies, which are driven by market forces. The other force is institutional arrangements, i.e., regulation and political priorities where policy informatics is anticipated to play a more prominent role in the future (Johnston 2015). These two forces influence the public organization which has to implement the technology. The agency's experience suggests that there is a conflict within the institutional arrangements. The rule-based case handling system received ample funding and support from central government. The system was completely aligned with the digitization visions outlined by the central government and represented the potential for a best-practice situation. However, other institutional arrangements (i.e., regulations) pulled in the opposite direction. The government attorney overruled the efficiency measures which were promised from the system and which aligned with the overall digitization policies. That led the agency to implement even more detailed procedures for discretion, resulting in an increase in the agency's legal staff.

This study suggests that digital nomos is still not established in this particular government agency. The socio-technical view on information systems posits that information technology influences the normative universe of any organization (Heeks and Bailur 2007). This view emphasizes that information technology is not a black box. Information technology influences the administrative nomos—and administrative nomos influence information technology. In an ideal world, the rule-based system should have been "unpacked" and supported the forming of the digital nomos, balancing tailored and individual decisions (Jorna and Wagenaar 2007). This experience however suggests that the implementation of the rule-based system was driven by a technology deterministic approach. The agency had to align its practices with the logic of the system. However, the system could not accommodate the detailed level of discretion necessary to produce a satisfying outcome for case handling. This experience suggests that technological determinism goes even further: the law needs to be streamlined to accommodate the rule-based system. The policy initiative (the digitization strategy), which triggered the implementation of the rule-based system, was based on recommendations or soft law. However, the outcome of the implementation efforts was that politicians had to issue revisions to the Danish Public Administration Act and devise a specific law to delegate competencies from the agency to a SSC. These specific policy initiatives were miles apart from the scenario of policy informatics stimulating involvement and data (Tsoukias et al. 2013; Johnston 2015).

Conclusion and Practical Recommendations

This chapter presented how e-government policy goals translated into the development efforts of a specific rule-based case handling system in a government agency. Primary and secondary data was used to illustrate how policy implementation of e-government strategies challenged the practices of the agency. The analysis mapped some of the organizational resistances and side effects of the policy implementation.

By using the lens of policy implementation and mapping the resistances and side effects of the policy (Tsoukias et al. 2013), this study found that the process of implementing a digitization agenda is not straightforward. E-government policy goals aimed at streamlining case handling processes involving high levels of discretion were met with legislative constraints.

It should not be a surprise that the implementation of a rule-based system is governed by the logic of technological determinism (Heeks and Bailur 2007). What is more interesting in the context of understanding the emergence of a digital nomos is that technology not only shapes the work processes but also the regulatory foundation. In order to make system-level bureaucracy work, the regulatory foundation of the bureaucratic processes has to be changed too. This confronts the policy makers with the dilemma highlighted by Jorna and Wagenaar (2007): the choice between operational discretion versus tailored decisions. Data from the interview with the shop steward of the agency clearly illustrates that the agency is in favor of tailored decisions. The preference of tailored decisions may be attributed to administrative nomos (Cover 1983) and professionalism (Susskind and Susskind 2015) or the more gloomy perspective of being redundant due to automation (Susskind and Susskind 2015; Zuboff 1988).

But the delegation of tasks to the SSC pulls in the opposite direction. Given that the employees are administering the regulation (Lipsky 1979), there may be more challenges than just getting a rule-based system to work. Legislative acts are applied locally and need organizational support or "willingness to implement" (Tummers and Bekkers 2014). Tummers and Bekkers found that the willingness to implement was strongly correlated to the perception of meaningfulness. The informant observed that the rule-based system had covered only a fraction of the processes involved in the agency's case handling and it was futile to struggle with the system. This suggests that the informant attributed little meaning to the system. Unless a more comprehensive rule-based system can be developed, there must be a fundamental reform of the legislation—not only a reform of the specific regulation of the agency but also a reform of the control mechanisms which are part of public administration for decades to come.

In conclusion the case illustrates that a move towards preparing rules and regulations for rule-based systems or other automated processes reinforces a need for more emphasis on policy analytics where the utilization of big-data and decision support systems play a central role (Tsoukias et al. 2013) in order to get a thorough understanding of the specific domain, along with involvement from stake-holders (Johnston 2015) in order to utilize the innovative capacity of technology in public institutions (Janssen et al. 2015; Krishnamurthy et al. 2013).

References

Andersen KV, Henriksen HZ, Rasmussen EB (2007) Re-organising government using IT. In: Nixon PG, Koutrakou VN (eds) E-government in Europe. London: Routledge, pp. 103–118
Angell IO, Samonas S (2009) The risk of computerised bureaucracy. J Inf Syst Secur 5(2):3–25
Barrett CL, Eubank S, Marathe A, Marathe MV, Pan Z, Swarup S (2011) Information integration to support model-based policy informatics. Innov J 16(1):v16i1a2
Barth TJ, Arnold E (1999) Artificial intelligence and administrative discretion implications for public administration. Am Rev Public Adm *29*(4):332–351
Belanger F, Carter L (2009) The impact of the digital divide on e-government. Commun ACM 52(4):132–135
Berger, J. B. (2014). Mandatory e-government has arrived: the silent protest from staff calls for the committed scholar–resistance must never be futile! In: The 25th Australasian conference on information systems. ACIS
Bench-Capon TJ (2015) Transition systems for designing and reasoning about norms. Artif Intell Law 23(4):345–366
Bovens M, Zouridis S (2002) From street-level to system-level bureaucracies: how information and communication technology is transforming administrative discretion and constitutional control. Public Adm Rev 62(2):174–184
Brynjolfsson E, McAfee A (2014) The second machine age: work, progress, and prosperity in a time of brilliant technologies. New York: WW Norton & Company
Buffat A (2015) Street-level bureaucracy and E-government. Public Manag Rev 17(1):149–161
Cordella A, Tempini N (2015) E-government and organizational change: reappraising the role of ICT and bureaucracy in public service delivery. Gov Inf Q 32(3):279–286
Cover RM (1983) Foreword: Nomos and narrative. Harv Law Rev 97(4):1–68
Dawes SS, Helbig N (2015) The value and limits of government information resources for policy informatics in Johnston. In: Erik W (ed) Governance in the information era: theory and practice of policy informatics. London: Routledge, pp. 25–34
Dworkin R (1978) Taking rights seriously, vol 136. Harvard University Press
Flyvbjerg B (2014) What you should know about megaprojects and why: an overview. Proj Manag J 45(2):6–19
Fountain JE (2001) Building the virtual state: Information technology and institutional change. Washington: Brookings Institution Press
Gil-Garcia JR, Pardo TA, Nam T (2015) What makes a city smart? Identifying core components and proposing an integrative and comprehensive conceptualization. Inf Polity 20(1):61–87
Greve C (2015) Ideas in public management reform for the 2010s. Digitalization, value creation and involvement. Public Org Rev 15(1):49–65
Heeks R, Bailur S (2007) Analyzing e-government research: Perspectives, philosophies, theories, methods, and practice. Gov Inf Q 24(2):243–265
Henriksen HZ (2016) Digital by default. In: Bjørn-Andersen (ed) Cases on IT leadership: CIO challenges for innovation and keeping the lights on. Copenhagen: Samfundslitteratur, pp. 199–213
Henriksen HZ (2015) Scrutinizing Open Government Data to Understand Patterns in eGovernment Uptake. In: Electronic government. Springer International Publishing, pp 144–155
Ingolfo S, Siena A, Mylopoulos J (2014) Modeling regulatory compliance in requirements engineering. In: Advances in conceptual modeling. Springer International Publishing, pp 127–132
Janssen M, Wimmer MA, Deljoo A (eds) (2015) Policy practice and digital science: integrating complex systems, social simulation and public administration in policy research, vol 10. Springer
Janssen M, Joha A, Weerakkody V (2007) Exploring relationships of shared service arrangements in local government. Transf Gov People Process Policy 1(3):271–284
Johnston EW (2015) Conceptualizing policy informatics. In: Johnston EW (ed) Governance in the information Era: theory and practice of policy informatics. London: Routledge

Jorna F, Wagenaar P (2007) The 'Iron Cage' Strengthened? Discretion and digital discipline. Public Adm 85(1):189–214
Krishnamurthy R, Bhagwatwar A, Johnston EW, Desouza KC (2013) A glimpse into policy informatics: the case of participatory platforms that generate synthetic empathy. Commun Assoc Inf Syst 33(21):365
Lipsky M (1979) Street level bureaucracy, vol 198(1). Russell Sage Foundation, New York
Markus ML (1983) Power, politics, and MIS implementation. Commun ACM 26(6):430–444
Nonaka I, Takeuchi H (1995) The knowledge-creating company: how Japanese companies create the dynamics of innovation. Oxford: Oxford University Press
Paulin A (2013) Towards self-service Government-A study on the computability of legal eligibilities. J Universal Comp Sci 19(12):1761–1791
Reddick CG (2005) Citizen interaction with e-government: from the streets to servers? Gov Inf Q 22(1):38–57
Reddick CG, Abdelsalam HM, Elkadi H (2011) The influence of e-government on administrative discretion: the case of local governments in Egypt. Public Adm Dev 31(5):390–407
Smith ML, Noorman ME, Martin AK (2010) Automating the public sector and organizing accountabilities. Commun Assoc Inf Syst 26(1)
Susskind R, Susskind D (2015) The future of the professions: how technology will transform the work of human experts. Oxford: Oxford University Press
Taylor JA, Lips AMB (2008) The citizen in the information polity: exposing the limits of the e-government paradigm. Inf Polity 13(3,4):139–152
Tsoukias A, Montibeller G, Lucertini G, Belton V (2013) Policy analytics: an agenda for research and practice. EURO J Decis Process 1(1–2):115–134
Tummers L, Rocco P (2015) Serving clients when the server crashes: how frontline workers cope with E-government challenges. Public Adm Rev 75(6):817–827
Tummers L, Bekkers V (2014) Policy implementation, street-level bureaucracy, and the importance of discretion. Public Manag Rev 16(4):527–547
West's Encyclopedia of American Law (2008) Second ed. The Gale Group, Inc
Ulbrich F (2010) Adopting shared services in a public-sector organization. Transf Gov People Process Policy 4(3):249–265
Zhang J, Luna-Reyes LF, Pardo TA, Sayogo DS (eds) (2016) Information, models, and sustainability: policy informatics in the age of big data and open government, vol 20. Springer
Zuboff S (1988) In the age of the smart machine: the future of work and power. Basic Books
Zuurmond A, Snellen ITM (1997) From Bureaucracy to infocracy: towards management through information architecture. In: Beyond BPR in Public Administration: an institutional transformation in an information age, Amsterdam: IOS Press, pp 205–224

Helle Zinner Henriksen is Associate Professor at Department of IT Management at Copenhagen Business School, Denmark and Adjunct Professor in the Department of Information Systems at University of Agder, Norway. Her main research interests are IT in public sector, including value of IT investments in public sector, diffusion of IT innovation, and consequences of automation in public sector. She has presented her research at several international conferences (European Conference on Information Systems, IFIP WG 8.5 (EGOV), IFIP WG 8.6, and the Hawaii International Conference on System Sciences) and published her work in the Journal of Strategic Information Systems, European Journal of Information Systems, Journal of Information Technology, Scandinavian Journal of Information Systems, Communications of the Association for Information Systems, Government Information Quarterly, Transforming Government: People, Process and Policy and International Journal of Public Administration.

Development and Use of Data-Centric Information Systems to Support Policymakers: Applied to Criminal Justice Systems

Susan van den Braak and Sunil Choenni

Abstract Reliable management information is invaluable for policymakers and advisers to make informed policy decisions. Depending on the information needs of the policymakers and the characteristics of the data required, different approaches exploiting different ways to process data from multiple sources, may be used. In this chapter, we describe three information systems currently in use in the Dutch criminal justice system. The first system is based on a dataspace approach and uses aggregate data to provide a view on the current state of the criminal justice system. This is particularly useful for evaluating current policy and monitoring the implementation of new policy. The second system utilizes a data warehouse to integrate individual level data and look back to older cases. Therefore, it is suitable for evaluating policy. Finally, the third system exploits time series data to forecast the capacity needed in the near future. This allows for planning new policy and monitoring its implementation. Based on our experience with developing and implementing such systems and their use in practice, we lay down a list of guidelines for developing management information systems in the public sector. These guidelines also address issues like data quality, misinterpretation, and privacy protection.

S. van den Braak (✉)
Research and Documentation Centre, Ministry of Security and Justice, The Hague, The Netherlands
e-mail: s.w.van.den.braak@minvenj.nl

S. Choenni
Research and Documentation Centre, Ministry of Security and Justice, The Hague, The Netherlands

Research Centre Creating 010, Rotterdam University of Applied Sciences, Rotterdam, The Netherlands
e-mail: r.choenni@minvenj.nl

J.R. Gil-Garcia et al. (eds.), *Policy Analytics, Modelling, and Informatics*, Public Administration and Information Technology 24,
DOI 10.1007/978-3-319-61762-6_5

Abbreviations

GUI graphical user interface
KPI key performance indicators
USB *Uitvoeringsketen Strafrechtelijke Beslissingen* (Enforcement of Criminal Convictions)

Introduction

The Dutch criminal justice system comprises an interdependent chain of linked organizations each covering a specific area of the enforcement of criminal law. Together they ensure that criminal offences are detected, prosecuted, and punished. Thus, the total output of the system is a collaborative effort of all organizations involved.

A well-functioning criminal justice system implies transparency, effectiveness and efficiency, correctness and carefulness, integrity, and other generally accepted principles. The organizations involved in the criminal justice system perform a wide range of tasks to meet and implement these objectives. While doing so they may benefit from innovations in the field of ICT, for instance, the shift from paper documents to digital records. Concrete ICT applications aim to improve specific processes at the operational level.

Also at a policy level, developments in ICT may be exploited, in particular to obtain reliable, valid, and consistent management information. Data-centric information systems make use of the increasing amount of data that is generated at the operational level and are of great value to policymakers and advisers. This is because management information provides a comprehensive overview of all relevant information they need to make decisions. It provides them with (descriptive, predictive, or comparative) insight into the functioning of the system as a whole, and can therefore be used to optimize procedures. When policymakers are able to detect bottlenecks, unexpected differences, and potential problems, they can take better and informed decisions on the effectivity of current policy and the necessity of new policy.

Generating management information typically involves collecting, analyzing, and integrating data (Laudon and Laudon 2004). However, while the organizations work closely together in the Dutch criminal justice system, this process is not as straightforward as it seems. The reason for this is that in the Netherlands there is no central organization for the governance of the system and, therefore, no single source of information. Relevant data are scattered over various different databases or information systems that are owned and managed by one of the organizations involved (Choenni and Leertouwer 2010). Each organization only registers data that are needed for its own processes and that fall within the scope of its responsibilities.

As a result, to monitor the criminal justice system, data from the separate collaborating organizations need to be integrated to some extent (van den Braak et al. 2013a).

In this chapter, we focus on information systems on the policy level. In these systems, data processing, which is subjected to a set of soft and hard requirements, plays a crucial role. Soft requirements may be derived from legal rules and regulations or ethical standards and values (Choenni et al. 2011b; Bargh et al. 2016). We analyze these systems along a number of dimensions, such as the information needs that the systems are intended to meet, the aggregation level of the stored data, and the data quality required. Based on this analysis, we lay down several guidelines for developing tools for (judicial) information management.

Above-mentioned dimensions are closely related. We note that stakeholders may have different information needs depending on their aim and focus, and that they require different types of data, both with respect to their aggregation level and quality (see van den Braak et al. 2012). For most tactical and strategic decisions taken by policymakers (e.g., optimizing processes and developing new policies), statistical data may suffice. For instance, when a policymaker wants to know whether certain types of crime require special attention (because they are on the rise), he only needs insight into the development of the number of committed crimes per type over time. When he wants to know which types of cases are settled out of court, he needs detailed statistical information on the settlements by the police and prosecution. These types of questions, relating to the output of the criminal justice system and the flows in it, can be answered using (combined) aggregated data from the organizations involved. However, for some applications it is necessary to integrate individual level data. This is for instance the case when policymakers want to follow individual cases through the system, for instance, to investigate why certain cases are not handled in time and at which stage they currently remain (van den Braak et al. 2013b). Thus, individual level data are used to create (statistical) management information. So, while individual level data are stored and processed, the results shown to the users are aggregated. Note that integrating data on an individual level is more costly and requires data with a higher quality. Several reasons for this include the expensive join operations that have to be performed and issues like inconsistency and redundancy that have to be resolved beforehand.

In this chapter we will describe three different tools for analyzing and integrating judicial data, each exploiting data of a different level of aggregation and providing a different view on the system. Each tool will be described in detail using illustrative examples taken from the Dutch criminal justice system. Moreover, it will be explained how they are currently being used by policymakers for evaluating policies and/or defining effective and sound policies. While our use cases focus on the judicial domain, the lessons learned are broadly applicable to other domains in the public sector. Therefore, based on the success factors of our examples, we provide a set of general guidelines for developing and implementing management information tools on a policy level. Together they offer a roadmap for the implementation of similar systems in other domains and/or countries.

Background

Research Questions and Methodology

The Research and Documentation Centre of the Dutch Ministry of Security and Justice systematically collects, analyses, and disseminates data about the Dutch criminal justice system. The data mainly originate from various departments and institutions of the ministry like the police, the Public Prosecution Service, and the judiciary, and are related to the judicial processes (e.g., temporal data corresponding to the progress of criminal cases). The research centre also obtains relevant information indirectly through external sources and partner organizations.

In 2005, with the arrival of the new head of the statistical data and policy analysis division of the research centre, a long-term strategy was formulated to exploit the collected data for the benefit of policymaking. Within the last 10 years, based on these data, a number of management information tools have been developed, implemented, and deployed to monitor the performance of the Dutch criminal justice system. Throughout the years, we have reported extensively about a number of these tools at several conferences and in several journals. Although these tools are developed for different purposes and are, therefore, based on different architectures and data processing principles and procedures, all of them contributed to a better insight into the performance of the Dutch criminal justice system for policymakers and advisers.

This paper is devoted to the following research questions:

- In what respects are the tools different?
- In what respects are the tools similar?
- What lessons can be learned from the development and use of the tools?
- What are the main advantages of the tools and the impact of their use?
- What were the main challenges when developing and implementing the tools?

To address these questions, we formed a working group of two researchers from our centre and carried out a multi-case study. We selected three monitoring tools that were developed at our research centre and are being used in practice for at least 4 years. Within a number of working sessions, the members of the working group analyzed and reflected on the tools and their use. For this analysis, the members studied several relevant publications regarding these tools. Furthermore, for some parts of their analysis, they exchanged and discussed their views with the developers and users of the tools. We note that since both members of the working group were involved in the development of at least one of the tools, the methodology used in this research can be marked in some sense as participatory research.

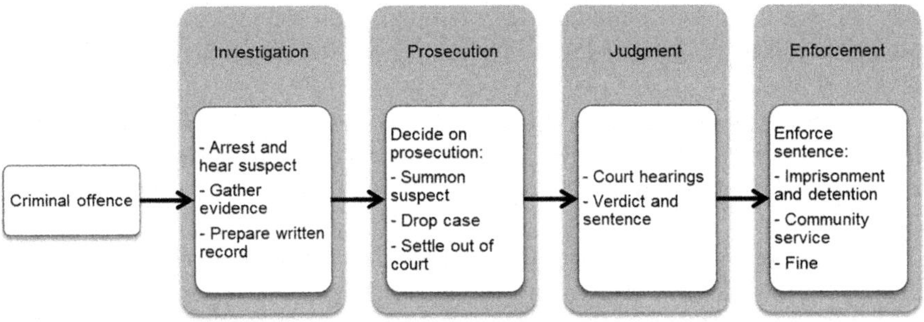

Fig. 1 Schematic overview of the Dutch criminal justice system

The Dutch Criminal Justice System

The Dutch criminal justice system (displayed in Fig. 1) consists of various organizations that work together to ensure the enforcement of criminal law: from investigation via prosecution and judgment to the enforcement of sentences (see also Tak 2008 for a more comprehensive overview).

The Dutch criminal justice system is often characterized as a chain: the organizations involved can be considered partners in a chain of events. This implies that the output of one organization is the input of another organization in the chain (e.g., Kalidien et al. 2010), which is true for the most part of the system. The chain of events in the criminal justice system is linear in the sense that a phase (e.g., investigation) must be concluded before the next phase may begin (e.g., prosecution) (see Netten et al. 2014).

Data in the Dutch Criminal Justice System

The organizations involved in the criminal justice system have in common that they register data about the criminal cases being worked on and the corresponding offenders. Therefore, each organization registers at least some attributes to identify the case and the suspect/convict involved. The different information systems thus partly store the same (or similar) information, and therefore, they are partially redundant. Redundancy poses a problem when it leads to inconsistencies. This occurs when some attributes are available in more than one source, while their values are not the same for the same entity. When integrating data, such inconsistencies have to be identified and resolved, but choosing one of the conflicting values is not straightforward and requires domain knowledge.

This is, however, not the only type of conflicts that occurs in the criminal justice domain: there are also issues with semantic interoperability. After all, each organization follows its own procedures and has its own goals and therefore, organizations may all use their own definition for seemingly similar entities. Differences in definitions occur on several levels, for instance, in the way in which cases are defined or labeled.

A related issue with the data in this domain is concerned with the quality of the data, as errors may occur. Some information about a case may not be registered correctly, or not be registered at all, by one of the organizations involved (Netten et al. 2014, 2016).

Above-mentioned data characteristics make data integration (in this domain) a challenging task involving domain knowledge. In the following subsection, two approaches for this task will be described that are currently employed in the Netherlands.

Approaches to Collecting and Integrating Data from Multiple Sources

While existing database management systems provide an extensive set of tools to manage data within a database, tools to manage data from different databases are lacking. Because database design primarily focuses on serving the information needs of a selected group of users, it is not common practice to establish relationships between different databases. A database is considered an independent and autonomous unit, in which the relationships with other databases are neglected and therefore not specified. However, to obtain adequate management information there is a practical need for tools to establish unambiguous relations between different sources. Although data integration can be automated for a large part, a significant amount of manual effort is sometimes still required. The main reason for this lies in the characteristics of the data used: issues such as redundancy and inconsistency make data integration harder and require that domain knowledge is used to ensure data quality and avoid undesired effects.

Generally, two solutions to relating or integrating data from different sources are distinguished: a data warehouse and a dataspace system (Kalidien et al. 2010; van den Braak et al. 2013a). On the one hand, integrating data in a *data warehouse* (Kimball and Ross 2011) is done on an individual level and involves data reconciliation, that is, the identification of data in different sources that refer to the same entity. This is relatively straightforward when a unique identifying key is present in all databases to be combined. On the other hand, integrating data in a *dataspace* (Franklin et al. 2005) is based on establishing rules using domain knowledge and may include aggregate data. Here, relating data means bringing (aggregated) data together that pertain the same entity and defining how the data should be interpreted in the context of other entities. Thus, this is more of a data co-existence approach than a data integration approach. The choice for one of the two approaches not only depends on the level of aggregation of the available data, but also on (1) the reasons why data are combined (the application) and (2) the issues with data quality (inconsistencies, incompleteness and inaccuracy) and semantic interoperability.

In the data warehouse approach, data are integrated rather tightly and on an individual level. This approach is preferred for applications in which it is necessary to follow cases through the criminal justice system (between the various phases). However, setting up a data warehouse requires that unique identifiers, or a set of common characteristics, are eligible for data integration (Choenni et al. 2010). Additionally, developing a data warehouse requires an enormous amount of upfront effort (Kalidien et al. 2010; van den Braak et al. 2013a), for instance, by performing expensive join operations (Choenni et al. 1993), and removing redundancy and inconsistencies. In contrast, in the dataspace approach, data are integrated rather loosely and on a higher level of aggregation. Dynamic (relationship) rules are established to link data and ensure data quality. Thus, a dataspace differs from a data warehouse in the sense that a common data model is not required and that there is no need to link data based on unique identifiers. This approach is more flexible and more effective for managing heterogeneous data. It is also useful for situations in which no individual level data are available (possibly due to privacy rules and regulations).

Thus, data integration can be executed in a variety of ways. Depending on the availability of the data, the application at hand, and the information needs of the users, one of the two solutions may be chosen. In the next section, it will be illustrated how they are used in practice in the Dutch criminal justice system.

Tools for Monitoring Performance in the Dutch Criminal Justice System

In this section, three information systems for providing insight into the functioning of the (Dutch) criminal justice system are introduced. Each system is described along different dimensions: its aim (and the information needs of the users), the data used (particularly their aggregation level), and its (current) implementation. Subsequently, their practical use in evaluating or defining policies is explained.

A Dataspace System for Monitoring Flows in the Criminal Justice System

In the Dutch criminal justice system, policy advisers periodically meet with experts and representatives of the organizations involved, to monitor developments in this system. These meetings are meant to evaluate existing policies and to identify possible problems in the system in time. Until recently, manually composed collections of written reports were used during these meetings that contained meaningful indicators for each of the organizations involved, for instance, their input and output per quarter. However, since no information was available on the interaction between organizations these meetings were considered less useful. A clear understanding of what happens with cases between organizations was missing and flows could not be compared in a structured way.

Therefore, to aid these policy evaluation meetings, a system that can periodically provide reliable, consistent, and valid management information was developed and implemented (van Dijk et al. 2016). This system allows policy advisers and experts to monitor flows within and between organizations and to display and visualize the data in a way that makes it easy for them to draw conclusions. For the purpose of this application it suffices to use aggregated data. This is because it aims to show the general (case) flows in the system and the relation between the input and output of different organizations.

Implementation

As depicted in Fig. 2, the system is based on the dataspace concept and therefore consists of three layers. Firstly, the bottom layer, the dataspace, contains source data of different types. Most sources are comma-separated files with aggregated data. Secondly, the space manager consists of two main components: (1) the variable database and (2) a relationship module. Both components will be described in more detail below. Thirdly, the interface layer serves as a communicator between the user and the space manager: it communicates with the space manager to get variables

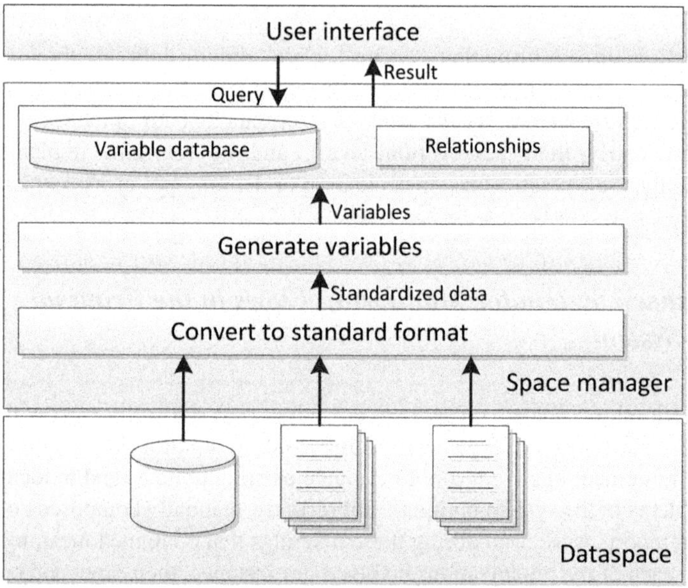

Fig. 2 Implementation of the dataspace system for monitoring flows

and present them to the users using tables or graphs. The implemented (web) interface presents the information in such a way that misinterpretations are minimized and unexpected deviations are highlighted (through alerts).

The variable database in the space manager contains a large set of hierarchical variables that provide meaningful notions to the end users. These variables were defined during several working sessions and interviews with potential users of the system (e.g., policy advisers and experts in the criminal justice domain). They are obtained by manipulating (database) attributes stored in the dataspace. For example, the output of the prosecution is the sum of the following three attributes as it can handle a case in three ways: (1) the number of summonses, (2) the number of settlements, and (3) the number of dismissals by the prosecution. Two auxiliary modules in the space manager are used to create variables: "Convert to standard format" and "Generate variables". In the former module, the data required to compute the value of a variable are extracted from the dataspace and converted to a standard format. In the latter module, the converted data are used to compute the value of a variable and the result is stored in the variable database.

The other main component of the space manager, the relationship module, contains rules, based on domain knowledge, to establish relationships between the different sources. Most relationships are modeled and implemented as if-then-else rules. There are two types of rules: to deal with (1) missing values and (2) semantic differences (for a detailed description of the use of such rules see Choenni et al. 2011a; van Dijk et al. 2016).

Use in Practice

The developed system allows policy advisers to monitor flows in the criminal justice system by providing insight into the input and output of the seperate organizations involved as well as the ratios between the output and input of collaborating, successive organizations. The interface helps them to visualize potential problems and facilitates the interpretation of the results. They are able to detect and investigate possible bottlenecks by using the implemented alert system. For example, when a policy adviser detects a large difference in the output of one of the organizations, he can search for possible causes by dividing the output into different subcategories. For instance, the output of the prosecution can be broken down into the number of dismissals, settlements, and summonses. Additionally, he can zoom in on several other dimensions, for instance, the output per type of crime or region. This may, for instance, reveal that there are bottlenecks in certain districts and the adviser can focus on those to solve the problem.

The system was evaluated using a set of questions relating to the requirements of the system (see van Dijk et al. 2016 for a detailed overview). Based on the feedback of the users, the interface layer as well as the underlying layers were gradually developed and improved. Afterwards, the interface was determined to be an adequate tool for the task at hand.

The present system (that has been updated a couple of times after the first prototype) incorporates 21 data sources coming from 18 different organizations. It contains almost 700 attributes from which more than 1300 variables are created. The system is currently being used to prepare the policy evaluation meetings mentioned above and is more flexible than the reports used in the past. It allows for more in-depth analyses, which is useful as the information need of the users is dependent on the current policy of the government. For instance, when the government wants to reduce street robberies, burglaries, and violent crimes, detailed and current information about these crimes is needed. Moreover, because the implemented system is updated quarterly, it contains information on recent developments in the system and allows policy advisers to take action promptly. In this way possible bottlenecks can be handled in an early stage. However, while the system provides useful insight into developments in the criminal justice system, expert knowledge is still indispensable to interpret the results.

In practice, policy advisers periodically view and investigate the results, and select surprising results for discussion in the policy evaluation meetings with experts. In these meetings, current policy is evaluated based on the recent results shown in the system pertaining to the previous quarters. This in turn may provide reasons to develop new policy. Because the results shown are fairly recent, the system is also very useful when implementing this policy as its effects can be monitored closely and policymakers and advisers can react rapidly when there are unwanted outcomes.

A Judicial Data Warehouse for Measuring Elapsed Times and Other Performance Indicators

In the Netherlands, the enforcement of sanctions is closely monitored by policy advisers of the Ministry of Security and Justice in a special program called *Uitvoeringsketen Strafrechtelijke Beslissingen* (USB, Enforcement of Criminal Convictions). This program aims to improve the processes in the enforcement phase, while strengthening the collaboration between the organizations involved. To do so, policy advisers from the program defined several key performance indicators (KPIs) in accordance with these organizations (van den Braak et al. 2013b). These KPIs measure elapsed times (see Netten et al. 2014) and pertain to different objectives of the USB program, such as promptness (the enforcement of the sanction begins quickly) and certainty (the convict is not able to avoid his punishment and completes it entirely). Thus, the KPIs examine the enforcement phase as a whole and measure whether the organizations work together properly. The performance on these KPIs is monitored in a system that brings together data from the different organizations.

To calculate the defined KPIs, data referring to the same case need to be integrated on an individual level. Aggregate data cannot be used, because a case must be recognizable or identifiable throughout the whole enforcement phase: from the verdict to the completion of the sentence. For the purpose of data reconciliation, the unique case number registered by the prosecution can be used, as it is also available in most registration systems in the enforcement phase. Therefore, a data warehouse approach was exploited.

Implementation

Figure 3 depicts a schematic overview of the developed system. It shows that first data are extracted from a set of sources (containing individual level data). These data are then encrypted, formatted, and cleaned. Finally, the relevant data are integrated and structured into a so-called event database from which the KPIs can be calculated.

The event database is obtained by performing a set of joins on the separate data received from the organizations involved. While doing so, it is assumed that data with the same case number are part of the same criminal case. The resulting database contains all relevant events that took place per case. These are sorted chronologically so that a detailed picture of the development of each case can be constructed.

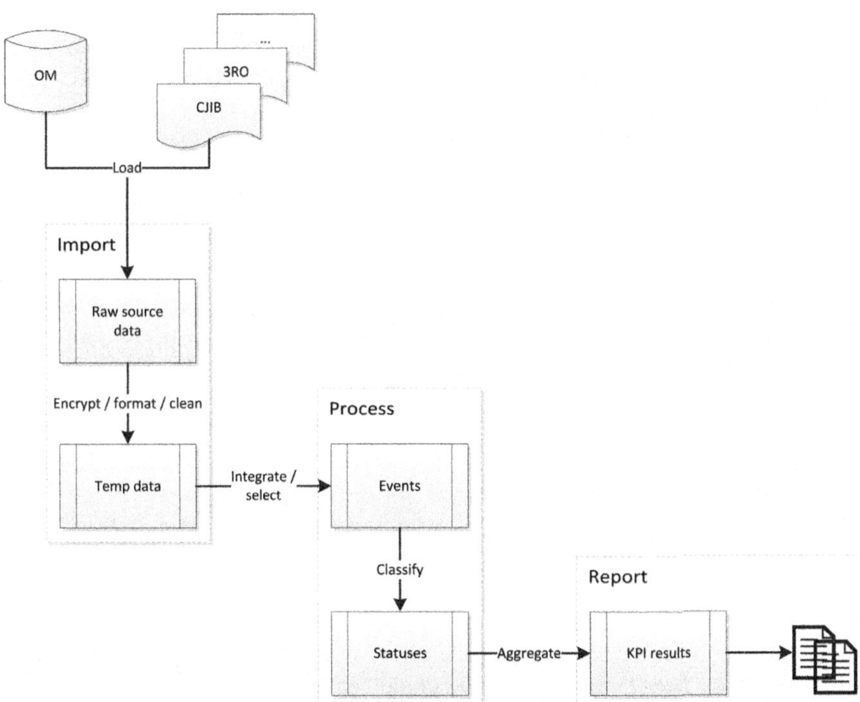

Fig. 3 Implementation of the data warehouse system for measuring performance indicators

This allows us to determine precisely which action was taken on which data by which organization. Thus, the structure of the event database resembles a timeline. From the timelines in the event database the KPI results are calculated using classification trees. This is done using cohort analysis. Here, a cohort is a set of cases (with the same sanction) that have their starting point (the prosecution's settlement or court's verdict) in the same particular period (usually a quarter). All these cases are followed for a certain predefined period (called the observation period). Based on the classification trees, for each case in the cohort its status at the end of the observation period is determined and each case is assigned to one of the categories. These results are then aggregated into KPI results per cohort.

All results are stored in the results table in the data warehouse. This table contains the aggregated statuses for all KPIs and sanctions given a number of observation periods (ranging from 1 week to 3 years) and a large number of cohorts (from 2008 onwards). From this table, KPI results can be calculated and presented in reports.

Use in Practice

Currently, four different KPIs have been defined, while 15 sanctions are distinguished. In total there are 38 classification trees as not all KPIs are relevant for all sanctions. The results are updated periodically (each trimester) and are currently published in written reports. These reports contain a selection of results: a limited number of cohorts is shown, while one observation period is chosen as the default. A web interface is being developed that provides access to all possible results and the complete history of results.

The written reports are discussed in scheduled feedback groups. Both policy advisers from the program and representatives from the organizations involved participate in these meetings. Together, they try to interpret the results and determine whether the measurements are correct or need to be altered. Each meeting focusses on one type of sanctions (i.e., custodial sentences, fines, community services, or conditional sentences). In the meetings, results of successive cohorts are compared, long-term trends are monitored, and large differences investigated.

To determine whether the performance is satisfactory or should be improved, norms were defined, relating to a certain chosen default observation period, per KPI and sanction. This was done in mentioned feedback groups in which additional domain experts participated. These norms (for the enforcement phase as a whole) are partly based on standing working arrangement between organizations, and (implicit) norms per organization. By making norms explicit, the organizations involved are forced to collaborate better. By relating the actual results to these norms, and watching whether the long-time trend is in line with or deviates away from it, policy advisers can use them to detect unwanted effects and evaluate current policy. This may help them to decide on implementing new policies to reduce any unnecessary delays.

There is one important reason why the KPI results are not very useful to monitor the implementation of new policies: given the method used, and depending on the chosen observation period, the results pertain to fairly old cases. For instance, an observation period of 3 years implies that the cases are at least 3 years old. In the judicial domain, where rules and regulations change frequently, this is not ideal. Additionally, policymakers have to wait a certain time (at least the duration of the observation period) before the policies' effects are truly reflected in the results. Thus, the KPIs do not provide a view on the current state of the enforcement phase, but look back to the performance in the past. When the results are disappointing, it is sometimes too late to take action, as the cases are relatively old and the convicts harder to find.

The system (and the method used) does have an important advantage: it allows for following individual cases. As a result, it provides a detailed view on (the speed of) the processes. This could not be obtained by using aggregate data only. The web interface under development aims to take away some of the drawbacks of looking back to older cases. Firstly, policy advisers are given the opportunity to look at more recent cohorts and compare their progress with older cohorts (for which the entire observation period already passed). Additionally, the results can be broken down per organizations and per region so that the cause of delays can be pinpointed precisely.

A Forecasting System for Predicting Workload

In the Netherlands, annual budgets are set for all ministries including the Ministry of Security and Justice. To establish a realistic and sound budget, policy advisers require reliable predictions of the workload of the different organizations involved. Such predictions may also help in developing effective and sound policies capable of handling the future workload.

Adequate estimates require large quantities of data, from inside and outside the justice system, as input. These data include historical time series of different relevant variables covering over 50 years or more. The predictions are partly based on a number of social, demographic, and economic developments, which are considered relevant for the justice system in the Netherlands. Examples of such developments are changes in unemployment, average income, and purchasing power. Thus, not all data needed pertain directly to the justice system.

To assist the budgeting process, a system was developed (Smit and Choenni 2014) to estimate the capacity needed in the near future by all organizations involved in all three fields of the Dutch justice system: criminal law, civil law, and administrative law. This forecasting system establishes relationships between the external developments and the various inputs in the justice system (such as crime rates or appeals to legal aid). Additionally, relationships within the justice system (i.e., how the input of one organization depends on the output of other organizations) are established. This is done on a macro level using regression equations. For this, aggregate data are used which are easier to obtain, while the error-prone and laborious process of integrating individual level data is avoided. In the remainder of this section, we focus on the part of the system that is related to criminal law, but the descriptions and findings for the other two fields are similar.

Implementation

In the implemented forecasting system, the criminal justice system is considered a network, where an organization gets its input from its predecessors and its output serves as input for its successors, except for some organizations in the network, that either get their input from external organizations or individuals (sources of the network) or are the endpoint of the flow of cases through the network (sinks).

As depicted in Fig. 4, the network system used for forecasting consists of several parts:

- a statistical model,
- a model generator,
- a database system,
- a variable management tool, and
- a graphical user interface.

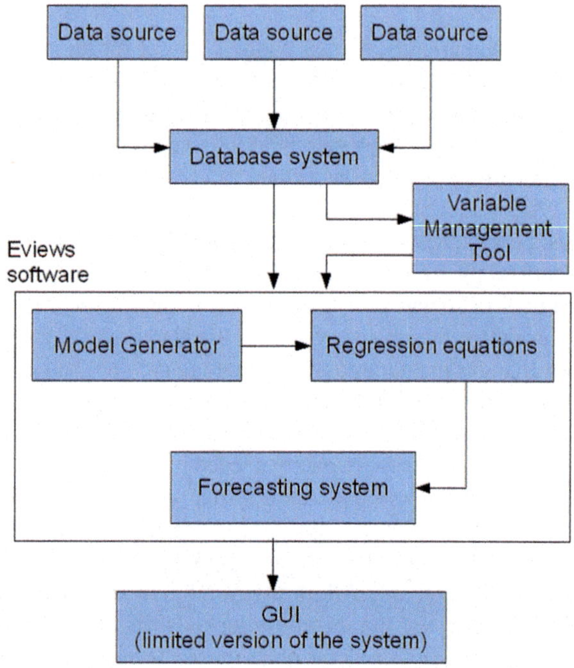

Fig. 4 Implementation of the forecasting system for predicting workload

The core of the system exploits regression analysis trying to fit a model to observed data. This fitted model is then used to describe the relationships between variables or to predict future values. The statistical model consists of hundreds of regression equations that explain a certain dependent variable in terms of an independent variable. These variables can either be external or internal. External variables describe outside developments such as demographic and economic changes. Internal variables are variables from within the criminal justice system, such as the number of registered crimes or the number of prisoners. All relevant variables are stored in the database system and are derived from various sources (e.g., the police, prosecution, and courts). For each variable the database contains a time series. Since one variable may be used in several regression equations, the variable management tool keeps track of all variables used. As a result, variables can be changed without the need to examine all equations.

The basic idea of the forecasting model is to relate the (forecasts of the) internal variables to the (forecasts of the) external variables. The model generator helps in automatically detecting which variables correlate. This separate tool searches through many possible regression equations and delivers a number of statistically "best models" (based on model selection criteria). Subsequently, the most suitable equations are selected manually using domain knowledge. The forecasting system then uses these equations to make forecasts over the desired period.

The graphical user interface (GUI) provides a limited version of the system, aimed at the policymakers and advisers who use its results. The GUI was developed to give them a better understanding of the vast and complex forecasting system. With this tool, a policymaker can change certain parameters to see the effect of this change further on in the network. It is limited in the sense that he can only alter one parameter at a time, while some parameters cannot be changed at all.

Use in Practice

The developed forecasting system has been in use for more than 10 years. The forecasts are updated every year and used for the budgeting plans of the Ministry of Security and Justice in the Netherlands. In year t forecasts are made for the years $t + 1$ until $t + 6$, generally based on data until $t - 1$. The updating process starts with collecting all data for all variables for the year $t - 1$ and the forecasts from year t onwards for the external variables. Using these data, a forecasting model is constructed to forecast the years t until $t + 6$. This results in preliminary forecasts that are refined in two ways. Firstly, the results are validated together with the organizations involved. Usually some mistakes in the underlying data are detected here. Secondly, the forecasts are corrected using updated data from year t that become available during the updating process.

These two steps result in so-called "policy-free forecasts". The results signify what would happen if there are no policy changes during the forecasting period. In this sense policymakers can use these forecasts as a warning about what happens if nothing changes. To create what is known as "policy-rich forecasts", outside of the forecasting model, the effects of all known and expected policy changes are added to the policy-free forecasts. The quantification of these changes can be done in different ways, for instance, using simulations or expert opinions. While doing so, the possible effects on all judicial organizations should be taken into account. Therefore, these policy changes and their consequences are discussed in plenary sessions in which all organizations involved take place.

This evidence-based approach has proven to be very useful. The estimates produced provide a sound underpinning for the annual budget and can be used in the negotiations with the Dutch Ministry of Finance. As a side effect, it brings together the many organizations in the justice system and thereby encourages better communication. As the system makes the dependencies and relations between organizations visible, it also provides insight into the functioning of the justice system and what could be done to improve it. Furthermore, the system can clearly show the consequences of changes, both internal (i.e., due to policy decisions or changes in the organization of the justice system) and external (e.g., demographic or economic changes).

Thus, the results help policymakers in planning and implementing effective and sound policies. Through the forecasting results, they have a reliable view on developments in the near future, while the forecasting system can also be used to evaluate alternative scenarios. Certain factors in the forecasts can be modified to see what the consequences are for the criminal justice system. For instance, policymakers can investigate what would happen if the criminal justice system becomes more repressive or more lenient. By using such scenario studies, policy advisers are better prepared to handle future developments. This is particularly useful in areas in which the potential costs of a capacity increase or decrease are high and it takes time to act on changing needs (e.g., building or closing prisons).

Concluding Summary

In this section, we described three management information tools currently in use in the Dutch criminal justice system in order to gain insight into the functioning of the system. In this section, the tools will be summarized along different dimensions relating to their functionalities, benefits, and impact. Subsequently, Table 1 provides a schematic overview of these dimensions.

The first tool utilizes a dataspace system to provide recent management information based on aggregate data. Heterogeneous data from different organizations are related in order to monitor (case) flows within and between organizations. The system allows users to compare the input and output of different organizations and alerts them when there are large differences. Because this approach uses fairly recent information, it is useful in both the implementation and evaluation stages of the policy process. The main advancement of this tool is that it brings together heterogeneous sources with a minimal amount of computational resources, thereby

Table 1 Schematic overview of the three information system described

	System		
Dimension	Dataspace	Data warehouse	Forecasting
View	Present	Past	Future
Data level	Aggregate	Individual	Aggregate and individual
Goal	Monitor flows (input and output)	Measure KPIs (elapsed time)	Predict workload and capacity
Policy stage	Implementation and evaluation	Evaluation	Planning and implementation
Data used	Hierarchical input and output variables	Chronological timelines with events	Historical time series, both internal and external
Situation before implementation	Written reports with incoherent numbers from separate organizations; no information on their interaction	No measurements available; performance was not monitored structurally	Isolated forecasts of the demand for sanction capacity (e.g., prisons)
Situation after implementation	Reliable, consistent, and valid numbers that are updated periodically; possibility to visualize data in a meaningful way and zoom in on results	Detailed insight into the (speed of) processes and the collaboration between organizations; possibility to follow cases through the system	Evidence-based and coherent forecasts for the entire justice system; possibility to evaluate alternative scenarios and the effects of internal and external changes
Impact	Better preparation of policy evaluation meetings; improved interpretation and understanding of results	Better ways for policy advisers to exercise control by well-founded norms; improved collaboration between organizations	Better position in budget negotiations, improved communication between policymakers and the organizations involved
Main challenges	Use domain knowledge to make relations between heterogeneous sources explicit	Operationalize definitions in consultation with parties involved, handle missing and inconsistent information	Collect reliable and correct historical time series

providing reliable and validated information on the interaction between the different organizations involved. Previously, before the implementation of the tool, such information was not available on a structural basis. Moreover, by making relations between sources and variables explicit using rules, it gives the users a deeper understanding of the data and their quality. This requires, however, thorough and detailed domain knowledge and is a challenging task.

In the second tool, individual level data is used for more advanced performance measures such as elapsed times. While the first tool provides a view on the current state of the system, this tool looks back to older cases, which makes it suitable for evaluation purposes only. Using the developed data warehouse it is possible to follow individual cases through the criminal justice system (between the different organizations) and monitor their progress. As a result, users can obtain a more detailed view of the processes and collaborations in it. Until now, there were no management information tools that tightly integrate data (coming from different sources) into an event database to follow cases. The developed method provides a blueprint for designing such tools and has already been used in other research projects. The main challenge when implementing such a tool is operationalizing the performance measures and choosing which attributes to use. This is not straightforward, because required relevant information is not always available, univocal, and correct (see also Netten et al. 2014).

The third tool exploits large quantities of data from various sources in order to predict the workload, and the capacity needed, for the different organizations in the system. Thus, this forecasting tool provides a view on the (probable) future state of the system and is, therefore, particularly helpful while planning and implementing policy. It is different from the other two tools in that it also uses data from external sources from outside the criminal justice system, for instance, demographic, social, and economic data. The main advantage of this tool is that it provides grounded estimates of the expected workload that were not available in the past. In this way, it makes the budgeting process more transparent and reliable. However, while the tool requires a large amount of historical data on both internal and external variables, data collection is a time-consuming and sometimes troublesome process. This is partly due to the fact that the required time series span 50 years or more, are not always complete, and may involve legacy data.

Discussion

In this chapter, we illustrated how information systems on the policy level can be implemented and used. While these approaches focus on the judicial domain, the lessons learned are broadly applicable to the public sector. Whenever management information is generated from large amounts of data, that need to be processed and integrated, the guidelines provided here are relevant and useful. Through our examples we have shown that choosing a suitable approach largely depends on the information needs of the users and the characteristics of the data required and/or available. Below we will summarize the lessons learned and success factors into a list of guidelines for developing and implementing such tools. These are divided into data-related and system-related guidelines. Together they offer a roadmap for the implementation of similar systems in other policy-related domains and in other countries.

Guidelines and Roadmap

Data-Related Guidelines

Identify Information Needs

The first step in developing information systems is eliciting the information needs of target users. Above examples illustrate that depending on the decisions they need to make and the policy stage they operate in, policymakers and advisers may require different types of information. Note that in some cases, the precise information needs are not know beforehand, and cannot be specified by the users, for instance, when policymakers want to exploit data, but do not know which insights they can gain from them. In such cases the approach is more gradual and participatory, and starts with collecting and analyzing available data.

Determine Information Requirements

After the information needs are specified, it has to be determined which data are needed to satisfy those needs. This involves establishing information requirements. In most cases, for information needs relating to tactical or strategic tasks on a policy level, it suffices to use aggregate or statistical data (van den Braak et al. 2012). However, as shown in one of the use cases, for some applications individual level data are required.

Collect Data Available

When the information needs and requirements are laid down, they have to be matched to the data available. This step thus involves collecting the data and assessing whether they can be used to answer the questions of the users. In the use cases, the gap between information needs and data collection was bridged top down: first it was established what the users want to know, then it was determined which data are required and how they can be obtained. In some cases, where the information needs are unknown, the approach has to be bottom up: start with the data and investigate what can be learned from them.

Assess Data Quality

The previous step of collecting data involves one important consideration: the quality and validity of the data. This is relevant for all governmental data as there are several issues with these data. Firstly, because registering relevant data often takes place on the operational level, and this may not be the core business of the organizations involved, incompleteness, inaccuracy, and inconsistency are not uncommon.

Secondly, some data were collected in the past, sometimes decades ago. Thirdly, the meaning of the data evolves over time as rules and regulations are constantly changing. These issues make it important to assess the quality of the data before they are used to generate management information. As explained above, when issues such as redundancy and inconsistency arise, additional measures and domain knowledge are needed to ensure data quality. For instance, in the first use case described, relationship rules are exploited to monitor data quality.

Manage Expectations

In some cases it is not possible to (completely) satisfy the information needs of target users, because the data required are not available. Sometimes data is available, but relevant attributes are missing or not reliable. Therefore, it is important to involve the customers in these first few steps of the development process and lower their expectations when necessary. In some cases, a bottom up approach can be used as an alternativeto try to find second best alternatives that meet their needs to a certain extent.

Operationalize Definitions

Most information needs can be translated into one or more meaningful indicators (for example, elapsed time is used in the second use case described). In order to measure them, their definitions need to be operationalized. This means defining unclear concepts into measureable variables related to the data available while setting down exact definitions of each of those variables. This is not always straightforward, as in some cases there are many alternative operationalizations (involving many choices) that do not necessarily yield the same results. In the second use case, such choices were modeled into classification trees that can be used to calculate indicators. Choosing an operationalization is also partly dependent on the availability of the data and their quality. In some cases, the required attributes are not available and alternatives have to be found that are strongly related to them (see the rules to handle missing values in our first use case).

Utilize Expert Knowledge

Our use cases have shown that domain knowledge from experts is invaluable for several reasons. Firstly, they can have a role in operationalizing definitions and mapping them onto information needs. Secondly, in complex domains such as criminal justice, in which errors frequently occur, domain knowledge plays an important role in interpreting the data and monitoring their quality. Thirdly, when data from different sources are combined, domain knowledge is need to establish explicit relations between sources and, for instance, solve inconsistencies. Therefore, typically, the development and implementation process should not only involve software developers and database specialists, but also domain experts.

System-Related Guidelines

Choose Data Integration Approach

Through our use cases, we have shown that there are at least two approaches to collecting and integrating data: one can choose between a data warehouse and a dataspace system. Which approach is most suitable, largely depends on the aggregation level of the data used. In a data warehouse, data are integrated explicitly on an individual level, while in a dataspace, data with a higher level of aggregation are related more loosely. Each approach has its advantages and disadvantages. The benefits of a data warehouse are that the data are fully structured and consistent, and can be queried easily and rapidly (using a single query). Its main drawback is that the development may be very costly, in particular when the source data are of low quality and highly redundant and inconsistent. In cases where data reconciliation is impossible or too time-consuming, the dataspace approach is a better alternative. A dataspace system is less costly to build and maintain. This is because using a dataspace, problems with the data can be managed more easily and effectively. A dataspace is also more flexible than a data warehouse in the sense that it is extendable: new data sources can be added without changing existing data sources. On the other hand, a dataspace may not suffice for situations in which a uniform view of the data is required.

Test in User Groups

User participation plays a crucial role in information system implementation. Obviously, their information needs are at the start of this process. Therefore, during the process, developers, domain experts, and end users should interact closely. Given the wishes of the end users, the developers and experts decide which functions should be implemented and how this can be done. As soon as a new feature is implemented, a user group should be asked to give their feedback so that suggestions for further improvement and development can be gathered.

User feedback is also valuable after the implementation of a management information tool is completed. This is because interpreting their results often require at least some domain knowledge. Therefore, the results of such tools need to be discussed and validated regularly in feedback groups. Typical members of these groups, besides policymakers or advisers, are experts and representatives from the organizations involved.

Take Into Account Security and Privacy

There are other issues that have to be taken into account when implementing management information systems and that influence above guidelines, namely privacy protection and information security. These issues entail that developed systems must be secured, data must be sanitized, and access to sensitive data must be controlled. With a view to this, aggregate data has a clear advantage over individual level data as data on a higher aggregation level do not provide personal information. Therefore, for privacy reasons, it is recommended to use aggregate data instead of individual level data whenever possible. Nevertheless, even applications that use aggregate data may pose a serious privacy threat (see van den Braak et al. 2013a).

Several precautions can be taken to respect the privacy of individuals and to minimize these risks (see Bargh et al. 2016), for instance, not storing privacy-sensitive attributes and hiding small numbers. When these precautions are followed, and data are published on a highly aggregated level, the risk of disclosing personal data and thereby violating the privacy of individuals is minimized.

References

Bargh MS, Choenni S, Meijer R (2016) On design and deployment of two privacy-preserving procedures for judicial-data dissemination. Gov Inf Q 33(3):481–493

Choenni S, Blanken HM, Chang T (1993) On the selection of secondary indices in relational databases. Data Knowl Eng 11(3):207–233

Choenni S, Kalidien S, Ariel A, Moolenaar D (2011a) A framework to monitor public safety based on a data space approach. Electronic Government and Electronic Participation. In: Joint proceedings of ongoing research and projects of IFIP EGOV and ePart, pp 196–202

Choenni S, Leertouwer E (2010) Public safety mashups to support policy makers. In: International conference on electronic government and the information systems perspective. Springer, Berlin Heidelberg, pp 234–248

Choenni S, van Dijk J, Leeuw F (2010) Preserving privacy whilst integrating data: applied to criminal justice. Inf Polity 15(1, 2):125–138

Choenni S, van Waart P, de Haan G (2011b) Embedding human values into information system engineering methodologies. In: European conference on information management and evaluation. Academic Conferences International Limited, p 101

Franklin M, Halevy A, Maier D (2005) From databases to dataspaces: a new abstraction for information management. ACM Sigmod Record 34(4):27–33

Kalidien S, Choenni S, Meijer R (2010) Crime statistics online: potentials and challenges. In: Proceedings of the 11th annual international digital government research conference on public administration online: challenges and opportunities. Digital Government Society of North America, pp 131–137

Kimball R, Ross M (2011) The data warehouse toolkit: the complete guide to dimensional modeling. Wiley

Laudon KC, Laudon JP (2004) Management information systems: managing the digital firm. Pearson Prentice Hall, Pearson Education, Inc., Upper Saddle River

Netten N, Bargh MS, van den Braak S, Choenni S, Leeuw F (2016) On enabling smart government: a legal logistics framework for future criminal justice systems. In: Proceedings of the 17th international digital government research conference on digital government research. ACM, pp 293–302

Netten N, van den Braak S, Choenni S, Leertouwer E (2014) Elapsed times in criminal justice systems. In: Proceedings of the 8th international conference on theory and practice of electronic governance. ACM, pp 99–108
Smit P, Choenni S (2014) On the interaction between forecasts and policy decisions. In: Proceedings of the 15th annual international conference on digital government research. ACM, pp 110–117
Tak PJ (2008) The Dutch criminal justice system. Wolf Legal Publishers
van den Braak S, Choenni S, Meijer R, Zuiderwijk A (2012) Trusted third parties for secure and privacy preserving data integration and sharing in the public sector. In: Proceedings of the 13th annual international conference on digital government research (DG.0'12), June 4–7, College Park, MD, USA, pp 135–144
van den Braak S, Choenni S, Verwer S (2013a) Combining and analyzing judicial databases. In: Discrimination and privacy in the information society. Springer, Berlin Heidelberg, pp 191–206
van den Braak SW, Netten CPM, Witzenburg RV (2013b) Inzicht in het presteren van de executieketen. WODC
van Dijk J, Kalidien S, Choenni S (2016) Smart monitoring of the criminal justice system. Gov Inf Q. ISSN 0740–624X, http://dx.doi.org/10.1016/j.giq.2015.11.005. (http://www.sciencedirect.com/science/article/pii/S0740624X15300204)

Susan van den Braak is a researcher at the Statistical Data and Policy Analysis Division of the Research and Documentation Centre (WODC) of the Ministry of Security and Justice in the Netherlands. She holds a PhD in Computer Science from Utrecht University and an MSc in Artificial Intelligence from Radboud University Nijmegen. Her doctoral dissertation focused on argument mapping software for crime analysis. For the Ministry of Security and Justice, she is currently working on various projects that involve combining and analyzing various large (judicial) databases. Her research interests include e-government, law enforcement, argumentation and visualization, databases and data science. She has published several papers in these fields.

Sunil Choenni holds a PhD in database technology from the University of Twente and an MSc in theoretical computer science from the Delft University of Technology. Currently, he heads the department of Statistical Information Management and Policy Analysis of the Research and Documentation Centre (WODC) of the Dutch Ministry of Security and Justice. Furthermore, he is a professor in the field of Smart and Inclusive Society at Rotterdam University of Applied Sciences in Rotterdam. Prior to joining WODC, he held several teaching and research positions at different universities and research centres. His research interests include data warehouses and data mining, open data, databases, e-government, cyber security, and human centered design. He publishes frequently papers in these fields and acts as a PC member for various conferences.

The Value of System Dynamics Modeling in Policy Analytics and Planning

Norman Wayne Porter

Abstract An efficacious policy and planning process must be focused on enhancing the ability of decision makers to make sense of an uncertain and complex environment. One tool that could prove useful in this process is system dynamics modeling, created by Jay Forrester at MIT. Use of small system dynamics models (with each module containing ten stocks or less) as a decision support tool has recently been explored in three areas of regional planning: modeling a regional economic and education strategy for Central Coast California; the modeling of U.S.-China relations; and, the modeling of violent extremist activity. In each case, an integrated system dynamics model was created or planned that included multiple modules that comprise a strategic system. The models allowed decision-makers to use a "flight control simulator" or "dashboard" to better understand potential, nonlinear, behavioral outcomes over time. When used in concert with other methods and tools of evaluation, system dynamics may provide enhanced understanding and key insights into problems previously thought too complex for this level of analysis and may encourage decision makers to examine a longer time horizon in overcoming policy resistance and establishing system stability.

List of Abbreviations

CLD	Causal loop diagram
K-12	Kindergarten through twelfth grade
MIT	Massachusetts Institute of Technology
NPS	Naval postgraduate school
S/I	Susceptibility and infectivity
U.S.	United States
USN	United States Navy
Yr	Year

N.W. Porter, PhD, CAPT, USN (Ret) (✉)
Naval Postgraduate School, Defense Analysis Department, 589 Dyer Road, Room 214, Monterey, CA 93940, USA
e-mail: capt.wporter@gmail.com; nwporter@nps.edu

J.R. Gil-Garcia et al. (eds.), *Policy Analytics, Modelling, and Informatics*, Public Administration and Information Technology 24,
DOI 10.1007/978-3-319-61762-6_6

Overview

Policy analytics are often intended to deconstruct complex challenges so that effective policy can address the individual elements that contribute to undesirable outcomes. An increasing recognition of the complex and probabilistic nature of our world and the human systems that populate it, leads to the inevitable conclusion that reductionist and deterministic approaches to policy analytics are unlikely to satisfactorily address today's problems. A primary objective of planning and policy analysis is to inform decision makers of the complexity of the environment in which they, and their competitors, operate and to broaden the horizon of their strategic thinking. Research in the areas of complexity and system thinking covers a spectrum of concepts that frame regional and global environments and has matured over time with our understanding of the laws that govern our physical universe. Common in much of this analysis is a focus on determining system boundaries, endogenous and exogenous impacts, identification and implementation of feedback loops, and an appreciation of the delays and time frames required to provide a sufficient understanding of relationships within and between systems. An efficacious planning and policy analysis process must be focused on enhancing the ability of decision makers to make sense of an uncertain and complex environment. One tool that could prove useful in this process is system dynamics modeling, created by Jay Forrester at MIT.

System dynamics practitioner, John Sterman, stated that, "…people use various cues to causality including temporal and spatial proximity of cause and effect, temporal presence of causes, covariation, and similarity of cause and effect…These heuristics lead to difficulty in complex systems…" (Sterman 2000, p. 28). This process of sense-making has a direct bearing on strategic thinking, planning, and analysis. System dynamics is grounded in non-linear dynamics and feedback control developed in mathematics, physics, and engineering. Because these tools are applied to the behavior of human-in-the-loop as well as physical and technical systems, system dynamics draws on cognitive and social psychology, economics, and other social sciences. In a bounded system, the application of system dynamics can provide both conceptual and qualitative insight. Systems that involve humans are driven by feedback loops in which both free choice and constraint are present.

In system dynamics modeling, causal loop diagrams (CLDs) are frequently employed as an initial step in the identification of the behavior of feedback structures at play within bounded systems and sub-systems. CLDs provide polarity notation to indicate whether feedback from one variable to another is positive or negative. Positive feedback indicates that as an independent, causal variable increases/decreases, a linked dependent variable increases/decreases in effect as a result (same effect, reinforcing the change). Negative feedback indicates that as a causal variable increases/decreases, a linked dependent variable decreases/increases in effect as a result (opposite effect, balancing the change). If these polarities are equal within the causal loop, they offset each other and result in reinforcing behavior. If the number of positive polarities in the loop does *not* equal the number of negative polarities, the result is balancing behavior.

A more quantifiable aspect of system dynamics is achieved through the mathematics of integral calculus in determining the rates of accumulation within a system that affect its overall behavior as a result of the feedback structures of the system. Accumulations of information, material, money, etc. are referred to as stocks, which can be measured through integration based on their rate of inflow or outflow (accumulation or depletion). These rates are known as flows, and they represent derivatives of integrals. In standardized system dynamics models, an inflow or outflow is denoted by a pipe and valve, and a stock is denoted by a rectangle (Sterman 2000). Traditionally, system dynamics has been used to bound an identified problem and to model the nonlinearities in an endogenous system. Working through an iterative process of system structure definition and feedback relationships, models are populated with data representing key parameters and independent and dependent variables in the system. The resultant model is then validated by comparing actual trends over a finite period of time with modeling results that are intended to closely replicate the actual trends demonstrated over the same period of time. Once the model structure is validated, current data is used to initiate modeling runs projected over a suitably chosen time horizon (selected to allow sufficient time for feedback within the system to be reflected in potential behavioral outcomes).

There is, however, another application of system dynamics modeling, in which plans or policies are modeled by incorporating proposed changes in the system structure going forward. A system's structure determines its behavior. While validation of this modeling can clearly not depend on the reflection of existing trends (since the plans or policies may not have been implemented yet), it can still provide decision makers key insights into possible courses of action through an enhanced understanding of the proposed structure's feedback mechanisms and potential behavioral outcomes over time. It is therefore not the specific predictability of the modeling and subsequent simulation that provide policy analytics and planning value, but rather the enhanced understanding of nonlinearities within the system and the ability for decision makers to easily evaluate changes in system variables and structure. The use of small system dynamics sub-models or modules (with each module containing ten stocks or less) as a decision support tool is currently being explored in both the public and private sectors.

This chapter will provide three case examples of exploratory work conducted by faculty and students at the Naval Postgraduate School (NPS): the modeling of a regional economic and education strategy for Central Coast California; the modeling of US-China relations in the Asia-Pacific region; and, the potential value of system dynamics modeling for the net assessment of counter terrorism policies.[1] In each case, an integrated system dynamics model was created or planned that included multiple modules comprising a strategic system. Each module was

[1] Sections of this chapter have borrowed from the author's unpublished dissertation, The Effects of System Dynamics Modeling on Systems Thinking in The Context of Regional Strategic Planning, (Porter 2014) that was approved for publication by the Naval Postgraduate School, Monterey, Ca. The views and opinions expressed are those of the author and do not reflect the official policy or position of the US Navy.

modeled endogenously, but was linked together in a system architecture intended to determine the collective effects of policy decisions implemented in one module on the others (the system of systems) in order to evaluate the potential behavior of the modules' inter-relationships over a prolonged time horizon. In two of the cases the models allowed decision-makers to use a "flight control simulator" or "dashboard" to perturb each system in order to better understand non-linear, potential outcomes over time.

While this modeling is non-predictive, it is intended to enhance foresight in a complex strategic environment by exploring both the risk and the opportunity space for planning and policy and the optimization and timing of the allocation of resources. When used in concert with other methods and tools of evaluation (e.g. agent-based modeling, wargaming and simulation), system dynamics may provide enhanced understanding and key insights into problems previously thought too complex for this level of analysis and may encourage decision makers to examine a longer time horizon, thereby enabling them to overcome policy resistance and establish system stability.

Framing the Problem: System Thinking for Planning and Policy Analysis

Although no single approach or methodology is sufficient for understanding, much less predicting, a complex and uncertain future, planners and analysts may indeed be better informed by understanding the implications and limitations of such methodologies despite the lack of a clearly predictive and "testable" outcome. The concepts of system dynamics provide for the setting of boundaries and the analysis of endogenous systems in terms of the stock (quantities of material), flow (the rates at which these stocks change), positive (self-reinforcing) and negative (self-correcting) feedback loops inherent in goal-seeking systems, and the delays associated with these interactions (Sterman 2000). By understanding the structure of these feedback loops, it may be possible to design policies that maintain the desired dynamic equilibrium of system behavior required to achieve stability (versus statis) amidst uncertainty.

Today, explanation and causation have taken on new meaning and the aim of design and predictability seem much less assured. A good theory for planning and policy analysis, then, involves a process that delivers a greater understanding of our complex environment but whose goal may simply be to provide a spectrum of desired outcomes rather than a single, designed course of action. Planners and analysts are now exploring the behavior of systems rather than focusing solely on their numeric by-products and traditional methods of variance and regression analysis. System dynamics was designed to provide this insight. In the face of global turbulence and uncertainty, less emphasis is being placed on developing specific long range plans of actions among some large multi-national corporations, focusing instead on shorter time horizons (Grant 2003). Policy analytics and planning have shifted their emphasis to enabling adaptability through increased environmental awareness and strategic thinking. This has resulted in less formal processes of planning and analysis with greater appreciation for creativity and innovation in the development of alternative future scenarios to enable flexibility in the face of uncertainty. In the end, both planning and policy analysis are seen as enabling better informed decision making.

Modeling to Offset Policy Resistance, Cognitive Biases, and Cross-Sectoral Dissonance

The value of system dynamics models would seem to lie in their ability to provide decision makers a better awareness of possible outcomes based on an accepted portrayal of system structures. System dynamics models may offer policy makers a tool that provides the means to better understand dynamic complexity, structure/behavior relationships, and to partially offset the effects of policy resistance in the planning and policy development process. System dynamics is a powerful method to gain useful insight into situations of dynamic complexity and policy resistance (Sterman 2000). Donella Meadows describes policy resistance as several actors working independently to achieve various goals within a system but finding their actions only exacerbate the problems they are attempting to address Meadows (1982).

Depending on their level of fidelity in modeling highly complex systems, many system dynamics models can overwhelm those unfamiliar with the science and methodology employed, thereby defeating the purpose of clarifying complex problems. One of the challenges facing system dynamicists, then, is to formulate models that are easily understood by decision makers. Ghaffarzadegan et al. (2011) propose the use of small system dynamics models (containing a limited number of stocks) to inform public policy decision making and to overcome policy resistance. Sterman explains that, "Policy resistance arises because we often do not understand the full range of feedbacks operating in the system" (Sterman 2000, p. 10).

For purposes of analysis, the specific numeric outcomes of the models may be seen as secondary to their value in providing stakeholders and decision makers with an enhanced understanding of the feedback mechanisms, non-linear behavioral trends, and systemic interrelationships within the strategy or proposed plan. Schaffernicht and Groesser found, "Research has demonstrated that more comprehensive and dynamic mental models seem to be at the foundation for improved policies and decisions" (2011, p. 57). Doyle et al. (1998) believed that the purpose of employing system methodologies, including the use of system dynamics modelling and flight simulators is to enhance those mental models by making them more complete and complex.

Sterman (2000) noted that system dynamics brings together many qualitative as well as quantitative disciplines, including cognitive and social psychology and economics. Kahneman and Tversky found that when confronted by uncertainty, people often use heuristic principles in order to simplify complex problems, but these overly simplistic heuristics can include various biases that may lead to significant systematic errors in judgment (Tversky and Kahneman 1974). Kurtz and Snowden (2003) concluded that decision makers and policy advisors fit reality into their existing mental models. This is a primary source of policy resistance and can result in short term fixes that only exacerbate systemic problems over time.

Boland and Tenkasi (1993) described research they conducted among diverse collaborators as perspective making and perspective taking in communities of knowledge. The authors use an open system, cybernetic model to explore feedback control mechanisms that impact communications among communities of knowing. In this context, perspective making represents the process of knowledge sharing that relies heavily on personal and shared narratives to enhance sense making. Perspective taking involves the presentation of diverse knowledge and making that unique knowledge available to others. The authors acknowledge that judgmental processes come into play on an individual level and the personal heuristics they apply often include cognitive biases, such as availability bias, that can lead to an over-estimation that their personal perspectives will be shared by others. Sterman asserted, "The heuristics we use to judge causal relations lead systematically to cognitive maps that ignore feedbacks, multiple interconnections, time delays, and the other elements of dynamic complexity" (2000, p. 28). It seems logical that using system dynamics models might also serve as a tool for expressing a shared narrative to enhance perspective taking and perspective making while diminishing cognitive biases.

Exposure to system dynamics modeling may, then, provide grounding for decision makers by increasing their understanding of the feedback mechanisms at play within and among the systems being modeled, causing policy analysts and planners to more rigorously evaluate assertions of cause and effect as well as some of their imagined outcomes. Better understanding the complexity and structure/behavior relationships within the systems of a strategy or plan provides clarity and a shared focus on longer term consequences that would aid in diminishing policy resistance.

In the examples that follow, students and faculty at the Naval Postgraduate School used system dynamics modeling to support decision makers planning an economic and educational development strategy in Central Coast California, to model US-China relations for the Department of Defense, and to explore a new approach to the net assessment of the effectiveness of US and partner counter terrorism strategies for the US government. These examples are intended to demonstrate planning and policy analysis applications of system dynamics in both the public and private sectors and to serve as an indication of how system dynamics might be used effectively to support decision makers.

Methodology

In order to answer the question, "How does system thinking and the use of system dynamics modeling inform regional strategic planning?" an action research approach was applied in case study analysis of the Steinbeck Innovation Cluster strategic planning effort in order to evaluate the impact of system dynamics modeling on the participants. In so doing, the core concepts of grounded theory—constant comparison and theoretical sampling—were employed. The value in this approach was the development of an emergent theory through both analysis of data gathered and sensitization derived from my own immersion in, and interpretation of, the data. The objective, therefore, was not to verify theory previously espoused by related literature, but to allow categories and concepts to emerge from the subjects themselves.

In contrasting grounded theory with logico-deductive theory, Glaser and Strauss (1967) extoll the benefits of theory that is inductively developed. Their strategy of comparative analysis for generating theory places the emphasis on theory as a *process* in which, "Qualitative research is often the most 'adequate' and 'efficient' way to obtain the type of information required to contend with the difficulties of an empirical situation" (p. 18). Dooley explains that when applied to theory building, case study research is a method that can help us understand complex issues while contributing to previous research. He adds that, "…case study research approaches the purpose and methodology of grounded study research—the conceptual development and operationalization of a new theory" (Dooley 2016, p. 350). Bradbury and Reason (2003) describe action research as being grounded in the experiences of the researcher who is working in partnership with the subjects of a study to address problems of significance to the community. The authors assert that various forms of action research, "have in common a commitment on the part of a group of organizational practitioners from diverse organizations, field-consultants/organizers and researchers to work together and share insights across the entire community and beyond" (p. 167). Action research was particularly relevant in my Case Study of the Steinbeck Innovation Cluster strategic planning process.

As part of the Steinbeck Innovation Project strategic planning team, I provided an overarching systemic vision for a sustainable strategy of economic growth in the region. I participated in every phase of planning, from long term vision to the assembly of key civic, academic, and private-sector stakeholders. I was able to attend all significant meetings and discussions and to introduce concepts of system thinking and system dynamics methodology. Further, I exchanged over 3500 e-mails and correspondence, had access to formal meeting minutes, and was consulted in the development of diagrams, presentations, and planning documents that would eventually constitute a formal regional strategic plan for sustainable economic development that came to be known as the Steinbeck Innovation Cluster.

In contrast to this action research/grounded theory approach, the Asia Pacific Region case study focused primarily on the system dynamics modeling of qualitative and quantitative research that had been conducted by several government and academic institutions. The resulting models were introduced to military decision makers as a potential tool for better evaluating complex strategic relationships. Similarly, the Net Assessment case study utilized extensive multi-disciplinary, collaborative, exploratory qualitative and quantitative research to develop a causal loop diagram and a system-dynamics counter terrorism modeling mock-up to help the sponsor develop both system dynamics and agent based models for a regional decision support tool.

Comparing the outcomes of these three case studies was not attempted here since little rigor was applied to qualitatively or quantitatively assess the value of the Asia-Pacific Region model and the Net Assessment model prototype. On the other hand, the Steinbeck Innovation Cluster research did involve the coding of pre- and post-intervention interviews to determine the model's effectiveness in increasing the participants' awareness of the non-linear feedback at play within the systems that constituted the strategy and the timelines over which these dynamics resulted in potential behavioral outcomes.

Modeling to Inform Strategy and Policy Decisions

Steinbeck Cluster Case Study

In early 2012, Salinas, California was confronted by a moribund local economy. Gang-related violence was deterring investment, there was insufficient employment opportunity for local residents, and this was exacerbated by a growing youth bulge. Water scarcity and restrictive water management policies made agricultural growth and sustainability a constant challenge, with farmers, ranchers, and vintners sharing diminishing sources of water. State and local regulations, taxes, and energy costs deterred new business development, and contributed to the stagnant job market, further decreasing area attractiveness for new home development and construction. The historic city center was characterized by empty store fronts and slumping small businesses. The kindergarten through twelfth grade (K-12) school system offered students little real hope of meaningful employment or academic futures. Local area Hispanic youth, recruited and pressured by gang members, often entered the sagging service sector, working in local hotels and restaurants, while many of their parents labored in the fields or struggled with local businesses. All of this, the mayor noted, was occurring just 20 miles from the wealth of the Monterey Peninsula, within reach of several renowned institutions of higher learning, and 60 miles from the largest engine of new business development and technology innovation in the nation, the Silicon Valley.

The Mayor of Salinas assembled a working group that consisted of prominent local area business leaders, city council members, shippers, growers, ranchers, and vintners, academics, bankers, and a technology investment consultant from Silicon Valley. Over time, an Executive Committee was formed that recognized the need for a systemic approach to growth and sustainability, and the non-profit Steinbeck Foundation was eventually formed to support the effort.

Agriculture production is the largest industry in the Salinas Valley and competition for scarce water resources is fierce among the large commercial growers of row crops. What became clear from early modeling and research was that the agriculture industry in Monterey County was already operating near the carrying capacity of the water system, so that attempting to develop a strategy of economic development dependent on adding agricultural acreage to increase production was not sustainable. Having studied the components and benefits of successful regional technology and manufacturing clusters (Porter 2003; Iammarino and McCann 2006; Bresnahan and Gambardella 2004; Gordon and McCann 2000), it seemed logical that by leveraging education, a strong youth demographic, the proximity of local research universities, and the innovation, technology and venture capital of nearby Silicon Valley, Salinas might focus its effort on manufacturing the components of "precision agriculture" (robotics, real-time sensors, IT infrastructure, field packaging, and product tracking) for export, while simultaneously improving local production and transitioning unskilled laborers to higher paying, more skilled jobs. Whether this commercial growth and emphasis on technology education was sustainable and sufficient to offset water shortages and to diminish the influence of gang activity would need to be understood over time.

Model Architecture

As the Steinbeck Innovation Cluster strategy took shape, it was clear that system dynamics modeling might assist city, county, and investment decision makers in understanding the systemic nature of their strategy for sustainable growth and help them prioritize their resources. Components of this strategy included education (agriculture technology curricula in high school, higher education, and vocational education); mitigating gang activity; improving area attractiveness for middle to high income earners and businesses; the integration of funding for operations, new business start-ups and research; and, a model of county-wide water supply and demand. Consistent with this strategic approach, with the help of area experts, five sub-models or modules were developed using iThink® (ISEE systems) modeling software: Education for Agriculture Technology Employment; Gang Membership and Programs; Water Management; Investment, Funds, Start-ups, and Research; and, Area Attractiveness/Middle to High Income Earners. Each module was initially modeled independently and then all modules were run as a single integrated system model with a color-coded system of systems model architecture that captures the elements of the Steinbeck Innovation Cluster strategy (Fig. 1).

The iThink® software also facilitated the creation of a control panel or decision flight simulator (Fig. 2) that allowed decision makers to manipulate those inputs they could reasonably control. Each time the model was run, all five modules ran simultaneously, providing input and output among modules as indicated by the

Fig. 1 Cluster model architecture

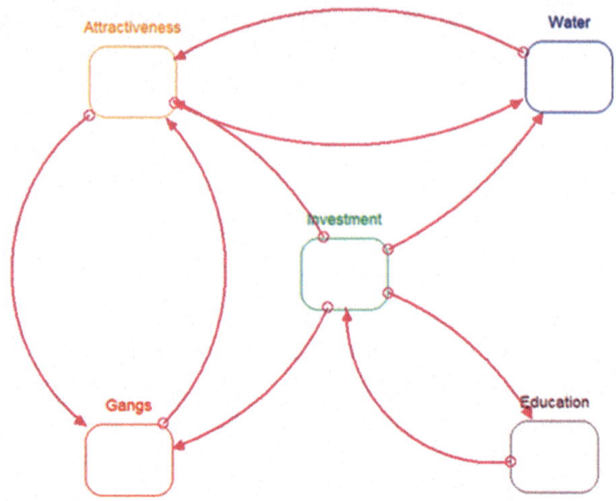

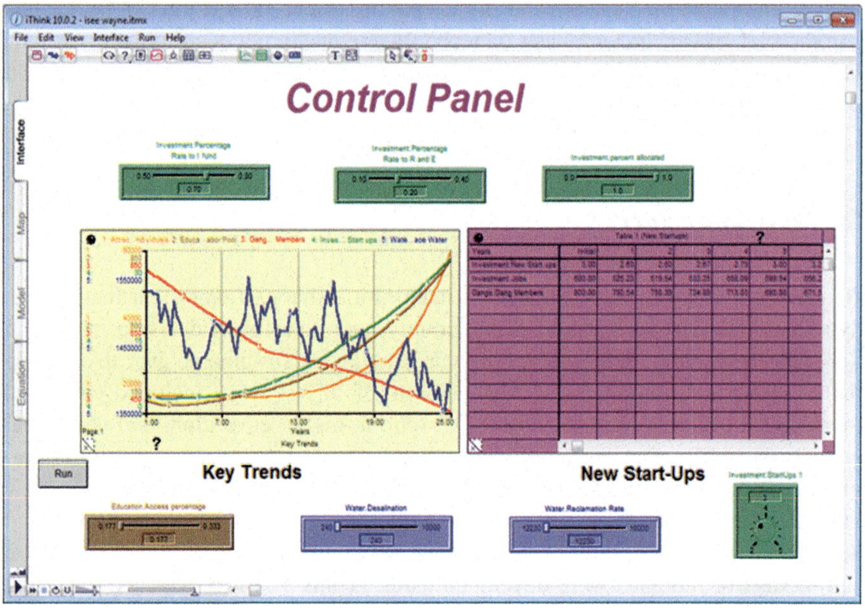

Fig. 2 Control panel/Flight simulator

system architecture diagram and noted within each stock and flow diagram. The resultant graph of system behavior was correspondingly displayed on the control panel/flight simulator dashboard. The purpose of the model and control panel was to enhance the Executive Committee members' understanding of interrelationships within the strategy and the impact policy decisions might have on the sustainability of the strategy over time.

Model Validation

Because system dynamics was used to model proposed structures and policies of the strategy rather than existing system structures, validating the model by its ability to replicate historic trends was, for the most part, not practical. Instead, key subject matter experts within the city of Salinas (e.g. Chief of Police, City Manager, Superintendent of High School Districts) were asked to validate the model structures and the data used to populate the models to ensure they reasonably reflected existing structures affected by the planned policy changes. Since many equations and parameters were based on estimates or assumptions, the specific numeric outcomes of the model were seen as secondary to their value in providing decision makers with an enhanced understanding of the feedback mechanisms, non-linear behavioral trends, and systemic interrelationships within the Steinbeck Innovation Cluster strategy.

The Modules

A screen capture of each module's stock and flow diagram is provided in Figs. 3, 4, 5, 6, 7. It should be noted that color coding corresponds to the colors used in the system architecture shown in Fig. 1. So, for example, in the stock and flow diagrams an orange convertor that appears in the Gang Membership, Violent Crime module (red), the Agriculture Technology Education module (brown), the Water module (blue), or the Investment module (green), indicates an input from, or output to, the Attractiveness (orange) module.

Investment for Education, Research, and New Business Start-ups/Jobs

Because this was a purely conjectural module based on projections of investment made as part of the Steinbeck Innovation Cluster strategy, there was no historical data upon which to base many of the equations. Assumptions were made based on research that had been done during the planning process. This module (Fig. 3) features six stocks that relate investment to education/research and new business start-ups.

Ag-Tech Education

This module (Fig. 4) illustrates the flow and feedback mechanisms among five stocks related to agricultural education, the labor pool, and unemployment/low income earners.

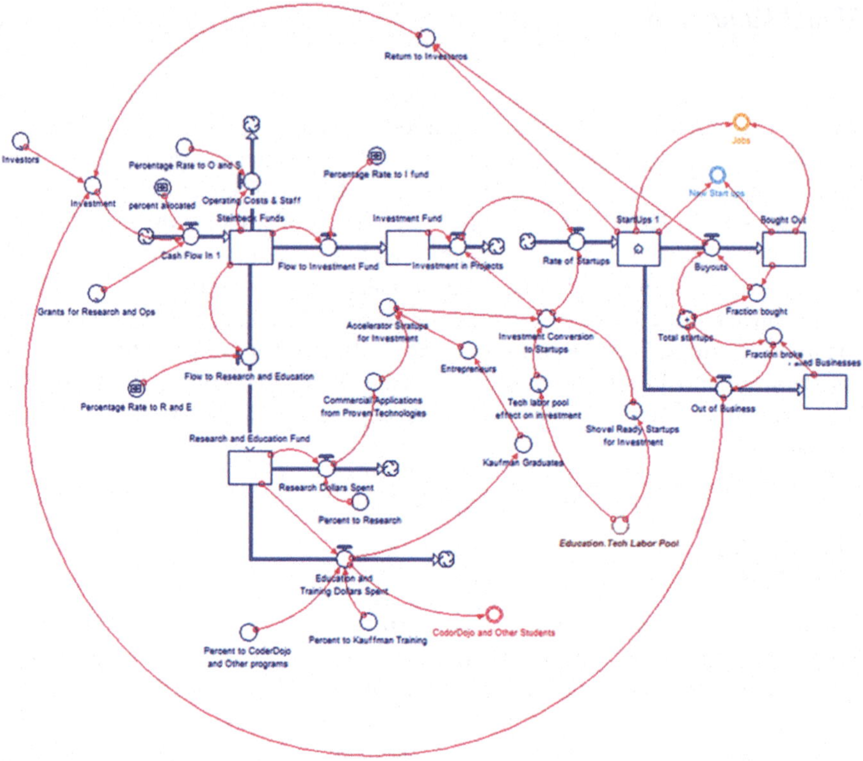

Fig. 3 Investment for research, education, start-ups/jobs

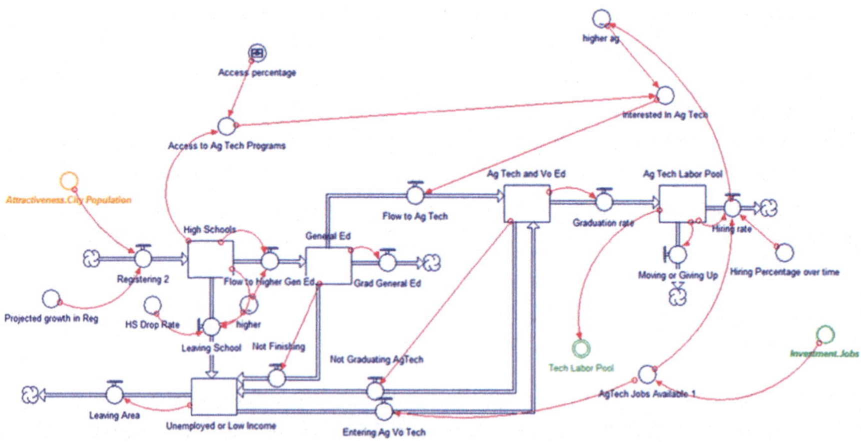

Fig. 4 Education for Ag-tech employment

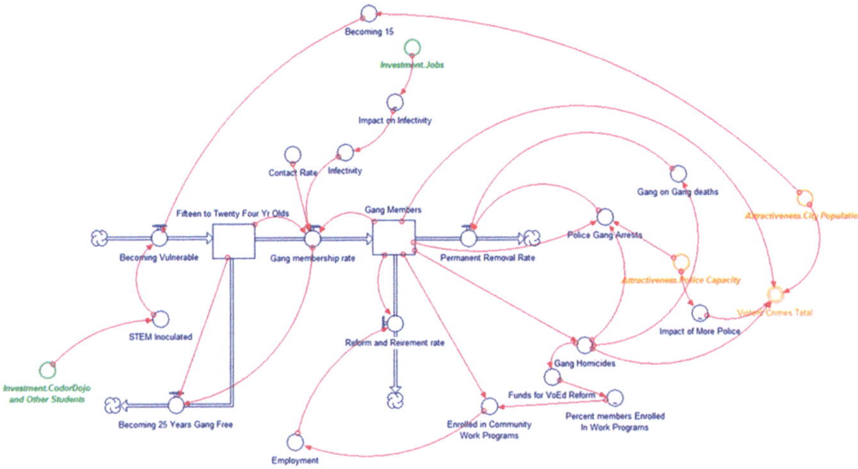

Fig. 5 Gang membership and programs

Gang Membership

The Gang Membership Module (Fig. 5) is based upon a susceptibility/infectivity (S/I), disease diffusion logistics model. The model features two stocks that relate the pool of susceptible youth to gang membership.

Water Management

All Monterey County water is supplied from within the county, largely through groundwater aquifers and some surface water capture. The Water Management Module (Fig. 6) features two stocks only of available water.

In-Flow of High Income Earners, Area Attractiveness

The Area Attractiveness Module (Fig. 7) features only one stock of high income earners in the agriculture technology sector and is intended to measure the in-flow of middle to high income agriculture technology professionals (incomes greater than $75,000/yr) into Salinas from a pool of potential in-migrants from five nearby cities.

Area Attractiveness was determined by comparing six factors, weighting each factor, and multiplying that by a competitiveness ratio (Salinas factor score/average score among the five). The scoring was based upon city by city comparisons in five of the six factor areas (water availability was not included in this).

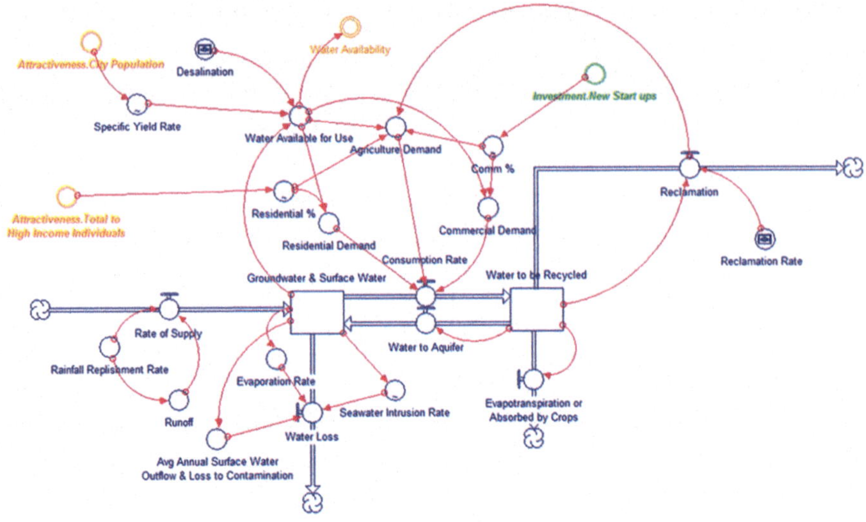

Fig. 6 Water management

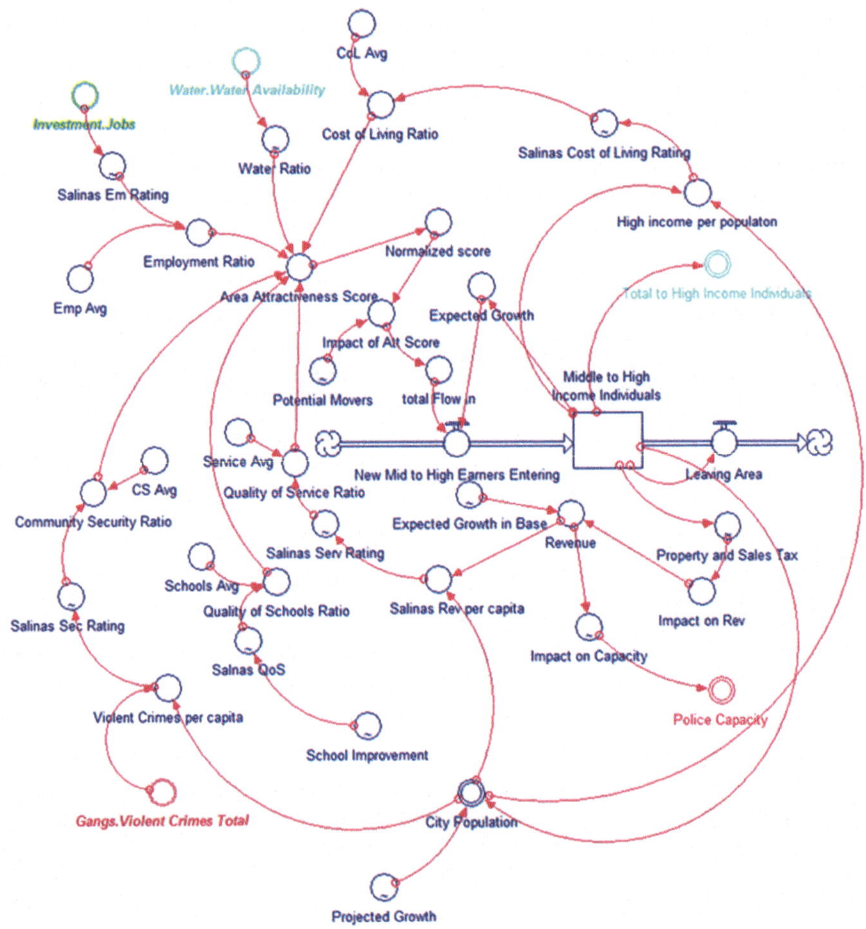

Fig. 7 High income earners, area attractiveness

Model Outcomes

As a system of systems that integrates all five of the modules described above, the Steinbeck Innovation Cluster model was intended to focus decision makers' attention on trends over time that resulted from the non-linear feedback mechanisms and inter-relational behavior of the system. The Control Panel allowed them to change values of only those variables/converters they could reasonably expect to modify. For example, stakeholders could not increase anticipated city revenue or diminish gang membership by simply changing the parameter values, any more than they could change anticipated rainfall. They could, however, manipulate the percentage of investment that they directed to the Investment Fund or to the Education and Research Fund; they could reduce the overall investment dollars they had anticipated they would receive; they could increase the number of high school students with access to agriculture technology curricula; and, they could increase (within system limits) the amount of water derived from desalination and from reclamation. The model could be run from any level of the hierarchy (from the module level to the architecture and Control Panel level), allowing more detailed analysis of each module's output, but the model always ran all modules simultaneously. When running the model for the Executive Committee, members were advised not to focus on specific numeric outcomes (based on assumptions made for modeling purposes), but to consider outcome trends in the context of the inter-relational feedback at play among the modules.

As a decision support tool, members were encouraged to change values of variables using the Control Panel, in order to judge the potential impact these policy changes would have on the behavior of the system over time. For example when participants increased the percentage of dollars invested in new start-ups from 70 to 80% and commensurately reduced the investment in research and education programs from 20 to 10%, fewer new start-ups were created over the 25 year run of the model, resulting in fewer jobs created and an increase in gang membership from the initialized run. In a subsequent run of the model, participants used the Control Panel to increase the percentage of high school students with access to agriculture technology programs from 17.7 to 25%. Surprisingly, this resulted in a significant increase in the number of start-ups and jobs created as well as a significant decrease in gang membership over the 25 year run of the model. This was largely due to the fact that a thick enough skilled labor pool was allowed to develop over time that could sustain the growth of new businesses and simultaneously incentivize youth at risk with job opportunities. Of note, while this increase in economic growth was the effect the Committee was trying to achieve, the model simulation indicated it had a significantly deleterious impact on the water supply, which forced committee members to recognize the need to address water shortages through increased desalination and water recovery reclamation in the near term to avoid an overshoot of water capacity as the economy grows.

The Value of the Model

The conclusion that was drawn from using system dynamics modeling to support the diversity of Executive Committee members, was that system thinking and the use of small, system dynamics models can enhance the awareness of decision and policy makers by clarifying dynamic complexity and structure/behavior relationships, and may contribute to collaborative, cross-sectional effort that diminishes the pitfalls of policy resistance in regional strategic planning. Specifically, small system dynamics models can:

- Help to dispel cognitive and judgmental biases within diverse, cross-sectoral planning teams
- Provide a decision tool for use during strategy development and implementation phases of planning to help prioritize objectives and resource allocations
- Provide an information tool to align cross-sectoral collaboration within the planning team
- Diminish cognitive biases by using quantitative and qualitative data to enhance the understanding of system behavior as a result of non-intuitive feedback mechanisms within the system structure

It is worth considering that the potential benefits of system dynamics modeling, in this case, did not come from the Executive Committee's direct involvement in building the models, the subject of much previous research (Ackermann et al. 2010; Rouwette et al. 2002; Rouwette et al. 2011; Snabe 2007). Although concepts of system thinking were introduced during the strategic planning process, the Executive Committee members were not exposed to the model until 15 months after the planning process had been concluded. Even then, each participant spoke of the clarity the model had provided and of the foreseen benefits of employing the model in taking the strategy forward.

US-China Relations in the Asia-Pacific Region Case Study

Having seen the system dynamics modeling that was done for the Steinbeck Innovation Cluster Executive Committee, members of the Department of Defense approached the Naval Postgraduate School and asked whether it would possible to apply the same methodology to explore the risks and opportunities (conflict and convergence) of US-China relations in the Asia-Pacific Region. As a first step in developing a comprehensive model, faculty and students identified the overall model boundary and pertinent key factors affecting regional geopolitical stability. Relationships between factors were initially postulated, and then refined as inputs from other subject matter experts, nationally, were received (Whitcomb et al. 2015). Figure 8 presents a high level overview of the initial modelling concept.

Each of the key factor areas became candidates for sub-models, and in fact resulted in five integrated modules or sub-models (Tension, Economic, Energy and Resources, Demographics and Stability, and Military Actions). The "Archetypes" referred to in

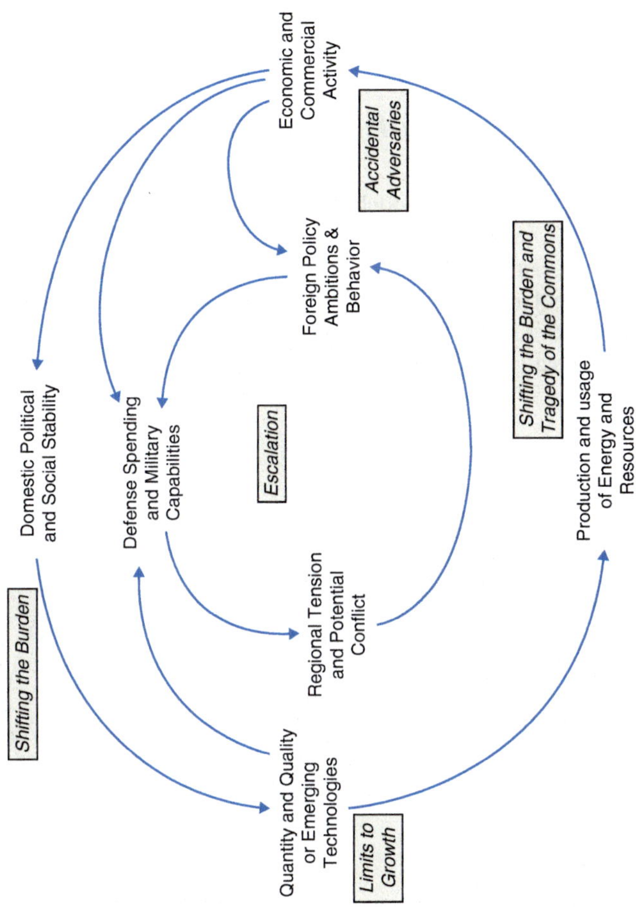

Fig. 8 Asia-Pacific conflict and convergence model (with archetypes). Source: [Whitcomb et al. 2015]

Fig. 8 (Accidental Adversaries, Shifting the Burden, Tragedy of the Commons, Limits to Growth, and Escalation) are system archetypes identified in Senge's *The Fifth Discipline* (1990, 2006). Each of these archetypes corresponds to a major component of the overall model. The first four archetypes are clearly associated with a model subcomponent (note that Shifting the Burden and Tragedy of the Commons are both utilized in the Energy and Resources sub-model). The Escalation archetype is at the center of three components (Defense Spending & Military Capabilities, Regional Tension & Potential Conflict, and Foreign Policy Ambitions & Behavior) which were aggregated into a more comprehensive Military Actions sub-model.

Modeling Tension in the Asia-Pacific Region

The NPS team began detailed model development by looking at the central Escalation loop shown in Fig. 8. The central loop depicted in Fig. 8 is not a complete causal loop diagram (CLD) in the classic sense of the term, since it lacks the standard "+" or "–" polarity notations that would define either a reinforcing loop or balancing loop. It does serve as a starting point for development of a more detailed CLD (Fig. 9) specifically addressing tension between the U.S. and China in the

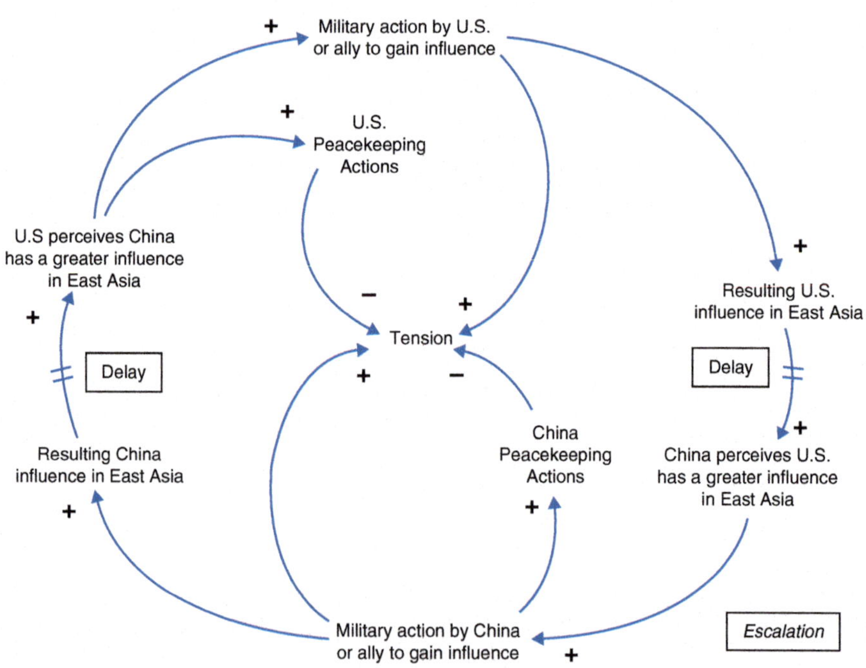

Fig. 9 Tension causal loop diagram. Source: [Whitcomb et al. 2015]

region. The methodology followed for developing the sub-model applies to the other areas as well, which resulted in an integrated model composed of modules or sub-models interconnected according to the overall scheme of Fig. 8. While detailed CLDs and sub-models are described in the project documentation (Whitcomb et al. 2015) this chapter will use the Tension sub-model as an exemplar for the application of the methodology to the other areas.

The tension sub-model is based on two competing actors (the U.S. and China) each taking actions to balance the actions of the other actor. China and the U.S. both have goals to influence the East Asia region. China's goals stem from their belief that ancient historical boundaries entitle them to exercise sovereignty over certain territories. The United States' goals stem from the many alliances that they have in the region which in some instances obligate them to military action. There are also economic interests for both the U.S. and China that drive the desire to maintain influence in the region, and these actions are modeled in the economic and energy sub-models. Each actor is assumed to have an internal picture of what their desired level of influence (political, economic, and military) is in the region, and what their desired level of influence in the region is for the other actor.

Each actor also has a perception of their own actual level of influence and the actual level of influence of the other actor. If China for example perceives that its actual level of influence is less than its desired level of influence, the assumption is that China will take actions to increase it. Examples of such actions could be the establishment of new diplomatic or economic relationships with other actors in the region, trade agreements with other actors, increased military equipment sales and training for third parties in the region, or increased presence through military exercises, operations, or new military basing agreements. Likewise, if China perceives that the U.S.'s level of influence is greater than China's desired level for U.S. influence in the Asia-Pacific region, China will take actions to limit or reduce U.S. influence in the region. Any action either the U.S. or China takes to correct what one or the other views as a delta in perceived versus desired influence thus creates tension between them. Without any tension-reducing actions on the part of China or the U.S., these reinforcing actions would continue to escalate until one or both of the parties is eliminated or change their goals.

As seen in Fig. 9, there is a delay between China's influence and the U.S. perception of China's influence. If the U.S. perceives China's actions are giving China greater influence in the region, then the U.S. is likely to take military actions such as sea and air "freedom of navigation" missions to counterbalance China's actions that would lead to increased tensions between China and the U.S., as illustrated in the CLD. Based on the CLD, tensions increase unless tension-easing actions are taken that allow the militaries of China and the U.S. to cooperate. Examples of tension-easing actions might be joint exercises conducted to combat piracy as well as joint exercises to aid nations that encounter natural disasters. These peacekeeping actions will lower tensions temporarily but the problem will not be solved com-

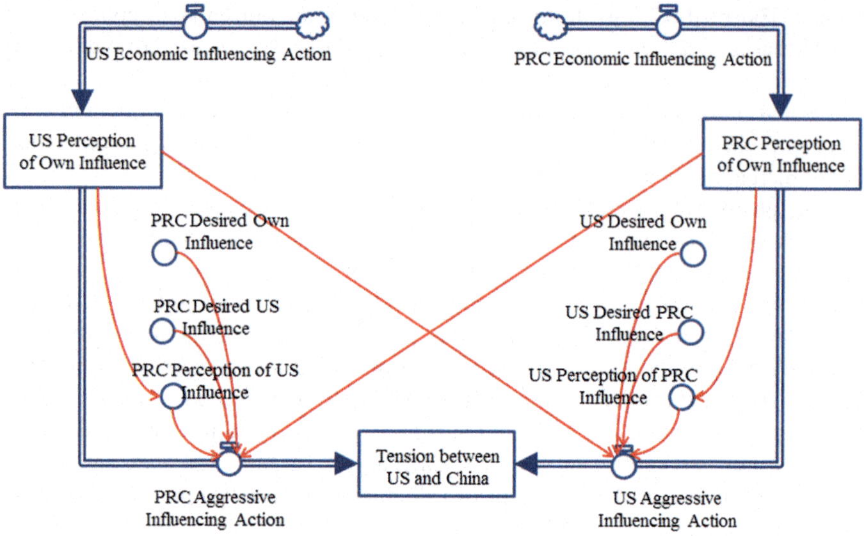

Fig. 10 Tension sub-model implemented using iThink® software

pletely through military means. There must be some economic or other contributions that lead to a consistent downward trend of tension, as was captured in the Tension Sub-Model in Fig. 10.

Examples of destabilizing actions that can be modeled by this tension sub-model are:

- Increase in defense spending above a certain threshold by China, the U.S., or ally.
- Disputes involving military vessels between China and the U.S., or between an ally of China and a U.S. ally.
- Declaration of an exclusion zone in international sea space using military vessels as enforcement.
- Introduction of new militarily significant technology.
- Military occupation of disputed territory.
- Nuclear proliferation issues.
- Official visits by high profile U.S. political figures to U.S. allies in the Pacific.
- Official visits by high profile Chinese political figures to China's allies in the Pacific.
- Cybersecurity

The Dashboard Flight Simulator

As noted earlier, iThink® software offers an interface layer in which user controls, model inputs, and model outputs can be arranged to give users a more intuitive interface without the need to delve into the details of the actual models.

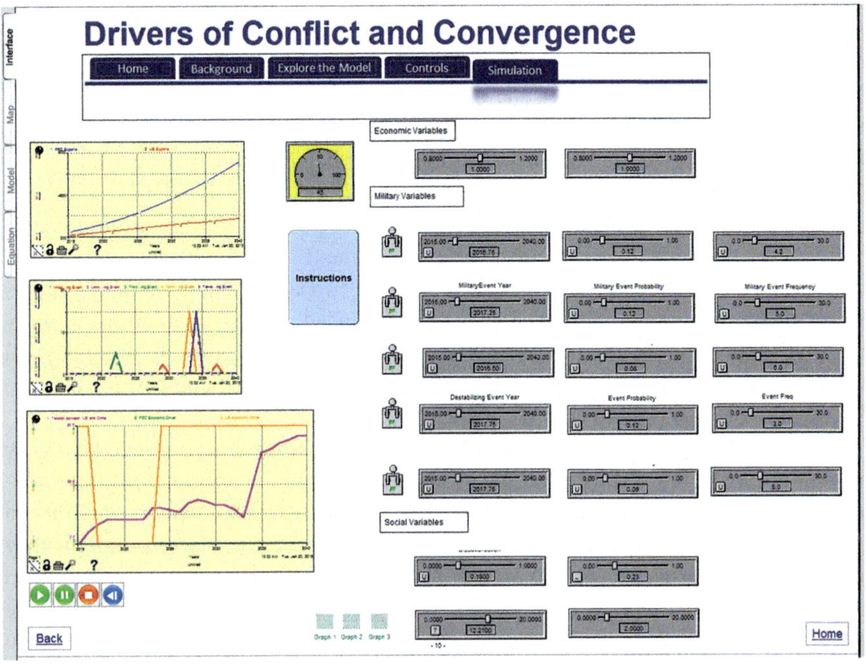

Fig. 11 US-China dashboard flight simulator interface

Simulation controls are provided to facilitate repetitive simulation runs, thus encouraging users to conduct "what if" analysis by varying the values of the inputs.

The Dashboard Flight Simulator (Fig. 11) is divided into three functional areas; inputs, outputs, and controls. The graphs on the left side display results (outputs) of the simulation runs. The larger graph on the bottom displays tension level over a period of 25 years. The user has the option to make any graph comparative by displaying the behavior of a single variable (a "variable of interest") over time in multiple simulation runs. Underneath the graphs are a series of simulation controls that let the user start, pause, stop, and reset simulation runs. On the right side of the screen is an area containing user input controls, in this case toggle switches and sliders that let the user change certain economic and social variable values, decide which potential destabilizing events will occur, and the likelihood and frequency of such destabilizing events occurring. The Dashboard Flight Simulator provided military planners a simplified simulation tool that encouraged experimentation and exploration of the underlying system dynamics model. By expanding the boundary of the system to include economic, demographic, and energy considerations as well as purely military aspects, planners gained insights into how interrelationships and feedback mechanisms might affect the range of potential actions the U.S. and its allies could employ in the pursuit of national objectives in the Asia-Pacific region over the next 25 years.

Modeling for Ongoing Net Assessment of Strategy and Policy

Discussions of net assessment generally begin with, or at least include, a reference to Andrew Marshall who, in the early 1970s brought to the National Security Council a concept for analyzing the balance of strategic power between the United States and the Soviet Union. In 1973, Marshall was made the Director of Net Assessment housed in the Pentagon, retaining that position until his retirement in 2015. Over the years, the subjects for net assessment shifted with the geopolitical and military dynamics of the times. But generally speaking, these net assessments were either geographically or functionally focused. Using a multidisciplinary approach, they applied qualitative and quantitative methodologies, looking beyond mere force on force metrics to the underlying complexities of competitive balance between two strategic powers (Skypek 2010).

Paul Bracken discusses the fact that Marshall's net assessments expanded the view of senior leaders beyond force on force considerations to the asymmetries that influence power and the extended time horizons over which strategies play out. Bracken makes the point that complex organizations tend to deconstruct large problems they encounter in order to make them more manageable then attempt to reconstruct them through coordinated policy. He asserts that this creates internal divisions within large organizations, and that net assessment was able to identify this phenomenon and to integrate Red and Blue (Soviet and U.S.) strategies in order to provide a net assessment of the competition between them (Bracken 2006). In other words, perceptions of a competitor's capabilities and intentions can drive one's own strategy and actions, and it is the net assessment of the resulting behavior that is important.

In 1999, the defense contractor Booz-Allen (1999) addressed the application of net assessment to network-centric warfare in a paper written for Marshall's Net Assessment office. The paper describes network-centric warfare as a complex adaptive system and asserts that focusing on networks "is also a shift from a closed to an open system in warfare where actors are no longer independent but part of a 'continuously adapting ecosystem'" (page iv). This is precisely what seems to have played out in the twenty-first century, when conflicts and adversaries are no longer clearly defined as "Blue" or "Red," but by the many shades of grey in which their interests and agents compete. Nowhere has this been more evident than in the confrontations between violent extremist, non-state actors and the alliances of disparate state partners determined to counter them.

Net Assessment Case Study

In 2014, the Defense Analysis Department of the Naval Postgraduate School was invited to participate in a multidisciplinary effort intended to develop a net assessment approach to U.S. government counter terrorism efforts. Over the course of the next year, subject matter experts from across academia and the public and private sectors were periodically convened to address various economic, sociological, cognitive, geopolitical, and anthropological aspects of domestic and foreign terrorism as well as evolutionary science, complexity and modeling implications. The aim was to develop a new-age, diagnostic approach to net assessment to better inform policy and decision makers of the effectiveness of U.S. and partner actions taken to offset and eliminate terrorist behavior and recruiting. As CAPT Todd G. Veazie, USN noted in the preface to a journal volume assembled by NPS dedicated to the net assessment of counter terrorism, "A good diagnosis will provide the necessary 'sense-making' to guide appropriate action in full cognizance of the long-term consequences of both action and inaction, in terms of potential threats and opportunities across a range of policy choices. A bad diagnosis can lead to policy choices that are inefficient, ineffective, and potentially tragic" (Veazie 2015).

Throughout this process, the complex and adaptive nature of organizations and movements involved in terrorism was a recurring theme, as was the actual complexity and evolutionary sciences that underpinned notions of self-organization and emergence among these organizations. Increasingly, emphasis was placed on the "ecosystem" within which terrorism thrives, and whether or not this was a suitable metaphor for conducting net assessment. The thrust of the effort initially attempted to define or bound the stakeholders in terms of traditional Red and Blue net assessment terminology, as well as defining Green. The questions of whether Blue should consist only of U.S. actions or those of a dynamic set of partners, and whether Red should consist only of named terrorist organizations or a loose affiliation of adversaries was hotly debated. While one concept of Green was as a representation of actors with tendencies toward either Blue or Red, a more intriguing approach was to think of Green as an ecosystem that represents the strategic environment and conditions in which both Blue and Red operate. Using this second conceptualization, the aspects of the Green ecosystem that seemed to have the most impact on Red and Blue actors were generally recognized as including economics, resources, technology and information, self-identity, security, and governance. This is portrayed in Fig. 12.

As discussed, net assessments are generally either functionally or geographically oriented. In what came to be referred to as Net Assessment 2.0, it was clear that while many instances of terrorism often shared similar motivations and broad objectives, the localized instantiation of violent extremism significantly varied from region to region, even while in some cases representing a global phenomenon and recruiting effort. After several months of collaborative thinking among the participants, it seemed necessary to capture a global portrayal of violent extremism with a simple causal loop diagram (Fig. 13).

The causal loop diagram in Fig. 13 captures only the highest level interaction among Blue stakeholders (intended here to represent U.S. and other nations, international organizations, non-governmental organizations, and private sector partners), Red stakeholders (intended here to represent violent Islamist extremist movements, organizations, individuals, and supporting adversaries to the U.S. and our partners), and the Green strategic environment (intended here to represent the conditions of the environment, identified in Fig. 12, in which Red and Green operate). As Blue and Red each perceives a gap between the conditions of Green and

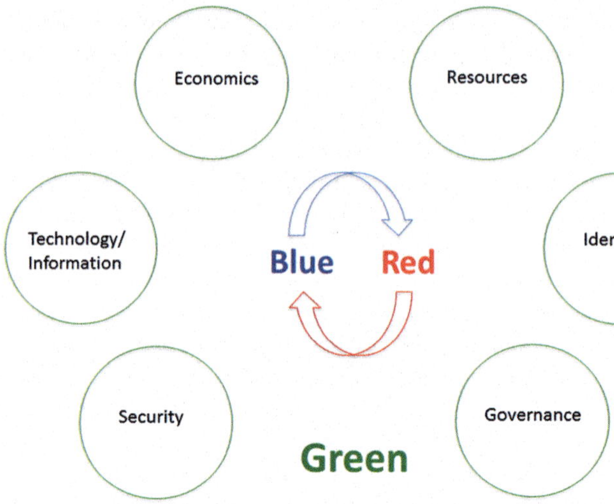

Fig. 12 Conditions of *Green*

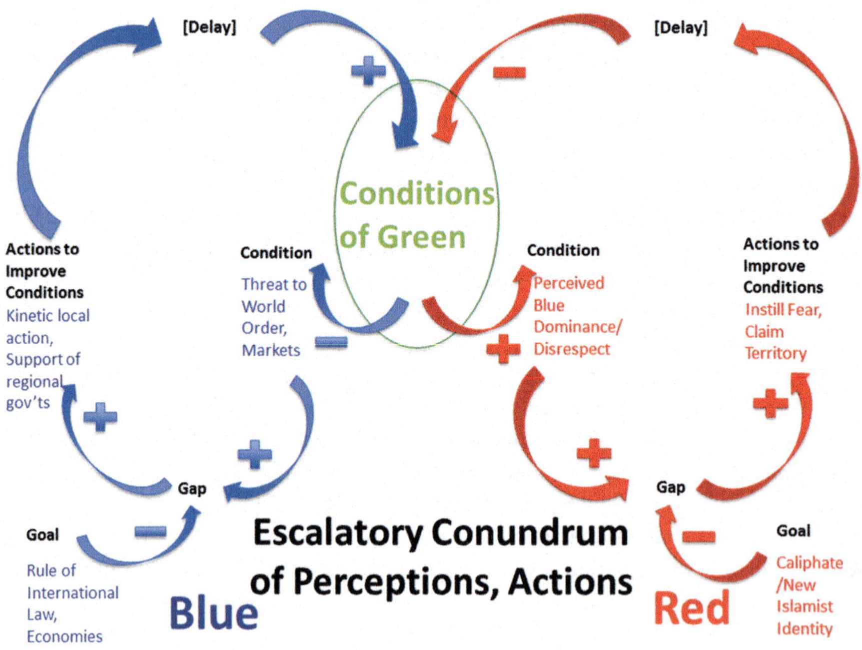

Fig. 13 Causal loop diagram of *Blue* and *Red* behavior in response to conditions of *Green*

their desired goal, they take actions to close the gap. But responsive adversarial actions only serve to maintain or increase the perceived gaps for both Red and Blue, resulting in an escalatory conundrum with no clear end in sight.

This was only a preliminary step toward developing the means for Net Assessment 2.0 to provide a more substantive understanding of the challenges that face us in the twenty-first Century than was provided by Marshall's Net Assessment office, particularly with regard to foreign and domestic terrorism. It has been written that Marshall's approach to net assessment was based upon four pillars: long term trend analysis; doctrine or concepts of operation that drive courses of action; the identification of strategic asymmetries; and, scenario-based wargaming for hypothesis testing (Skypek 2010). While these pillars remain relevant, advances in non-linear mathematics and the corresponding increase in computing power that is now available have opened the door for deeper analysis. For instance, Cioffi-Revilla asserts that computational social science, based on an information-processing paradigm that recognizes society as a complex adaptive system changing states in response to conditions within its environment has much to offer (Cioffi-Revilla 2014). This recognition argues for a broader approach to net assessment.

The need for a better quantitative and qualitative understanding of the interrelationship of Red and Blue behavior in today's complex and uncertain environment may represent the essence of Net Assessment 2.0. Planners and analysts need methods and tools to conduct longitudinal net assessments as policy and actions change over time and are reflected in behavioral changes, greatly accelerated by social media and internet connectivity. One approach would be to develop system dynamics models to capture unique local or regional manifestations of the escalatory conundrum illustrated by the causal loop diagram above. Figure 14 shows a sample stock and flow model mock-up of localized competition between Blue and Red under the conditions of Green discussed earlier.

This stock and flow diagram is, again, based on the susceptibility and infectivity disease diffusion model discussed earlier. It is intended to illustrate some of the interactions among Blue and Red entities (both local and global) within the conditions of Green (as shown in the CLD Fig. 13) in a local context. This is only a mock-up and is shown here to provide a rough conceptualization as to how different systems might impact the rate of radicalization of a susceptible population (both locally and globally) in light of local governance, Blue responses to Red violence, and outside Red funding and recruitment. If such a system dynamics model was to be developed as a system of systems representation, it seems feasible that this could inform the design and development of an agent based model. Both models would be populated by actual data, with genetic algorithms used in the agent based modeling to simulate the evolution of agent identities, motivations, and behaviors in response to the actions of Red and Blue and the environmental conditions of Green. This approach might eventually provide a longitudinal Net Assessment 2.0 that could support a more flexible and efficacious strategy over time. Gartner stated, "It is vital for the security of the world's nations to be able to determine which of these groups are more capable of threatening, which areas are likely to see the greatest emergence of violent non-state actors, and how environmental factors influence the

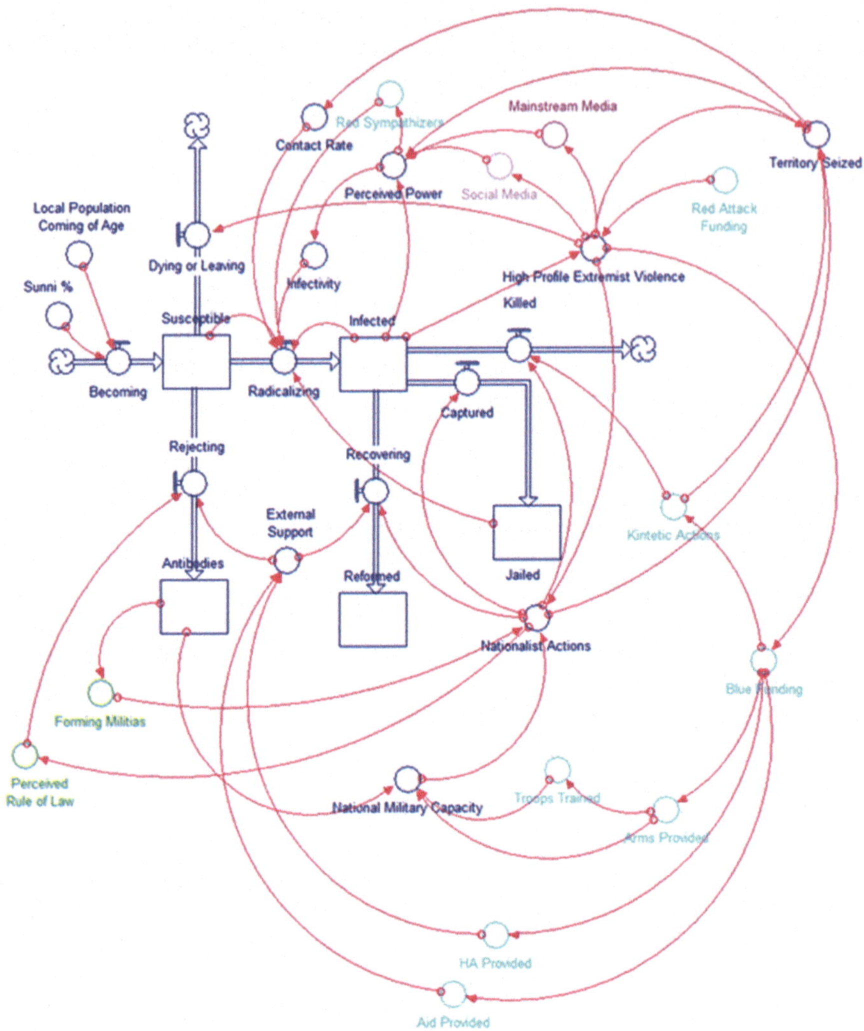

Fig. 14 Stock and flow mock-up of *Red* and *Blue* impact on radicalization

insurgent-counterinsurgent dynamic. Net Assessment 2.0 thus represents an essential global concern" (Gartner 2015).

Conclusion

The focus of this chapter has been on the use of simple system dynamics modeling as one tool available to assist policy and planning analysts in order to better inform decision makers and, ultimately, to overcome the pitfalls of policy resistance, cognitive biases, and dissonance among communities of interest seeking to find common analytic ground. The value of system dynamics modeling lies in its ability to bound a specific systemic problem, frame its structure and the feedback mechanisms inherent in that structure, populate the model with actual or estimated data, and to run simulations over a time horizon sufficient to provide insights into the potential, non-linear behavior of the system. Further, these models can provide decision makers with simple tools and graphics to immediately test variations of structure or data. While these simulations aren't intended to be predictive in the most literal sense, they can provide a better understanding of possible non-intuitive and unintended consequences of policy decisions or actions under consideration as well as much greater awareness of the feedback relationships that affect system—and system of systems—behavior. System dynamics can also provide key insights into system behavior that may then be explored through other analytical tools such as wargaming, agent-based modeling, Delphi groups, or more traditional tools of operations research, including optimization. Whether system dynamics is applied to analyze planning, policies related to state on state or other forms of competition, or to support net assessment efforts, it is clear that its applications may be of great value for both planning and policy analytics.

Acknowledgement Gary Parker, a faculty member of Naval Postgraduate School (NPS), contributed to this chapter by providing his first-hand insights and technical expertise regarding the modeling that was done at NPS for US-China relations in the Asia-Pacific region. His assistance was greatly appreciated and his work is cited as (Whitcomb et al. 2015).

References

Ackermann F, Andersen DF, Eden C, Richardson GP (2010) Using a group decision support system to add value to group model building. Syst Dyn Rev 26(4):335–346

Boland RJ, Jr., Tenkasi RV (1993) Perspective making and perspective taking in communities of knowing. CEO Publication, Los Angeles, T 93-18(238)

Booz-Allen H (1999) Measuring the effects of network-centric warfare, Office of the Secretary of Defense, Net Assessment, Washington, D: OMB No. 074–0188

Bracken P (2006) Net assessment: a practical guide. Parameters Spring 2006:90–100

Bradbury R, Reason P (2003) Action research, an opportunity for revitalizing research purpose and practices. Qual Soc Work 2(2):155–175

Bresnahan T, Gambardella A (2004) Building high-tech clusters: Silicon Valley and beyond. Cambridge University Press, Cambridge

Cioffi-Revilla C (2014) Introduction to computational social science: principles and applications. Springer, London

Dooley LM (2016) Case study research and theory building. Adv Dev Hum Resour 4(3):335–354

Doyle JK, Radzicki MJ, Trees WS (1998) Measuring change in mental models of dynamic systems: an exploratory study. Unpublished manuscript. Department of Social Science and Policy Studies. Worcester Polytechnic Institute, Worcester

Gartner S (2015) Introduction to the special issue. CTX J 5(3):11–13

Ghaffarzadegan N, Lyneis J, Richardson GP (2011) How small system dynamics models can help the public policy process. Syst Dyn Rev 27(1):22–44

Glaser BG, Strauss AL (1967) The discovery of grounded theory: strategies for qualitative research. Aldine Transaction, Piscataway

Gordon IR, McCann P (2000) Industrial clusters: complexes, agglomeration and/or social networks? Urban Stud 37(3):513–532

Grant RM (2003) Strategic planning in a turbulent environment: evidence from the oil majors. Strateg Manag J 24(6):491–517

Iammarino S, McCann P (2006) The structure and evolution of industrial clusters; transactions, technology and knowledge spillovers. Res Policy 35:1018–1036

Kurtz CF, Snowden DJ (2003) The new dynamics of strategy: sense-making in a complex and complicated world. IBM Syst J 42(3):462–483

Meadows D (1982) Whole earth models & systems. Coevolution Q:153–163

Porter ME (2003) The economic performance of regions. Reg Stud (37.6&7):549–578

Porter NW (2014) The effects of system dynamics modeling on systems thinking in the context of regional strategic planning. Naval Postgraduate School, Monterey

Rouwette EAJA, Korzilius H, Vennix JAM, Jacobs E (2011) Modeling as persuasion: the impact of group model building on attitudes and behavior. Syst Dyn Rev 27(1):1-–21

Rouwette EAJA, Vennix JAM, van Mullekom T (2002) Group model building effectiveness: a review of assessment studies. Syst Dyn Rev 18(1):5–45

Schaffernicht M, Groesser SN (2011) A comprehensive method for comparing mental models of dynamic systems. Eur J Oper Res 210:57–67

Senge PM (1990, 2006) The fifth discipline: the art and practice of the learning organization. Currency Doubleday, New York

Skypek TM (2010) Evaluating military balances through the lens of net assessment: history and application. J Mil Strateg Stud 12(2):1–25

Snabe B (2007) The usage of system dynamics in organizational interventions. Deutscher Universitats-Verlag/GWV Fachverl, Wiesbaden

Sterman JD (2000) Business dynamics: systems thinking and modeling for a complex world. Irwin McGraw-Hill, Boston, MA

Tversky A, Kahneman D (1974) Judgment under uncertainty: heuristics and biases. Sci New Series 185(4157):1124–1131

Veazie TG (2015) Preface to the special edition. CTX J 5(3):1–2

Whitcomb CA, Abdel-Hamid T, CAPT Wayne Porter, USN (Ret), Beery PT, Wolfgeher CA, Parker GW, CDR Michael Szczerbinski, USN, and Major Chike Robertson, USA (2015) A systems approach to modeling drivers of conflict and convergence in the asia-pacific region in the next 5–25 years. NPS Technical Report SE-15-001, May 2015. Naval Postgraduate School, Monterey, CA

Norman Wayne Porter holds a PhD in Information Sciences and two Masters of Science degrees—in Computer Science and Joint C4I Systems Technology—from the Naval Postgraduate School. His military service included three tours on the personal staff of ADM Mike Mullen, including Special Assistant for Strategy to both the Chief of Naval Operations and Chairman of the Joint Chiefs. He is a "Walton Fellow" at Arizona State University's Global Institute of Sustainability and recently served as Chair, Systemic Strategy and Complexity at Naval Postgraduate School (NPS) in Monterey, Calif. He retired from the Navy in July 2014 after 28 years of military service and now works at NPS as Director of the Littoral Operations Center, Executive Director of the Common Operating Environment Laboratory, and Senior Lecturer in Defense Analysis. He is also an independent systems and strategy consultant.

Deconstructing Social Policy Innovation Through the Use of Complex Systems Theory: A Methodology for Modelling and Simulation of the Impact of ICT-Enabled Social Innovation

Gianluca Misuraca, Luigi Geppert, and Csaba Kucsera

Abstract The recent financial crisis has put the economy and society of several European Union Member States under enormous pressure, at a time when the demand for social services is growing due to the ageing of societies. State budgets were decreased during the crisis, and considerable cuts in social services were made. This situation calls for the adoption of innovative, long-term social policy strategies and for modernized welfare systems, which foster more efficient, responsive and appropriate social services. However, though many initiatives have been launched and funds allocated, evidence on the results obtained is as yet lacking.

Abbreviations

AAI	Active Ageing Index
ABMS	Agent-Based Modelling Simulation
CGE	Computable General Equilibrium
CLD	Causal Loop Diagrams
DG EMPL	European Commission, Directorate-General for Employment, Social Affairs and Inclusion

G. Misuraca (✉)
European Commission, Joint Research Centre, Directorate Growth & Innovation, Human Capital and Employment Unit, Edificio EXPO, c/Inca Garcilaso, 3, 41092 Seville, Spain
e-mail: gianluca.misuraca@ec.europa.eu

L. Geppert
Fair Dynamics Srl, Via Carlo Farini, 5, 20154 Milan, Italy

Department of Economic Sciences and Business Management, Faculty of Economics, Catholic University of Milan, Via Necchi 7, Milano 20123, Italy

C. Kucsera
Freelance researcher and consultant, Budapest, Hungary

J.R. Gil-Garcia et al. (eds.), *Policy Analytics, Modelling, and Informatics*, Public Administration and Information Technology 24,
DOI 10.1007/978-3-319-61762-6_7

DRHE	Dublin Region Homeless Executive
DS-HM	Dynamic Simulation—Hybrid Model
ICT	Information and Communication Technologies
IESI	ICT-Enabled Social Innovation in support to the Implementation of the Social Investment Package
i-FRAME	Impact Framework for Real and Meaningful Evaluation
JRC	European Commission, Joint Research Centre
PASS	Pathways Accommodation and Support System
RCT	Randomized Control Trial
ROI	Return on Investment
SD	Systems Dynamics
SIP	Social Investment Package
SNA	Social Network Analysis
SROI	Social Return on Investment

This chapter presents the ongoing development of a methodological framework to analyze the economic and social return on investment of initiatives in social policies and services. Their key component is ICT-enabled social innovation, and they are defined as social policy innovations. This framework—called i-FRAME—proposes a novel approach that combines System Dynamics and Agent-Based Modelling Simulation in a hybrid model. The inspiration for this approach lies in different strands of the literature, such as the debate on the crisis of evidence-based policy-making, complexity theory and system thinking and is informed by a critical appraisal of traditional economic modelling approaches. After a brief overview of the policy context and research background, the chapter illustrates the method, sources and key findings of the literature and practice review, identifying the knowledge gaps. This helped us to develop the rationale underpinning the design of the i-FRAME modelling and the simulation approach proposed. The proposed innovative toolkit for policy modelling and informatics which underpins the i-FRAME is then outlined and its validity is discussed with an example of how the modelling approach was applied to the case of the Pathways Accommodation and Support System (PASS) initiative. This online system collects and provides access to real-time data on homelessness and bed occupancy across Ireland. The application of the i-FRAME to the case of PASS indicates that the approach proposed could be used in ex-ante, in-itinere, and ex-post simulation of the impact and the functioning of a given policy action involving ICTs as enablers in social services planning, design or provision. The chapter concludes with suggestions as to possible directions for further research and outlines the policy implications.

Introduction: A Social Investment Perspective to Address the Social Crisis in Europe

The financial crisis and the great recession that followed have brought to the fore dire social problems and also the need to find innovative ways to boost growth in socially-sustainable ways. Hence, across Europe and in most OECD countries, policies have increasingly focused on innovation, mainly the traditional form of economic and industrial innovation. Lately, however, policies have also focused on the concept of 'social innovation' (BEPA 2010). In the European Union, torn between austerity and budget constraints and its traditional social model, social innovation—fuelled by the enabling role of Information and Communication Technologies (ICTs)—is seen as a valuable means of modernizing social protection systems while keeping public budgets under control. ICT-Enabled Social Innovation can help make existing services more efficient and effective, but also produce new services as a result of integration between different policy silos. ICTs can also promote conceptual innovation by changing the base (empirical and analytical) upon which policies are first designed and later evaluated (Misuraca et al. 2011).

In 2013, the European Commission launched a Communication on the Social Investment Package (SIP) urging Member States to prioritize social investment and the modernization of their welfare systems. This would address unemployment, poverty, and social exclusion challenges brought about by the economic crisis and sustainability challenges posed by an ageing population. ICT-enabled social innovations have come to be considered a crucial element for implementing the SIP. Many initiatives have been launched and funds allocated, yet evidence on the results obtained is lacking. In addition, it seems that some of these initiatives were designed and funded without due impact assessment. In this context, the European Commission's Joint Research Centre in Seville, in collaboration with Directorate General Employment, Social Affairs and Inclusion, launched a project, called IESI,[1] to better understand how ICT-enabled social innovation can support the implementation of the SIP, and in turn have an impact on social policy reform in EU Member States. A key component of this research is the elaboration of a methodological framework (called i-FRAME) to assess the social and economic impacts of ICT-enabled social innovation initiatives. These promote social investment and are aimed at fostering social protection modernization, while maintaining and improving service access and quality.

i-FRAME has two objectives:

- first, to develop a framework of analysis of the initiatives surveyed in the IESI research, and guidance for conducting in-depth thematic analysis through case studies. This provides a structured approach to analyze such initiatives and insights for their replicability and transferability at policy/practice level across the EU.
- Second, it will serve as a framework for conducting the analysis of economic and social returns on investment initiatives, the key component of which is ICT-enabled social innovation.[2]

[1] IESI stands for 'ICT-Enabled Social Innovation to support the implementation of the Social Investment Package'. For more information see: https://ec.europa.eu/jrc/en/iesi.

[2] The IESI research proposed the following definition for ICT-Enabled Social Innovation: 'A new configuration or combination of social practices providing new or better answers to social protection system challenges and needs of individuals throughout their lives, which emerges from the innovative use of Information and Communication Technologies (ICTs) to establish new relationships or strengthen collaborations among stakeholders and foster open processes of co-creation and/or re-allocation of public value.' (see JRC Science for Policy Report, Misuraca et al. 2015a).

This will allow us to devise recommendations on how the European Commission and Member States could model and simulate the impact of ICT-enabled social innovations initiatives which promote social investment.

This chapter presents the results of the work conducted to create the i-FRAME and, particularly, that part of the framework that consists of an empirically-informed Agent-Based System Dynamics modelling simulation of ICT-enabled social innovation for the modernization of welfare services. The inspiration for this approach lies in different strands of literature, such as complexity theory and system thinking, and a critical appraisal of modelling approaches. It also includes an innovative toolkit for policy modelling and informatics.

After a brief overview of the policy context and research background in this introduction, Section "Method and Sources Behind the Development of the i-FRAME" describes the literature and practice review that has informed our selection and design of the modelling approach. In Section "Review of the State of the Art: Key Findings and Rationale Underpinning the i-FRAME", the main findings of this review are briefly discussed. These allowed us to identify current knowledge gaps and challenges that provide the rationale for the application of the i-FRAME modelling and simulation approach. Section "Proposed approach for Modelling and Simulating the Impact of ICT-Enabled Social Innovation in the Modernization of Social Policies in the EU" presents: (a) the preliminary proposal for the application of a dynamic simulation modelling approach to ICT-Enabled Social Innovation for the modernization of welfare services; and (b) gives an example of the application of the modelling approach to test its validity. In Section "Conclusions and Future Directions: Towards i-FRAME2.0" we offer our conclusions and discuss possible directions for further scientific development and policy implications.

Method and Sources Behind the Development of the i-FRAME

The methodological approach used to develop the i-FRAME comprised a number of sequential activities, which unfolded over the duration of the IESI project (2014–2016). It also received input from various validation activities which were conducted in parallel with the research. We based this iterative approach on desk research to review the state of the art on policy impact assessment; conceptual work to outline and structure the proposal of the i-FRAME meta-framework and related operational components; and consultation with experts from different research, practice and policy communities, including representatives of key relevant stakeholders and policy-makers at local, regional, national and international level.

More specifically, we conducted a review of the state of the art and conceptualization work during the first phase of the IESI research. This allowed us to suggest a first outline for the i-FRAME and propose an approach and operational components. These were discussed in various IESI Expert and Stakeholder Consultation Workshops. Based on the feedback we received at these events, and the results of the analysis of the review of the state of the art and further conceptualization work, we developed the initial version of the **i-FRAME (V1.0)** (Misuraca et al. 2015b). A revised version of the **i-FRAME (V1.5)** (Misuraca and Kucsera 2016) followed, based on a structured theoretical framework of a simulation model for social impact assessment, which was presented to experts and stakeholders for discussion. This version was validated and enhanced by applying it to selected case studies and through expert review and consultation with stakeholders. This will allow us to develop the final **i-FRAME (V2.0)** and policy recommendations for its further use, implement it as a computer-based simulation model and promote its scaling-up at EU level. We expect that the consolidated version of the i-FRAME will provide a common methodological framework made up of the following components: typologies of initiatives, dimensions of impact, indicators, methods and tools used to analyze the economic and social return on investment of initiatives (which are based on ICT-enabled social innovation and promote social investment to support the modernization of social protection systems in EU Member States).

We undertook a **systematic review of the scientific literature** and of relevant projects and initiatives to identify approaches used—or proposed—for assessing the social impacts of ICT-enabled social innovations promoting social investment. In this review, we aimed to collect suitable evidence for comparing different methodological approaches to the modelling and simulation of complex systems. We sought to provide justification for the selection of complex systems approaches as the most suitable for modelling and simulating the impacts of ICT-enabled social innovation initiatives, which are characterized by nonlinearity and retroactions that cannot be easily represented by other methods. We aimed also to review existing examples of applications of policy modelling approaches that use complex systems, in order to identify those key characteristics that could be used to shape the operational components of the i-FRAME.

The literature reviewed included a cross-analysis of the following methods: General Equilibrium Models; System Dynamic Modelling; Markov Modelling; Agent-Based Modelling; Discrete Event Modelling. They represent the most common methodological approaches used for policy modelling and simulation (Di Comite and Kancs 2015). For this review, we decided to concentrate search efforts on recent papers, discussing the advantages and disadvantages of the different methods. However, for the review of cases of policy modelling approaches based on complex systems, and in particular Systems Dynamics (SD) and Agent-Based Modelling Simulation (ABMS), we carried out an extensive and systematic search. We wanted to identify examples where modelling and simulation approaches were used to assess impacts of ICT-enabled social innovation initiatives which promoted the modernization of social protection systems. Initially, we came up with 953 papers which potentially addressed the use of dynamic simulation in the fields of analysis. We then examined the abstracts of these papers and retained only those which explicitly contained a description of dynamic simulation models in the field of analysis. This gave us a sub set of 137 papers, which was reduced to 65 after reading them in full.

Review of the State of the Art: Key Findings and Rationale Underpinning the i-FRAME

Limitations of Existing Approaches for Assessing Impacts of Social Policy Innovation

The main findings from the review of the scientific literature and evidence emerging from practice confirm that there is no assessment framework that has been developed ad-hoc and used specifically for evaluating impacts of social policy innovations. It also shows that social impact assessment is under-researched and evaluations undertaken so far are methodologically weak. It is still largely perceived as a 'nice to have' but is generally not included in policy intervention design. There is also a lack of accepted and tested methods, tools and indicators to assess the social and economic impact of policy interventions to improve welfare systems and related services. However, social impact assessment is gaining momentum both from research and policy perspective.

Though we found no specific methodology for social policy innovation per se, a few emerging trends can be identified (for recent policy and practice-oriented reviews see Misuraca et al. 2013; Epstein and Yuthas 2014; Misuraca et al. 2014; Misuraca and Viscusi 2015) in the broader social and public sector innovation field. Performance measurement concepts and tools have been transposed to the domain of social investment, social entrepreneurship and venture philanthropy. Concepts such as the Return on Investment (ROI) are taken from the private sector and developed into SROI (Social Return on Investment). Within this first trend one could also include attempts to transpose 'impact assessment tools' derived from economics and accounting to measure social impacts. These are mainly based on social accounting methods, benchmarking or rating systems, cost-benefit/effectiveness analysis and so on. The second trend follows the long-standing tradition in evaluation research of focussing on outcome measurement. Third, there have been different efforts to look into the metrics of innovation and the role of ICTs in public sector organizations, such as the Innovation Union Scoreboard (European Union), the Global Innovation Index (Cornell & INSEAD), or the approach for measuring innovation in Public Sector Organizations (NESTA). Other metrics focus on social, normative or environmental dimensions, such as the OECD Better Life Index, the Active Ageing Index (AAI), or the European System of Social Indicators (GESIS).

However, the level of complexity of our domain of interest puts it beyond the reach of standard micro-level impact assessments and traditional economic models. Social investments, and especially innovation in social policies and related services, add various sources of system complexity: (a) policy interventions have life course effects (i.e. they should represent a continuum of measures that accompany people through the key stages of their lives: childhood, working-age, parenthood, and old age); (b) integrated intervention means that measures in the various policy areas overlap and mutually reinforce each other: in other words, institutional complementarity adds another layer of complexity; c) heterogeneous behavior of agents (intervention on human activity, which in turn influences the interventions and the entire system).

In this respect, the review of the literature and practice demonstrates that traditional evaluation approaches are not equipped to deal with such complexity. Therefore, in line with the interdisciplinary aspects that characterize social policies, alternative methods should be used to complement more traditional evaluation techniques, such as scenario building, modelling and simulation to support stakeholders in coping with innovation-related uncertainties. These methods contribute to our understanding of the various factors which influence the evolutionary process of social policies and their innovation. They also allow us to define favorable conditions by considering alternative development paths and outcomes. When complexity is one of the dominant dimensions of a phenomenon (or a series of interlinked phenomena), the most appropriate way to analyze them as part of a 'complex system' is by using modelling first and then possibly simulation in order to understand the system's behavior and predict its evolution.

Recognizing and Explicitly Dealing with Complexity: Need for New Policy Modelling Approaches

Addressing the ecosystem dynamics of ICT-enabled social innovation which promote social investment definitely reveals new and emerging properties that do not belong to the individual units that make up the system. While the properties, interacting rules and contextual constraints of each unit are usually known before (ex-ante), the emergent properties of a complex system are unknown before and manifest themselves only later (ex-post). Modelling and simulation can give first indications of the need for resources, the possible transition times needed to achieve an objective, the duration of an initiative, and so on. This then helps the analysis and assessment of alternative policy options and consequently can provide evidence to assess impacts of the policy interventions by understanding the structural, operational and organizational changes that can result from the policy.

A complexity-acknowledging policy modelling approach is thus considered best suited to 'wicked problems' (Klijn 2008) and an important means of analyzing systems' self-organizing processes (Butler and Allen 2008). Nevertheless, system complexity poses challenges to classical evidence-based policy making in terms of structure and, especially, of dynamics. Complex systems include many system subcomponents which interact with each other. How the system works cannot be understood in terms of its parts but must be analyzed as a whole. However, it is the dynamic aspect of complexity that most challenges policy-making and classical evidence-based policy approaches.

As pointed out by Forrester in 1971, complex social systems often show counterintuitive behaviors as a result of the interaction of agents over time. The most troubling elements of dynamic complexity with respect to the classical linear policy-making thinking are feedback loops, time delays, stocks, and flows. Dynamic complexity is the source of low predictability as a small input can have large effects, so that equilibrium points are by definition unstable. Dynamic complex systems can only be characterized by punctuated equilibria whereby long periods of stability are interrupted by short bursts of change (Bovaird 2008). Order is emergent in the sense that it results from local interaction rather than central direction. As policy interventions concern structurally and dynamically complex systems, their effects can be dampened or taken up as a result of intricate relations and feedback loops among agents. In these contexts, experiments and replication are either not feasible or ineffective, especially when the policy, taken to affect part A of a system, ripples out to other parts. In addition, substantial time lags can mean that the consequences of action cannot be observed within the short span of the policy cycle.

This critique of traditional evidence based policy-making, which comes from system thinking and complexity theory, does not mean we have to renounce it. Complexity theorists do not subscribe to a perspective that we may define as 'ontological unpredictability', according to which socio-economic systems with human activity are simply unpredictable and can be neither studied empirically nor modelled or simulated. On the contrary, complexity theorists recognize epistemic uncertainty and propose the application of better and innovative modelling approaches as a solution. Examples of these are: system thinking and dynamics (Sterman et al. 2014), or the integrated application of Social Network Analysis (SNA) with Agent-Based Modelling (ABM) as illustrated by Ormerod (2010) and applied in Ormerod and Wiltshire (2009).

Traditional modelling includes all computation-based simulations, which are based on a stylized representation of the agents. They can be called 'representative agents' or 'variable-based' models (Gilbert 2008). They deal with aggregates of individual units (i.e. households) and not with each individual agent within the aggregate. Households can enter the models only through quantitative data that characterizes all of them, as average representations rather than heterogeneous agents. The quintessential example of classical modelling are Computable General Equilibrium (CGE) models, whose reliance on the assumption and requirement of macro-level equilibrium are at odds with the unstable and punctuated equilibrium that complex theory posits. Both micro and macro traditional modelling can also be characterized as 'discrete event' models in that they capture well-defined interventions that lead from one state to another, but they cannot capture involved process-based evolution.

Meanwhile, System Dynamics (i.e. Forrester 1971) is a modelling approach that substantially differs from classical micro and macro modelling but also relies on stylized agents and variables. SD models basically function with a set of simultaneous differential equations. Each of these equations calculates the value of a variable at the next time step, given the values of other causal variables at the current time step. They also start from cross-section and longitudinal sets of data, but rely on more sophisticated causal models, which include feedback loops and time lags (Borshchev and Filippov 2004).

Like SD, ABM is not entirely new as it appeared first in the early 1990s. However, the amount of relevant literature has only grown recently, proposing its application in various disciplinary domains (Degushi 2004; Gilbert and Terna 2000; Gilbert and Troitzsch 2005; Grimm and Railsback 2005; Railsback et al. 2006; Liebrand et al.

1998; Macy and Willer 2002; Schweitzer 2003; Tesfatsion and Judd 2006). Yet ABM applications remain limited mostly to industry-specific issues, and they have not been used to tackle broader and longer-terms issues such as those related to social policy and welfare systems.

In this context, a promising hybrid approach has emerged in literature and practice in which Systems Dynamics and Agent Based Modelling are integrated. This is part of attempts already underway to develop more sophisticated and integrated ICT-enabled models, which are emerging in both complex systems and policy informatics. The 'hybridization' with SD is expected to both explain the dynamics of a policy intervention and exploit the computational power of the ABM method to create, analyze and simulate models including individual agents interacting within an environment (Borshchev 2013).

Policy modelling based on Systems Dynamics and ABMS has the advantage of not being constrained by equilibrium assumptions, as other modelling approaches are. It can include agents with diverging preferences and market behaviors, and accounts for the greater degree of unpredictability that springs from complex social interaction (of specific interest when dealing with policy-making). However, much more effort is needed to develop such an approach and integrate ABM with other methods and insights from, for example, behavioral economics, evolutionary economics (Dosi et al. 2008).

Therefore, we consider the hybridization of system dynamics with agent-based models as a promising avenue for complex policy issues and for building the i-FRAME, as the literature review confirmed. In this regard, system-thinking modelling has already proven to be very relevant and suitable for addressing health care and social care interventions (Piniewski et al. 2011; McKelvie 2012), In terms of application, however, the IESI research can be considered pioneering in the field of ICT-enabled social innovation.

However, applying an Agent-Based System dynamic modelling entails a number of challenges that can be grouped into three main areas:

- **Analytical**: the definition of the key variables and the cause-effect relationships amongst them requires an extensive review of the literature and the involvement of experts in group modelling sessions;
- **Empirical**: the lack data available to test the model may hamper its application, unless a gathering data approach based on the development of scenarios of use that are close to reality or real-case studies is deployed from the initial design of the model.
- **Methodological**: the validation of the theoretical and methodological approach developed for designing a model based on systems thinking and simulation requires consensus on the characteristics of the scenarios of use to be selected and analyzed during the empirical activity. This can be achieved by involving experts and stakeholders and using consolidated indicators or proxy variables when available as baseline. In this respect, an added complication relates to the need to address different levels of aggregation, where different behaviors arise and prevail.

The next section presents how we propose to address these challenges by describing the overall design of the i-FRAME. We also describe the methodology underpinning the application of a hybrid dynamic simulation modelling approach in the domain of ICT-enabled social innovation initiatives promoting social investment.

Proposed Approach for Modelling and Simulating the Impact of ICT-Enabled Social Innovation in the Modernization of Social Policies in the EU

Outlining the i-FRAME as a Meta-Framework for Social Policy Impact Assessment

In order to assess social and economic impacts generated at micro, meso and macro level, it was decided to set up a specific methodological framework as current approaches are limited in scope. Nevertheless, the i-FRAME can benefit from existing methodologies and approaches that can be combined according to the specific needs and levels of analysis. The i-FRAME should therefore be considered as a methodological framework which provides guidelines on how to operationalize different methodologies and build a comprehensive (computer-based) simulation model that can provide specific inputs to devise appropriate macro/policy indicators.

To do so, the i-FRAME approach must capture the direct and indirect effects, and the intended and unintended consequences of policy interventions. It must understand how these affect beneficiaries, as well as the social innovation eco-system and the welfare system in which these initiatives are embedded. This means we must study the socio-economic effects on both individuals and the context of reference and link them to the social service delivery models and social systems in which they operate. The specific role of ICT-enabled innovations and the social nature of the initiatives under investigation should also be factored in the analysis, possibly through quantified (possibly monetized) indicators and variables. For this reason 'proxy-indicators' or value perception of stakeholders and beneficiaries may be used when data is lacking.

However, while the broader i-FRAME 'meta-framework' is still highly abstract, this phase of the research aimed to propose a concrete methodological approach. This has been done by defining the operational components that may be used to assess the impacts of ICT-enabled social innovation initiatives that promote social investment at micro, meso and macro level.

At the micro and meso levels, the analysis of the diffusion and net-contribution of ICT-enabled social innovation in social services should allow us to estimate—through an aggregated measurement of outcomes for beneficiaries and social service providers—the direct perceived impact on social value and on the performance of the social innovation ecosystem in which an initiative is implemented. For this purpose, a number of methodologies and evaluation techniques are available. First, baseline data and benchmark indicators need to be identified (e.g. through reviews of the literature and thematic analyses). In addition, the limited availability of quality data to measure performance and impacts of initiatives in both social services and social innovation requires a systematic micro-data collection process. This data is used to analyze impacts at micro level and also feed the modelling and simulation tools to assess impacts at meso and macro level.

In order to elicit specific data and unknown variables of direct and indirect effect and also to demonstrate causality scientifically and indisputably, specific methodologies based on counterfactual approaches to policy evaluation can be used, such as Randomized Control Trial (RCT). Another option is to run Social Policy Experimentations to test a policy intervention on a small population and evaluate its effectiveness before deciding on whether to scale up.[3] In this regard, one key difficulty, besides identifying the causal relationships between interventions and outcomes, relates to the complexity of linking effects of interventions at the micro or meso level to the broader effects at macro level. The i-FRAME assesses these effects indirectly by estimating indicators of social value impact on welfare systems. It applies a dynamic simulation model to estimate the effects of key variables of impact through forecast methods—such as Systems Dynamics and Agent-Based Modelling Simulation. These forecasting methods are validated through concrete, real-life scenarios of use and should complement other modelling approaches, which can be applied as appropriate to study social policy impact.

Addressing Complexity Through Hybrid-Dynamic Simulation Modelling: The i-FRAME Approach

The main assumption for using dynamic simulation modelling is that it can overcome the key problem of modelling and simulating policy decisions in complex social systems by reducing 'overconfidence' in the effects expected from a policy initiative. **This is particularly important when analyzing 'downstream' and 'upstream' interventions, i.e. two opposite policy options that can be considered when addressing social problems and designing policy relevant interventions**, like the introduction (or not) of elements of ICT-enabled social innovation. 'Downstream' measures target the consequences of harm—the effect—after it has occurred to try and stop things getting worse, while 'Upstream' measures aim to prevent harm before it occurs, by addressing the 'causes' of the problem (Coote 2012).

When resources are scarce for social policy reform, policy-makers try to design policy initiatives that offer the best trade-off between these two types of interven-

[3] Social policy experimentations require both designing a policy-relevant intervention and measuring its actual impact. They bring innovative answers to social needs; are small-scale probing interventions made in conditions where impact can be measured; and can be scaled up if results are convincing. See: http://ec.europa.eu/social/main.jsp?catId=1022.

tions in order to minimize costs and maximize potential outcomes. However, the balance between the two options depends on the behavior of the stakeholders involved. The dynamic of the decisions is difficult to model and simulate, especially ex-ante, before a policy initiative is implemented.

This is a typical situation where dynamic simulation is useful since it helps us to understand the consequences of decisions before they happen. However, it requires combining:

- different layers of analysis (micro-meso-macro);
- individual perspectives and behaviors;
- stakeholders' perspectives and behaviors at different layers of interaction;
- effects that can be achieved over different periods of time.

In addition, assessing the impact of policy interventions based on ICT-enabled social innovation on the modernization of social protection systems is difficult because of:

- the complexity of the context in which the policy interventions and the related ICT-enabled social innovation initiatives are conceived. Assessing their impacts requires in-depth knowledge of the dynamics of causality relationships among relevant variables and their negative and positive interactions;
- the cost of setting up a robust counterfactual approach that measures the causality relationships of all the relevant variables which can directly measure the impact of the social policies and the related ICT-enabled social innovation initiatives;
- the need to reach wide consensus among the relevant stakeholders on the results achieved with the impact evaluation process.

A possible representation of all the above is shown in Fig. 1 below. The horizontal arrow represents the sphere of social-related behavior and actions of an individual over time (i.e. from birth to death). Throughout his/her lifespan, the individual interacts with a multidimensional space representing his or her characteristics (i.e. lifestyle and related social and health risk factors) and the external context in which the individual lives from the micro, to the meso and macro levels of interaction.

If we apply the conceptual model described above to life-long learning and the employment status of a person (as shown in Fig. 2), this person's knowledge capital evolves throughout his/her life and can be influenced by the micro/meso/macro characteristics of the context in which he/she lives. Cultural norms and the structure of the socio-cultural offer in a given context, the conditions of the area where he/she was born, and the psychological hazards he/she is subject to determine the initial accumulation of knowledge capital of the person. At the same time, the psycho-physical and health characteristics of the person are influenced by the context and, in turn these influence the person's behavior and his/her choices and capability to seize the opportunities offered by the context in which he/she lives.

The example above illustrates clearly the complexity of the context which underpins the i-FRAME. Social policy innovations can directly affect the individual by modifying his/her psycho-physical and health conditions, which in turn can influ-

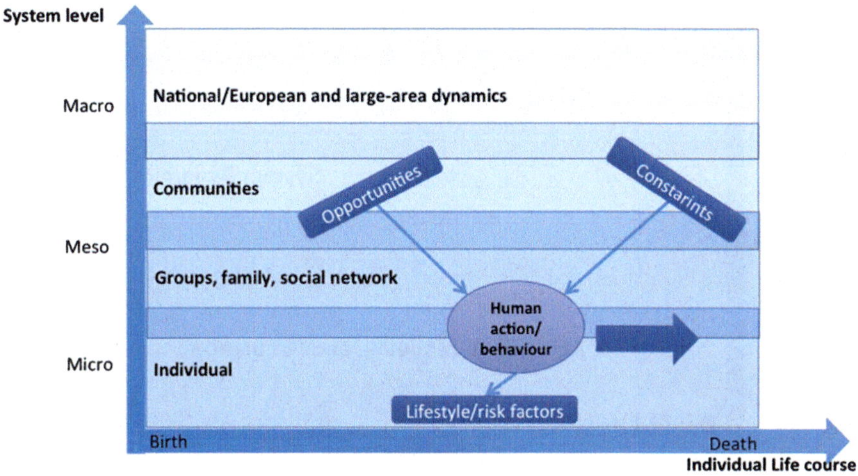

Fig. 1 Conceptual model underpinning the i-FRAME methodological approach. Source: Adaptation from the work of Glass and McAtee (2006)

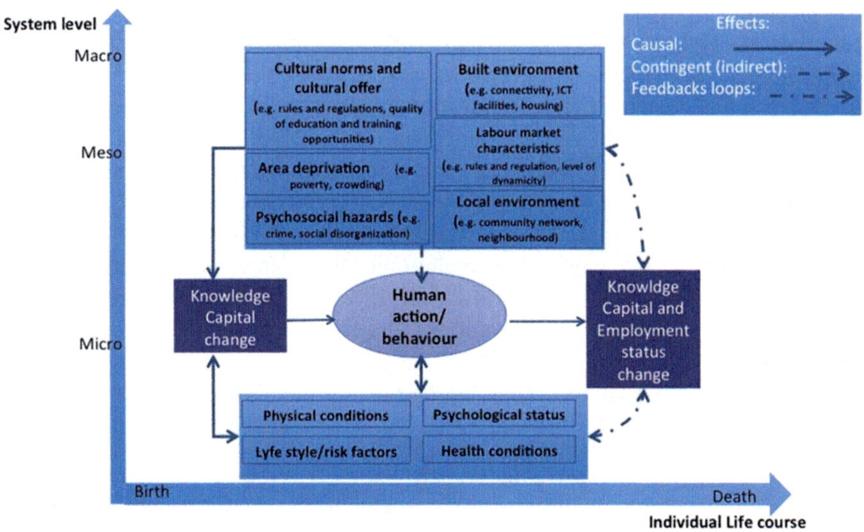

Fig. 2 Example of application of the conceptual model underpinning the i-FRAME. Source: Own elaboration

ence his/her behavior and actions. At the same time, they can modify the context in which the person lives, by changing opportunities and constraints which, in turn, can influence his/her actions and behavior throughout his/her life.

Modelling and simulating such complexity with numerous causality relationships and feedbacks loops, non-linearity and a time dimension can be done using a

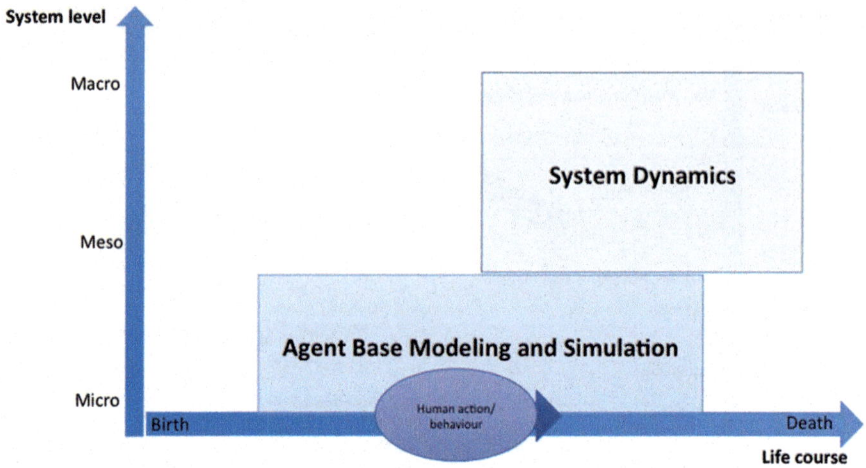

Fig. 3 Level of abstraction in the conceptualization of hybrid simulation modelling. Source: Own elaboration

combination of dynamic simulation models like SD and ABMS. The '**Dynamic Simulation - Hybrid Model**' or **DS-HM** we propose provides a powerful methodological approach which addresses all the specificity and complexity of evaluating the impact of ICT-enabled social innovation initiatives.

As shown in Fig. 3, the ABMS component of the DS-HM can easily model the dynamic characteristics of individuals in the target population and it allows us to simulate their behavior throughout their lives. At the same time, the SD component can capture the complexity and dynamic of the context in which the individual operates. It can also capture the evolution of the context's characteristics over time due to the interaction among different layers of the system represented by causality relationships and feedback loops that can interplay with the behavior of the individuals.

We prefer the DS-HM over other approaches, for the following reasons:

- The structure of SDs represents a well-defined paradigm that gives a high-level abstraction of the problem. The representation of stocks and flows shows at a glance the general logic of the system, making it easy to understand and depicting possible counter-intuitive situations.
- Yet, a pure SD approach does not immediately show the behavior of the single classes of objects (people, means, sub-structures) involved in the problem. In other words, the effort needed to take into account possible behaviors of single classes of objects that contribute to the problem requires the introduction of other structures (arrays of stocks and flows) that tend to complicate the overall model.
- A pure ABMS model can be difficult to understand in terms of macro general behaviors. As the ABMS paradigm is based on a bottom-up approach, it focuses on the detailed aspects of each single class of agents. The macro properties and

dynamics of the whole system depicting the problem, are often lost when glancing at the structure of the model but become evident 'ex post', as a result of the simulation.

In brief, the DS-HM offers the following advantages:

- it catches both high-level system structures and single agent behavior;
- the macro properties of the system emerge from the interaction of agents; this limits the use of 'heroic assumptions' to the general structure of the problem and the related system;
- it takes into account both agents' behavior and system's structures;
- it uses both stocks and events as sources of the problem's dynamics.

We propose therefore that the i-FRAME methodological approach be operationalized using a DS-HM, in order for it to:

- constitute a meta-framework that allows policy makers and stakeholders to better understand impacts;
- drive the scaling-up of the modelling and simulation of impacts in a given socio-economic context;
- support the comparison of the scaling-up effects in different welfare system contexts.

This is particularly important since the i-FRAME aims to support policy makers at European Commission and Member State level, by suggesting ways to assess impacts of ICT-enabled social innovations initiatives promoting social investment. At the same time, it should also become a useful practical resource for various stakeholders involved in the implementation of ICT-enabled social innovation initiatives to design their social policy intervention logic and understand how to measure outcomes and maximize the social impacts of their initiatives.

Operationalizing the Methodological Approach: The i-FRAME Decalogue and Its Application

Following the rationale presented above, we propose that the i-FRAME should be operationalized using dynamic simulation modelling. This is the most effective technique to help us understand and analyze problems of complex systems, because of: (1) its ability to represent virtually any dynamic problem (and the related system) without using sophisticated mathematics; and (2) its potential for running 'experiments' in a virtual, risk-free computer-based environment. Indeed, a model can be defined as a tool that helps us to understand and predict a system's behavior, and simulation as the execution (either statistically or over time) of a model. The process of modelling and simulation thus encompasses rigorous conceptualization, development and testing of a model, and the execution of that model to study its behavior.

It should be underlined that no model can include all the details of any given situation. Any attempt to build a general simulation model to study and evaluate the effects (i.e. the outcomes and impacts at micro, meso and macro level) of a specific policy would therefore not produce valuable results. Hence, any simulation model should start from a well-defined problem with its specific characteristics and properties to model the system itself: the system should represent the set of components that have to be modelled in order to understand the dynamics of the surrounding environment where the problem forms, evolves and generates other effects.

Within this context, the hybrid approach underpinning the i-FRAME based on the integration of SD and ABMS can be used for high and/or low level abstraction of a policy problem. It uses SD to describe the general behavior of the system, while ABMS identifies the behavior of the system's units through appropriate state-chart representations. This allows the i-FRAME methodology to capture both the general, high-level aspects and the unit-specific characteristics of the whole system. This approach thus allows us to address the aspects of a policy problem that are considered particularly meaningful, avoiding over-detailing. In a simplified frame, only the most important facts, properties and behaviors are considered and analyzed in order to understand how the system adapts and evolves over time.

To operationalize the approach proposed, we use the SD modelling technique to represent the general structure of the problem, i.e. the structure that models the relationship between the variables included in the problem (in terms of 'attributes' and relationships). We use ABMS to describe the behaviors of the single units (groups of stakeholders involved) that contribute to the problem. The combined modelling of SD and ABMS allows us to identify the contribution of each single group to the figure which characterizes the SD attribute 'population'. The resulting model is created by aggregating different domain-specific sub-models.

It is important to note that the proposed methodological approach is not sequential, but circular and reiterative, and requires the strong involvement of domain experts and stakeholders, as well as policymakers. No model is robust and/or realistic enough on its own without the suggestions, ideas and 'mental models' of domain-experts (Vennix 1999; Zeigler et al. 2000; Van den belt 2004). These should be involved particularly during the development phase, in workshops and discussions to help the developer shape the structure and details of the model step-by-step.

Building on the above, the practical steps for implementing the i-FRAME methodology can be summarized by what we have called the i-FRAME Decalogue:

1. **Start from the definition of a case/problem/need**, and reconstruct the logic model which represents how the case/problem/need is addressed by the ICT-enabled social innovation initiative.
2. **Define the levers for output, outcome and impact assessment** in accordance with the logic model identified, and define the impact, outcome and output assessment indicators in accordance with specific agreed 'levers'.
3. **Identify those areas in the case/problem that have an impact or are affected, and how they are addressed by the ICT-enabled social innovation initia-**

tive. Developing Causal Loop Diagrams (CLD) can help us understand the main cause-effect relationships of the problem.
4. **Check for existing similar dynamic simulation models** to identify domain-specific sub-models that are already available.
5. **Look for and check the attributes and methods for each domain-specific sub-model** in existing dynamic simulation models, and adapt them to the case/problem addressed.
6. **Improve the dynamic simulation model by adding domain-specific sub-models as they are identified**, and complete the logical representation of the case/problem addressed. To this end, develop a methodological pathway in dynamic model development that combines variables and indicators from qualitative (Causal Loop Diagram) and quantitative analysis (stocks and flows and agent-based models).
7. **Adapt and improve each domain-specific quantitative sub-model** (stocks and flows diagram and/or state charts with analytical description of the state transitions) also using Group Model Building Approach **and combine the sub-models in the final dynamic simulation model which represents the case/problem addressed**. To achieve this, use aggregate approaches (i.e. hybrid models) to build consensus around difficult policy problems and facilitate the presentation of results, leaving room for policymakers and stakeholders to concentrate on feedback and develop an endogenous perspective of the policy actions.
8. **Define the conditions** (initial data/information) **for each scenario to be studied.**
9. **Analyze the scenario through different simulations** (by changing the internal levers of the model).
10. **Compare the scenarios and devise policy recommendations.**

Clearly the above steps are not the 'perfect scheme' for achieving the ambitious objective of assessing impact and proposing policy recommendations. However, they offer valuable guidance for stakeholders and policymakers—with the help of domain experts. This allows them to achieve realistic results when assessing the impact of a policy initiative, and to identify all the elements required to describe the problem in its 'entire complexity'.

To demonstrate the validity of the proposed methodology and test quantitatively the approach underpinning the i-FRAME, we applied its DS-HM component to the **Pathways Accommodation and Support System** (PASS) initiative. This online system collects and provides access to real-time data on homelessness and bed occupancy across Ireland. It improves the planning, delivery, monitoring, and coordination of services across various public and third sector agencies. The data collected are linked to profile data, housing and support needs assessment, ongoing support planning; engagement with accommodation, outreach and day services, and reasons for departure.

PASS was launched in the Dublin region in 2011 to: prevent homelessness, reduce the duration of homelessness to less than 6 months and ensure the delivery

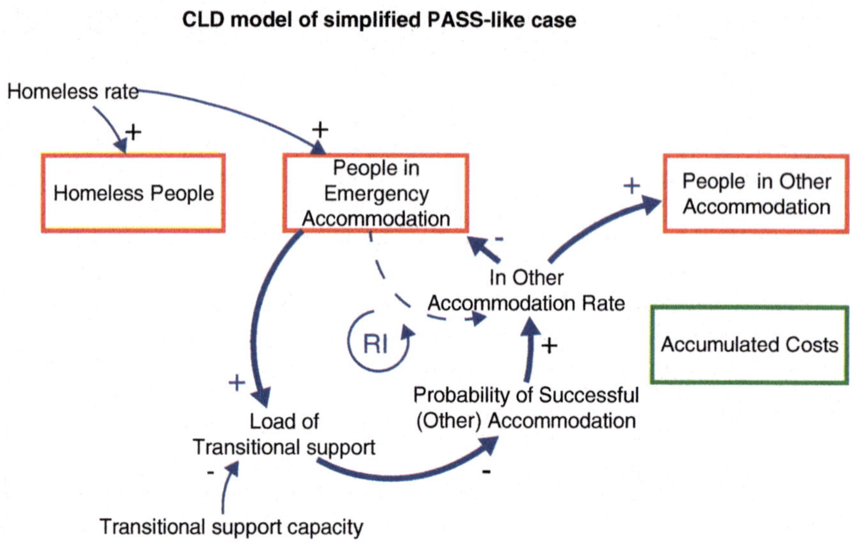

Fig. 4 CLD model of simplified PASS-like case. Source: Fair dynamics elaboration for JRC

of services for homeless people that meet their needs. All funded services addressing homelessness are required to use PASS under Service Level Agreements.

A general model for this case is very complex, as it implies numerous situations which interact with one another. Nevertheless, to test the i-FRAME methodology, a simple model was prepared to capture how the PASS system manages the transition of a homeless person from initial 'emergency accommodation' to more stable accommodation.[4] The model developed is coherent and reproduces over time data officially published by the Dublin Region Homeless Executive (DRHE).

As described in Fig. 4, the Causal Loop Diagram features three main stocks in the model: homeless people, people in emergency accommodation, people in other accommodation. The flow rate is based on the support capacity of the system variable—the transitional support capacity—which influences the rate at which people go from emergency accommodation to other accommodation.

The whole dynamic mechanism of the model is based on variations in the probability of success when transferring people from emergency to other accommodations. These variations may depend on the load of support that, in turn, depends on PASS's Transitional Support Capacity. In other words, the PASS system influences transitional capacity, i.e. the number of 'figurative' operators providing the service and the 'load' of the support offered.

[4]The model includes impacted domain-specific sub-models (i.e. relevant population, social housing delivery, care service delivery, financing of the intervention and labour market).

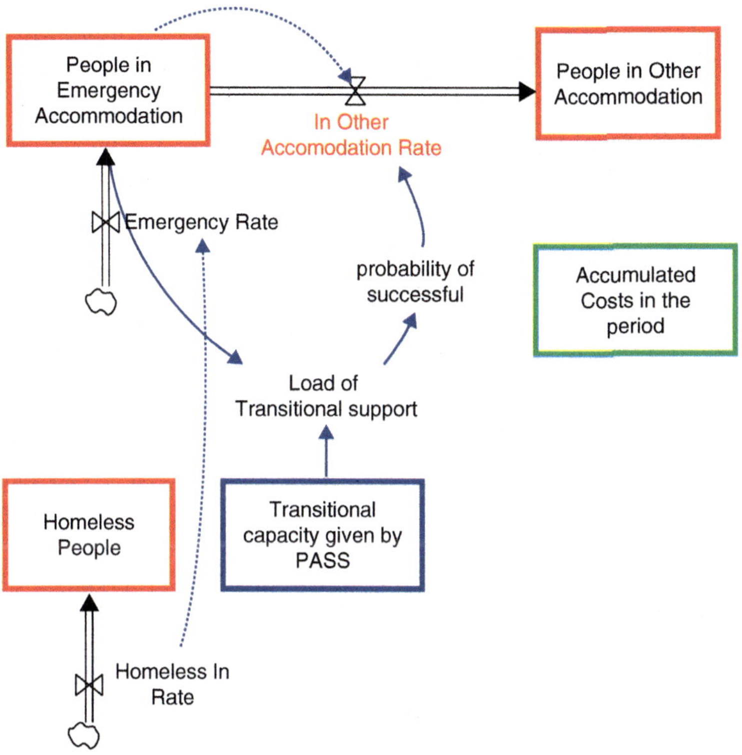

Fig. 5 Stock and flow diagram of PASS-like model. Source: Fair dynamics elaboration for JRC

The model captures the dynamic relationship between the support capacity of the services offered by the PASS system (transitional) and the rate of homeless people going into each type of service. The logic of the reinforcing loop is:

People in emergency accommodation → load of transitional support (affected by the PASS system through the change of the support capacity) → probability of success → in-transitional rate → people in other accommodation.

The support capacity of the service underlies the loop.

In order to validate the model quantitatively, the simulation modelling of the PASS-like model used data from the Homeless Agency Partnership (2008). A Group Model Building Approach was also used to present the dynamic simulation model of the PASS case at an Experts Workshop at JRC-Seville in July 2015.

Figure 5 below shows the stock and flow diagram used for the PASS-like simulation modelling application. The first stock represents homeless people in the Dublin area. Thanks to PASS, the end beneficiaries of the service can move to the stock of people in emergency accommodation, the first type of accommodation provided. The transitional service capacity, fed by the PASS system, influences the number of homeless who can enter into the other accommodation type. The accumulated costs represent the budget allocated to provide the services. The flows represent the rates of people going from one stock to another. 'Homeless people in' rates indicate the share of the population that becomes homeless. The emergency rate regulates the flow of people going into emergency accommodation. The other accommodation rate regulates the flow of people going from emergency to other accommodation. ICTs act in this case as enablers that increase the number of figurative operators who provide and coordinate the services. This, in turn, increases the probability of success: more people move to housing from emergency accommodations.

The levers that regulate the flows used to calibrate the model are:

- (Q1-2014) New homeless people in emergency accommodation at baseline (437).
- Existing homeless people in emergency accommodation at baseline (1869).
- Transitional support capacity, measured by the number of figurative operators affected by the PASS system (15).
- Homeless rate, measured by the number of homeless people per week (from the case documentation).
- Unit cost, measured by the cost per homeless person per week (2.100€) adjusted to a coherent value for the delivery of the service following discussion with experts and including the direct costs of the service delivered. For the calibration, the cost is taken into account only for the new homeless entering in the PASS system, leaving out those already existing in emergency accommodation.

The time period used to run the model is 78 weeks from Q1-2014 to Q3-2015.

By using the levers with the above values, in this timespan, there were 3383 homeless people, 3441 people in emergency accommodation, and 1379 people in other accommodation (assuming 0 at baseline, Q1-2014). The output of this exemplificative simulation is reported in Fig. 6.

The quantitative validation of the PASS scenario indicates that this approach could be used in ex-ante, in-itinere, and ex-post simulation of the impact and the functioning of a given policy action involving ICTs as enablers in services planning, design or provision. Furthermore, the proposed approach is also useful for simulating the impacts of different policy options. For instance, experimenting with the levers used for the simulation model shows that even if the number of operators in the system is changed, the probability of success reaches a cap over time where the number of beneficiaries is no longer affected.

In this simplified simulation, the effect of ICTs is represented by an increase in the figurative number of operators which leads to an increase in both the efficiency and effectiveness (probability of success) of the initiative. By decreasing the 'load' or increasing the number of figurative operators, the 'probability of success' increases and the 'time taken to move to housing' decreases, making the social intervention more efficient and effective.

Applying the i-FRAME approach to PASS shows the effectiveness and efficiency of this intervention in planning and executing the services. The real-time tracking of homeless people and the availability of beds made it possible to increase bed occupancy to 99%, thanks to the sharing of information between all agencies and other stakeholders, including volunteer organizations. This ensured efficient use of available resources and reduced duplication of efforts.

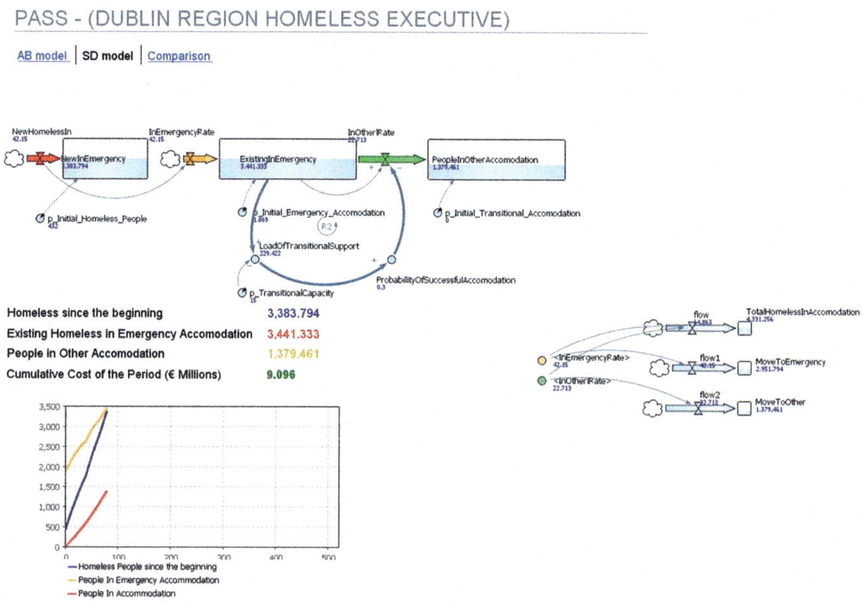

Fig. 6 Output of simulation modelling of PASS. Source: Fair dynamics elaboration for JRC

The introduction of PASS allowed the timely use by local government of the generated information to manage, monitor and evaluate the services efficiently. PASS provides statistical information on the homeless population's profile and usage of services that can be used to assess the effectiveness of the strategy, identify emerging trends and monitor service delivery. Every record is unique and allows local government to track each homeless person's progress, and assess his/her income, employment, training, education and health needs.

Service delivery is improved by the shared information system that allows the agencies and other stakeholders involved in service provision to track and share tasks and provide a care continuum and integrated service delivery. The statistics provided can be used to plan future service developments and monitor the quality of the services delivered. They also provide insights into the challenges homeless people face, profiling the characteristics of those who use the services. The statistics also help to reconfigure service provision to fulfil long-term strategic policy objectives.

The initiative helps optimize the use of financial and human resources by reducing the overall costs of homelessness for society by delivering the essential required service to any household experiencing homelessness. This response was much more preferable to early approaches which aimed to get people 'housing ready'. Statistics show a significant improvement in the number of individuals who moved into independent living and became integrated in society with full employment and better health.

Conclusions and Future Directions: Towards i-FRAME2.0

The systematic review carried out of the state of the art in modelling and simulation of social policy impacts shows that traditional econometric and innovation models do not provide a comprehensive approach that integrates social and economic concerns and heterogeneous agents' behaviors, roles and relationships. This review also shows that there is limited application of dynamic simulation modelling approaches, especially in relation to social policies and ICT-enabled innovation initiatives.

To fill the gap identified and respond to the need to design alternative methods and innovative computer-based tools to complement more traditional evaluation techniques, we have proposed in this chapter a new methodological framework for policy modelling and informatics called i-FRAME. This framework can address the complexity of social policies and related systems by using modelling and simulation techniques to support stakeholders in tackling innovation-related uncertainties and their dynamics.

The difficulty of modelling complex phenomena in social policies is further increased by the presence of human activities and their impact on the whole system. Therefore, the i-FRAME methodology 'factors in' agents' behavior and its dynamics. It takes a hybrid dynamic modelling and simulation approach, integrating Systems Dynamics and Agent-Based Modelling. We suggest that this is the most suitable way of addressing the ecosystem dynamics of ICT-enabled social innovation promoting social investment.

The application of the i-FRAME to the PASS case shows that dynamic simulation modelling is a powerful tool for understanding complex social systems and a promising option for supporting modelling and simulating policies promoting social investment. Without a structured methodological framework like the one proposed in the IESI research (the i-FRAME methodology), the dynamic simulation approach is difficult to implement because of the high level of expertise required for its adoption and the cost of applying it to any given problem.

The i-FRAME methodology aims to help policy makers understand better the dynamic relationships and possible impacts of alternative policy options in highly complex and uncertain situations. The ten steps which make up the i-FRAME Decalogue allow us to apply the proposed dynamic simulation model to evaluate ICT-enabled social innovation policy interventions. This model has been shown to have a high degree of replicability for social policy impact simulation. The methodology allows us to represent problem-solving as a circular process from the definition of the problem to the simulation of the impacts of given policy measures, to the direct and active participation of all the policy actors and domain experts. It allows them to shape the problem into a dynamic simulation model and then define the most suitable policy instruments to maximize the positive impacts of the planned policy actions on the target beneficiaries.

Thus, the positive results of the preliminary application of the i-FRAME approach open the door to a more extensive and systematic implementation of the proposed methodology. This approach can support policy actors in simulating ex-ante, in-itinere and ex-post impacts of initiatives where ICT-enabled social innovation plays an important role in social services delivery. Nevertheless, further research is required to review and improve the theoretical and methodological approach that underpins the i-FRAME. The overall aim of this approach is to provide a comprehensive framework for the evaluation of social and economic impacts of ICT-enabled social innovation initiatives which promote social investment at micro-meso and macro level. The focus of the approach could also be broadened beyond the ICT realm. This would allow us to develop and further validate the operational components of the improved methodological approach of the i-FRAME (v2.0). We would apply it to real-life cases and build a computer-based simulation model which would factor in possible counter-intuitive behaviors. A flexible approach must be taken to re-calibrating the model as a consequence of ex-post analysis or changes in the theoretical assumptions/causal relationships and/or dynamics hypothesis which underlies the model and its components.

At policy level, the application and use of the i-FRAME computer-based simulation model would support relevant policy institutions at various levels. It would give them impact estimates for their ex-ante analysis of potential impacts of new social policy initiatives based on ICT-enabled social innovation. It would also provide in-itinere and ex-post analysis of the returns of policy initiatives throughout the policy cycle, thus contributing to effective evidence-based policy-making. This would help the European Commission and EU Member States to strengthen the social dimension of the European Monetary Union, and thus shape a better future for Europe.

Acknowledgments This chapter is based on research conducted by the authors as part of the project 'ICT-Enabled Social Innovation in support to the Implementation of the Social Investment Package' (IESI) undertaken by the European Commission's Joint Research Centre (JRC) in Seville in collaboration with DG Employment, Social Affairs and Inclusion (DG EMPL) and led by Gianluca Misuraca. We are grateful to all colleagues that contributed to the activities of the IESI research and experts who participated in the Workshops organized by JRC to validate the i-FRAME methodological approach.

Disclaimer The views expressed in this chapter are purely those of the authors and may not in any circumstances be regarded as stating an official position of the European Commission.

References

BEPA (2010) Empowering people, driving change: social innovation in the European Union. European Commission. http://ec.europa.eu/bepa/pdf/publications_pdf/social_innovation.pdf

Borshchev A (2013) The big book of simulation modeling: multimethod modeling with AnyLogic 6. AnyLogic North America

Borshchev A, Filippov A (2004) From system dynamics and discrete event to practical agent based modelling: reasons, techniques, tools. In: Proceedings of the 22nd international conference of the system dynamics Society, vol 22

Bovaird T (2008) Emergent strategic management and planning mechanisms in complex adaptive systems. Public Manag Rev 10(3):319–340

Butler M, Allen P (2008) Understanding policy implementation processes as self-organizing systems. Public Manag Rev 10(3):421–440

Coote A (2012) The wisdom of prevention: long-term planning, upstream investment and early action to prevent harm. New Economics Foundation, London. www.neweconomics.org/publications/entry/the-wisdom-of-prevention
Degushi H (2004) Economics as an agent-based complex system. Springer, Tokyo
DI Comite F, Kancs D (2015) Macro-economic models for R&D and innovation policies. JRC Technical Reports Series
Dosi G, Fagiolo G, Roventini A (2008) Shumpeter Meeting Keynes: a policy-friendly model of endogenous growth and business cycles, Working paper Series, 10/2008, Laboratory of Economics and Management Sant'Anna School of Advanced Studies. Available at: http://www2.econ.iastate.edu/tesfatsi/SchumpeterMeetsKeynes.GDosiEtAl2008.pdf
Epstein MJ, Yuthas K (2014) Measuring and improving social impacts: a guide for nonprofits, companies, and impact investors. Berrett-Koehler Publishers, San Francisco
Forrester JW (1971) Counterintuitive behavior of social systems. Technol Rev 73:53–68
Gilbert N (2008) Agent-based models. Sage Publications, Los Angeles
Gilbert N, Terna P (2000) How to build and use agent-based models in social science. Mind Soc 1(1):57–72
Gilbert N, Troitzsch KG (2005) Simulation fort the social scientist, 2nd edn. Open University Press, Milton Keynes
Glass TA, McAtee MJ (2006) Behavioral science at the crossroads in public health: extending horizons, envisioning the future. Soc Sci Med 62(7):1650–1671
Grimm V, Railsback SF (2005) Individual-based modeling and ecology. Princeton University Press, Princeton
Homeless Agency (2008) Counted in, 2008 – a report on the extent of homelessness in Dublin. Homeless Agency, SPSS, Dublin
Klijn E (2008) Complexity theory and public administration: what's new? Public Manag Rev 10(3):299–317
Liebrand BG, Nowak A, Hegselmann R (1998) Computer modeling of social processes. Sage, London
Macy M, Willer R (2002) From factors to actors: computational sociology and agent-based modeling. Annu Rev Sociol 28:143–166
McKelvie D (2012) Envisaging social impact bonds modelling 'what might be' a whole system collaborative approach. The Symmetric Partneship LLP
Misuraca G, Kucsera C (eds) (2016) Proposed methodological framework to assess the social and economic impact of ICT-enabled social innovation initiatives promoting social investment in the EU - i-FRAME V1.5. JRC Technical Reports Series
Misuraca G, Viscusi G (2015) Shaping public sector innovation theory: an interpretative framework for assessing ICT-enabled innovation for governance and policy-making, in Electronic Commerce Research, Special Issue on Electronic Government: Investment in Communities, Firms, Technologies and Infrastructure, Springer
Misuraca G, Codagnone C, Rossel P, (2011) Foresight and policy modelling on ICT for governance: exploring the next frontiers. In proceedings of the 4th international Seville conference on future oriented technology analysis (FTA), 12 and 13 May 2011 - FTA and grand societal challenges: shaping and driving structural and systemic transformations. foresight.jrc.ec.europa.eu/fta_2011/intro.html
Misuraca G, Codagnone C, Rossel P (2013) From practice to theory and back to practice: reflexivity in measurement and evaluation for evidence-based policy making in the information society. Govern Inf Q 30:S68–S82. doi:10.1016/j.giq.2012.07.011
Misuraca G, Centeno C, Torrecillas C (2014) Measuring the impact of einclusion actors: impact assessment framework. JRC Science for Policy Reports Series
Misuraca G, Colombo C, Kucsera C, Carretero S, Bacigalupo M, Radescu R (2015a) Mapping and analysis of ICT-enabled social innovation initiatives promoting social investment through integrated approaches to the provision of social services. JRC Science & Policy Report Series
Misuraca G, Colombo C, Carretero S, Kucsera C (2015b) i-FRAME: initial proposal of methodological framework to assess the social and economic impact of ICT-enabled social innovation initiatives promoting social investment. JRC Technical Reports Series

Ormerod P (2010) N squared: public policy and the power of networks. RSA, London. http://www.thersa.org/__data/assets/pdf_file/0003/330258/RSA_Pamphlet-publicpolicy.pdf
Ormerod P, Wiltshire G (2009) Binge drinking the UK: a social network phenomenon. Mind Soc 8(1):135–152
Piniewski B, Codagnone C, Osimo D (2011) Using ICT to co-produce better health outcomes: a paradigm shift towards Health 2.0. European Commission, JRC-IPTS Scientific & Technical Report
Railsback SF, Lytinen SL, Jackson SK (2006) Agent-based simulation platforms: Review and development recommendations. Simulation 82(9):609–623
Schweitzer F (2003) Brownian agents and active particles. Springer, New York
Sterman J, Franck T, Fiddaman T, Jones A, McCauley S, Rice P, Rooney-Varga JN (2014) World climate: a role-play simulation of climate negotiations. Simul Gaming 46:348
Tesfatsion L, Judd K (2006) Handbook of computational economics, vol vol 2. North-Holland, Amsterdam
Van den Belt M (2004) Mediated modeling: a system dynamics approach to environmental consensus. Island Press, Washington
Vennix JA (1999) Group model-building: tackling messy problems. Syst Dyn Rev 15(4):379–401
Zeigler BP, Praehofer H, Kim TG (2000) Theory of modelling and simulation: integrating discrete event and continuous complex dynamic systems. Academic, San Diego

Gianluca Misuraca is Senior Scientist and Team Leader on Social Innovation and Governance at the European Commission´s Joint Research Centre, Directorate for Growth & Innovation where since 2009 leads resarch on the impat of the digital transformation on governance and society. Before joining the European Commission, Gianluca was the Managing Director of the Executive Master in eGovernance at the Ecole Polytechnique Fédérale de Lausanne (EPFL). During his career he held several positions as policy advisor for different International Organisations, cooperation agencies and consulting in the area of eGovernment, regional development, research and innovation. Gianluca's background is economics with focus on the interface between ICTs and public sector innovation. He holds a Diploma of Specialisation in EU Economics and Law, a diploma in Security Management, an Executive Master in eGovernance and a Ph. D in Management of Technology.

Luigi Geppert Since the early 1990s he's been using modelling approaches (Systems Thinking, System Dynamics, Discrete Events, Agent based) and tools to analyse and support strategic issues. He developed an innovative and successful methodology for the analysis and the optimisation of strategic resources, applied for several companies in different industries. He has also been acting as an Adjunct Professor of 'Strategy and Dynamic Systems' and 'Business Management' at the Catholic University of Milan for several years and has been collaborating with several European Universities. He's commonly used to working with international groups, and to be involved in management and scientific projects of different nature and subject, from general strategy to change management and organisational issues, by using simulation modelling as a pervasive approach to depict, analyse and simulate the behaviours of complex systems.

Csaba Kucsera is currently a freelance researcher and consultant in Budapest, Hungary. At the time of writing this chapter, he was working for the European Commission's Joint Research Centre, Human Capital and Employment Unit on the socio-economic analysis of the efficiency, usability and impact of ICT-enabled services on independent living and long-term care of older people. Earlier, at the Centre for Social Sciences, Hungarian Academy of Sciences he has been involved in numerous cross-country comparative studies as a researcher on various subtopics of ageing: quality of life of elderly people living alone, their subjective health, loneliness and social participation, social support, and in a project that developed a quality-of-life-supporting ICT device's prototype. Csaba earned both his MA and PhD in sociology at the Eötvös Loránd University in Budapest, Hungary.

Part III
Democracy and Governance

Citizen-Sourcing for Public Policy Making: Theoretical Foundations, Methods and Evaluation

Euripidis N. Loukis

Abstract The public sector, motivated by multiple success stories of the 'crowd-sourcing' in the private sector, and also by the increasing complexity of social problems and needs, has started moving in this direction as well, and this gives rise to the gradual development of the 'citizen-sourcing'. It is important to develop appropriate policy informatics for this purpose, and in particular theoretically sound effective ICT-based citizen-sourcing methods, which enable the efficient retrieval of policy-relevant information, knowledge and ideas from citizens, and then the advanced processing of them in order to calculate useful policy analytics, which can provide substantial support for public policy making. This chapter initially provides an overview of the research that has been conducted in this area by the research group of the author in the last decade, as part of several European projects, concerning the application in the public sector of crowd-sourcing ideas and the development of ICT-based methods for this purpose. We present briefly four such ICT-based methods that we have developed for the 'active' as well the 'passive' citizen-sourcing, initially aiming at the general public and latter focusing on the experts. Then leveraging the experience gained from the development and some first pilot applications of these methods we propose some theoretical foundations from previous political and management sciences research, which can be used for the future development of effective ICT-based citizen-sourcing methods for supporting public policy making, as well as for their evaluation. Next, based on them an evaluation framework is developed for the multi-perspective evaluation of such methods. Finally, an outline of the evaluations of these ICT-based citizen-sourcing methods is provided, based on parts of this evaluation framework, as well as a comparison of them.

E.N. Loukis (✉)
Department of Information and Communication Systems Engineering, University of the Aegean, Gorgyras and Palama 2, Karlovassi 83200, Samos, Greece
e-mail: eloukis@aegean.gr

J.R. Gil-Garcia et al. (eds.), *Policy Analytics, Modelling, and Informatics*, Public Administration and Information Technology 24,
DOI 10.1007/978-3-319-61762-6_8

Abbreviations

DOI Diffusion of Innovation
ICT Information and Communication Technologies
TAM Technology Acceptance Model

Introduction

Private sector firms have started shifting from the established 'closed innovation' paradigm, which is based on their own internal information, knowledge and ideas, towards the 'open innovation' paradigm, which exploits to a large extent also external information, knowledge and ideas, possessed by other firms (e.g. suppliers, customers, business partners, research centers, universities, etc.), as well as by 'crowds' of individuals, this latter being termed 'crowd-sourcing' (Chesbrough 2003a, b, 2006; Howe 2006, 2008; Brabham 2008, 2012, 2013; Huizingh 2011; West et al. 2014). In particular, crowdsourcing is defined as 'a new web-based business model that harnesses the creative solutions of a distributed network of individuals, in order to exploit 'collective wisdom' and mine fresh ideas from large numbers of individuals' (Brabham 2008). It can be highly valuable because it allows taking advantage of the 'wisdom of the crowds': previous management research and practice has revealed the high potential of a diverse crowd of individuals to provide a wealth of information and knowledge, as well as innovative solutions to problems and ideas for innovations, which can be comparable or even better than those provided by 'internal' firms' experts; 'external' crowds can be much more diverse in expertise and experience that the 'internal' R&D and experts, and this can be the source of a larger quantity and variety of insights and approaches (Surowiecki 2004; Majchrzak and Malhotra 2013; Brabham 2013). There has been extensive research for the identification and development of effective and efficient private sector practices of open innovation, and especially of crowd-sourcing, the assessment and evaluation of them, and also the discovery of the specific contexts for which each of them is more appropriate; a short discussion of this research is provided by Hetmank (2013) and Rechenberger et al. (2015).

The public sector, motivated by the multiple 'success stories' of private sector crowd-sourcing, and also by the increasing complexity of social problems and needs, has also started moving in this direction as well, and this gives rise to the gradual development of the 'citizen-sourcing' (Linders 2012; Nam 2012; Mergel and Desouza 2013; Ferro et al. 2013; Prpić et al. 2015). In particular, government agencies increasingly attempt to use ICT in order to exploit the extensive information and knowledge possessed by citizens, and also their creativity and ideas, in order to develop highly efficient and effective new innovative public policies, and also improvements to existing ones, or adaptations of them to new needs and conditions in society. However, there has been much less research and practice in the area of public sector citizen-sourcing in comparison with the area of private sector crowd-sourcing, so there is much less theoretical basis, methods, experience, and maturity in the former area than in the latter. It is, therefore, important to develop appropriate policy informatics for this purpose, and in particular effective and theoretically sound ICT-based methods for citizen-sourcing, which enable the efficient retrieval of policy-relevant information, knowledge and ideas from citizens, and then the advanced processing of them in order to calculate useful policy analytics that can provide substantial support for public policy making. In general, it is important to conduct extensive research in order to reach in the area of public sector citizen-sourcing a level of theoretical foundation, and also effectiveness and maturity, comparable to those of the private sector crowd-sourcing area.

This chapter aims to contribute in this direction; in particular it makes the following contributions:

1. It initially provides an overview of the research that has been conducted in this area by the research group of the author in the last decade as part of several European projects, concerning the application in the public sector of crowd-sourcing ideas and the development of ICT-based methods for this purpose. Most of this research has been conducted as part of the FP7 'ICT for Governance and Policy Modeling' research initiative of the European Commission (for more details about it see http://cordis.europa.eu/fp7/ict/programme/challenge7-governance_en.html), which had as main objective the development of novel advanced ICT tools in order to engage different types of societal groups and communities in public policy making processes, and support the creation, collection and sharing of policy relevant group knowledge of them, and the incorporation and exploitation of this knowledge by government agencies for improving governance. In particular we outline four ICT-based citizen-sourcing methods that we have developed for the 'active' as well the 'passive' citizen-sourcing, initially aiming at the general public and latter focusing on the experts.
2. Based on the experience gained from the development and some first pilot applications of these methods we propose some theoretical foundations from previous political and management sciences research, which can be used for the future development of effective ICT-based citizen-sourcing methods aiming to support public policy making, as well as for their evaluation.
3. Then based on these theoretical foundations an evaluation framework has been developed for the multi-perspective evaluation of such ICT-based citizen-sourcing methods.
4. Finally, we provide an outline of the evaluations of these ICT-based citizen-sourcing methods, based on parts of this framework, and a comparison of them.

Chapter Methodology

The methodology we followed in order to prepare this chapter includes five steps, which correspond to the next five Sections "Literature Review" to "Citizen-Sourcing Methods Evaluation and Comparison" of our chapter:

(a) Initially previous literature on private sector crowd-sourcing and public sector citizen-sourcing has been reviewed; in Section "Literature Review" are outlined some important and representative studies from the latter research stream, as it is more relevant to our research (it is beyond the scope and the length limitations of this chapter to provide full literature reviews of these two streams).
(b) Then in Section "Four ICT-Based Citizen-Sourcing Methods" we outline four ICT-based citizen-sourcing methods (in Sections "Active Web 1.0 Expert: Sourcing" to "Passive Web 2.0 Expert: Sourcing" respectively) developed by the research group of the author in the last decade, which are complementary and enable both 'active' and 'passive' citizen-sourcing, targeting the general public as well as the experts. For each method we present briefly its main structure, and then the techniques used for the collection of external policy relevant content, as well as for its processing in order to calculate from it useful policy analytics. All four methods have been developed through a common methodology, which consists of five stages: (1) development of the basic idea through co-operation among research project partners; (2) detailed analysis of requirements of potential users of the method (participating as partners in the corresponding European project); (3) detailed development of the method and the corresponding ICT platform; (4) pilot applications of the method and based on them evaluation of it; (5) taking into account the conclusions of the evaluation final formulation of the method and the corresponding ICT platform. Furthermore, for each method are provided references with more detailed descriptions of it and its development methodology.
(c) Since there is a lack of a sound theoretical background for the development of the public sector citizen-sourcing area, we leveraged the experience gained from the development and some first pilot applications of these methods for the identification of some theoretical foundations from previous political and management sciences research, which can be useful for the development of more ICT-based citizen-sourcing methods in the future, as well as for their evaluation. In particular, we identified useful theoretical foundations from previous research in the areas of wicked social problems, crowdsourcing risks, absorptive capacity and technology acceptance models/diffusion of innovation, which are presented in Sections "Wicked Social Problems Theory" to "Technology Acceptance Model/Diffusion of Innovation Theory" respectively.
(d) Based on these theoretical foundations a framework has been developed for the multi-perspective evaluation of such methods of ICT-based citizen-sourcing from political, crowd-sourcing, absorptive capacity and diffusion perspectives, which is presented in Section "A Multi-Perspective Evaluation Framework".

(e) Finally, in Section "Citizen-Sourcing Methods Evaluation and Comparison" an outline is provided of the evaluations of these ICT-based citizen-sourcing methods, which have been conducted using parts of this evaluation framework, and then a comparison of them; furthermore, references are provided with more detailed descriptions of the methodologies and the findings of these evaluations.

Section "Conclusions" summarizes the conclusions and proposes future research directions.

Literature Review

Considerable research has been conducted for the identification and development of efficient and effective crowd-sourcing practices; reviews of this research are provided by Hetmank (2013) and Rechenberger et al. (2015). However, for public sector citizen-sourcing, there is a lack of research similar to the one conducted for private sector crowd-sourcing (i.e. having similar levels of breadth and depth), probably because the former is a more recent phenomenon than the latter. Limited research has been conducted concerning the application of crowd-sourcing ideas in the public sector, the development of efficient and effective methods and practices for this purpose, and the evaluation of them from various perspectives (Lukensmeyer and Torres 2008; Hilgers and Ihl 2010; Linders 2012; Nam 2012; Mergel and Desouza 2013; Ferro et al. 2013; Prpić et al. 2015). In the following, we focus on two important and representative studies in this area. Nam (2012) developed a framework for the description and analysis of government agencies citizen-sourcing initiatives, which includes four main types of them: (a) contest (= competition-driven citizen-sourcing, with material (usually monetary) incentives (e.g. cash, prizes) or/and career opportunities); (b) wiki (= collaborative website that can be edited directly using a web browser by anyone with access to it, with non-monetary reasons motivating participation, such as amateurism (commitment to hobbies) and altruism (voluntary contribution to society)); (c) social networking (= forum for discussion and interaction, which motivates participation primarily through the desire and expectation of forming new relationships and strengthening existing ones); (d) social voting (= it allows citizens to post their own ideas, make comments on others' ideas, and rate them; they provide a unique motivator for engagement: citizens can make their voices be heard by other citizens and by the government). One of the few studies analyzing citizen-sourcing initiatives following practices similar to the ones of private sector crowd-sourcing is the one of Mergel and Desouza (2013). They analyze the case of USA Challenge.gov platform, which enables federal agencies to host contests on problems and challenges they face, create awareness for them and bring citizens together in a competitive scenario to solve them; this platform allows citizens to provide solutions, and also review and evaluate solutions provided by other, vote on solutions, and even get involved in the implementation of solutions and subsequent evaluation of new policies or other types of public sector innovations. This study provides interesting insights into the implementation process of such private sector models for addressing public sector problems, and also the role that public managers play in such initiatives.

Extensive further research is required in the area of public sector citizen-sourcing, in order to develop theoretical foundations of it, and efficient and effective methods and practices, and also to understand better and assess their value, and in general create a higher level of maturity in this area, comparable to that of the private sector crowd-sourcing area. Our research makes a contribution in this direction.

Four ICT-Based Citizen-Sourcing Methods

In this section are presented four ICT-based citizen-sourcing methods, which have been developed by the research group of the author in the last 10 years, as part of several European projects. They support both 'active' and 'passive' citizen-sourcing, aiming at the efficient retrieval of policy-relevant information, knowledge, and ideas from citizens, both from the general public and also from knowledgeable experts, using initially web 1.0 oriented tools, and latter web 2.0 oriented ones that exploit highly popular social media platforms (e.g. Facebook, Twitter, YouTube, Picasa, Blogger).

Active Web 1.0 Expert Sourcing

The first method involves 'active expert-sourcing' using web 1.0 oriented government operated structured e-consultation fora; this term denotes an electronic web space hosting structured policy related consultations on topics defined by government, in which participants can enter specific types of semantically-annotated postings, which should be associated to previous postings according to some predefined rules, based on a "discussion ontology". This structure aims to stimulate more effective electronic consultations, with more mentally-processed, focused, and, therefore, higher-quality contributions of the participants, which are properly associated with the ones of other participants. The design of this method has been based on the 'Issue Based Information Systems' (IBIS) framework (Kunz and Rittel 1979; Conklin and Begeman 1989; Conklin 2003) (see Section "Wicked Social Problems Theory"), so the primary model of structured e-consultation forum supported allows each participant to enter five types of postings: issues, alternatives, pro arguments, contra argument and comments. Also, a number of possible associations between them have been defined in accordance with IBIS: for each issue participants are allowed to enter alternatives or comments, for each alternative they can enter relevant pro arguments, contra arguments or comments, for each argument (pro or contra) can enter other arguments (pro or contra) and for each comment other comments. In the following Fig. 1 we can see part of such a structured e-consultation conducted

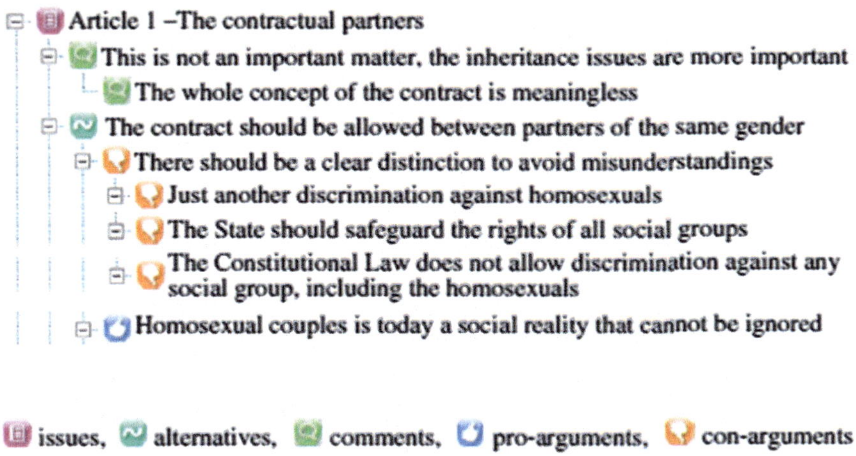

Fig. 1 Active web 1.0 expert-sourcing' using a structured e-consultation forum

by the Greek Parliament about a bill under formation concerning the 'Contracts of Voluntary Co-habitation'. This method has been developed as part of the LEX-IS project ('Enabling Participation of the Youth in the Public Debate of Legislation among Parliaments, Citizens and Businesses in the European Union') (www.lex-is.eu), which has been partially funded by the 'eParticipation' Preparatory Action of the European Commission. More information about this method is provided by Loukis and Wimmer (2010, 2012).

Active Web 2.0 Citizen Sourcing

This second method involves web 2.0 oriented 'active citizen-sourcing' making use of government agencies' own social media accounts. In this method a government agency has a strong active role, posing a particular social problem or public policy direction in its own multiple social media accounts, and soliciting relevant information, knowledge, opinions and ideas from citizens. The relevant content generated by citizens is automatically retrieved from these multiple social media accounts and then processed in order to produce useful policy analytics. The practical application of this method is based on a central ICT platform that automatically publishes multimedia content (e.g. a short text, a longer description, images, videos, etc.) concerning a social problem of interest or a public policy under formulation to multiple accounts of a government agency in various social media (e.g. Facebook, Twitter, YouTube, Picasa, Blogger), using their application programming interfaces (API), in order to actively stimulate discussions on it. During these social media consultations we continuously retrieve and monitor various types of citizens' interactions with the content we have posted (e.g. views, likes, ratings, comments, retweets), and

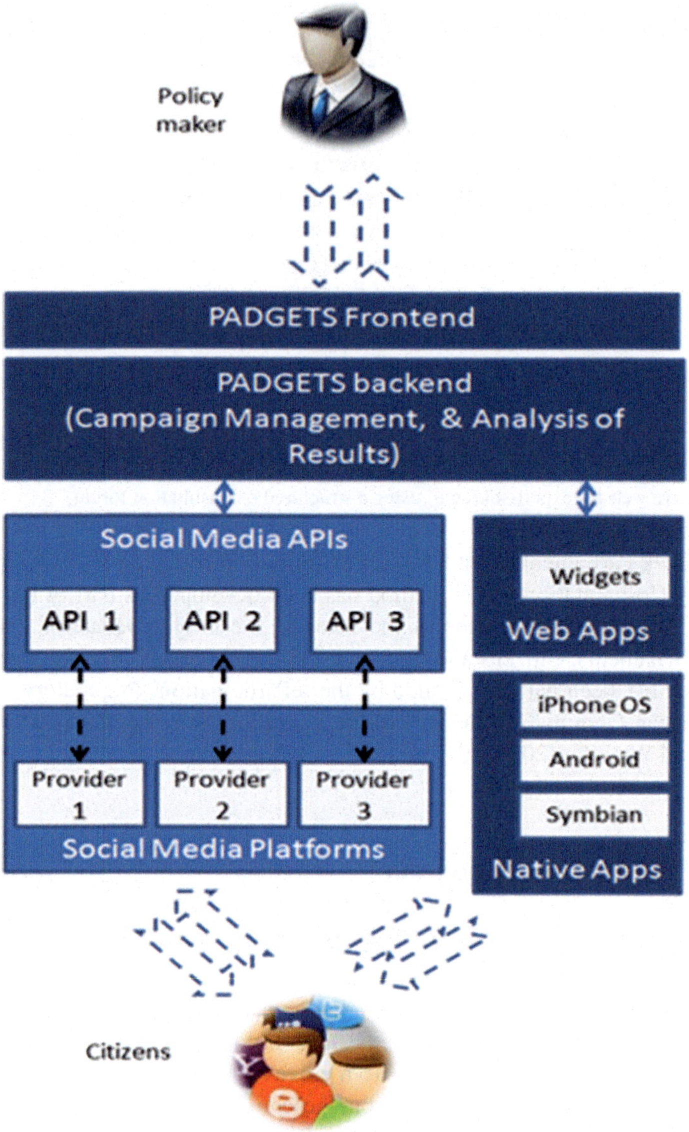

Fig. 2 Active web 2.0 citizen-sourcing using multiple government social media accounts

possibly make interventions in them whenever necessary. Finally, we automatically retrieve these interactions, again using the API of these social media, and then make advanced processing of them (e.g. calculation of various web analytics, opinion mining, issues extraction, sentiment analysis) in the above central ICT platform, in order to calculate policy analytics that allow gaining interesting insights and drawing conclusions. This method has been developed as part of the European research project PADGETS ('Policy Gadgets Mashing Underlying Group Knowledge in Web 2.0 Media') (www.padgets.eu). In Fig. 2 this active crowdsourcing method is illustrated. More information about it is provided by Ferro et al. (2013) and Charalabidis et al. (2014a).

Passive Web 2.0 Citizen Sourcing

The third method involves web 2.0 oriented 'passive citizen-sourcing' using again social media, however going beyond the accounts of government agencies in various social media, and exploiting political content developed by citizens in various external social media accounts not belonging to the government (e.g. various external political blogs, fora, news websites, and also external Facebook, Twitter, etc. accounts). In this method, government has a less active and more passive role, collecting and analyzing content on a specific topic or public policy of interest, which has been freely generated by citizens, without direct stimulation and direction by the government, in various external social media sources, and then making advanced processing of it and calculating useful policy analytics. The first step of the practical application of this method is to build the 'domain model', which is an ontology-based representation of the objects of the thematic domain in which we intend to intervene through a public policy (e.g. energy domain, education domain, health domain). Then the second step is to build the 'policy model', which is a representation of the public policy we want to collect relevant content about from the social media, consisting of a number of 'policy statements' associated with one or more nodes of the abovementioned domain model. The third step includes the definition by the user of a list of social media sources (e.g. external political blogs, news websites, and also Twitter, Facebook, etc. accounts) which are going to be crawled, in order to find relevant content about the topic or public policy of interest. These sources are searched regularly against the abovementioned domain and policy models, and the collected content undergoes sophisticated processing using opinion mining techniques: initially opinions and arguments are extracted, and then sentiment analysis of them is performed. The results are presented to the user in visualized form; a typical results' visualization screen is shown in the following Fig. 3. This method has been developed as part of the European research project NOMAD ("Policy Formulation and Validation through Non-moderated Crowdsourcing") (www.nomad-project.eu/). More information about it is provided by Loukis et al. (2015) and Androutsopoulou et al. (2015).

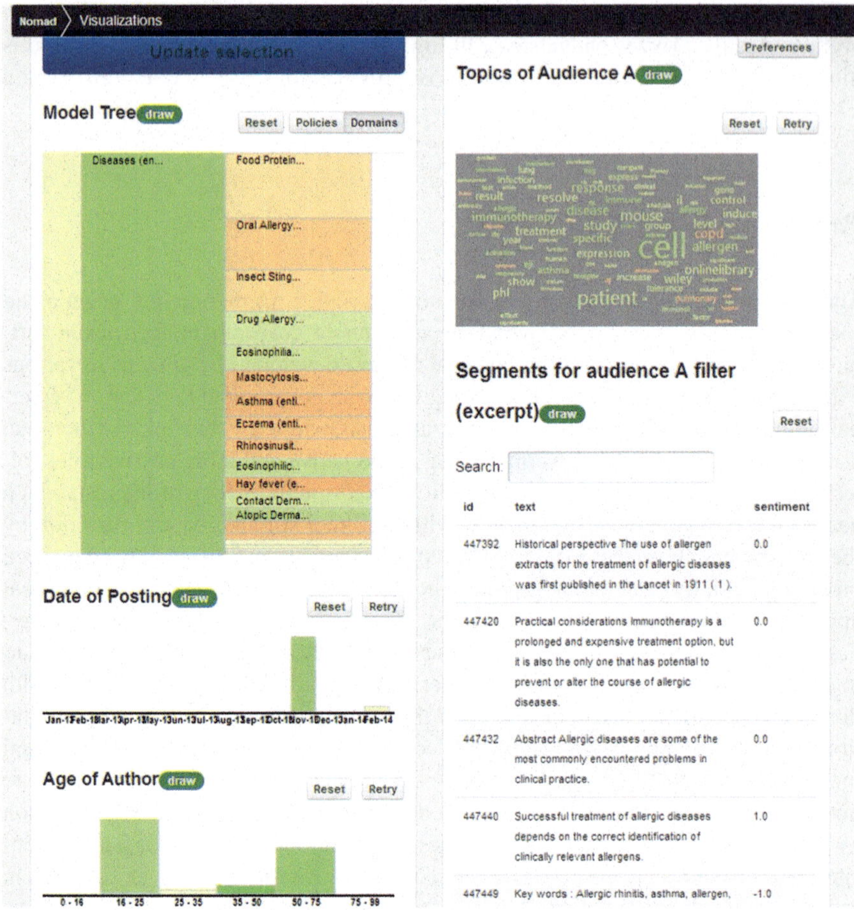

Fig. 3 Passive web 2.0 citizen-sourcing—typical results' visualization screen

Passive Web 2.0 Expert Sourcing

The fourth method involves web 2.0 oriented 'passive expert-sourcing', aiming at collecting policy relevant information, knowledge, and ideas published by experts in various external social media accounts not belonging to government (e.g. political blogs, fora, news websites, and also external Facebook, Twitter, etc. accounts). It adopts a 'selective' approach, focusing on the most knowledgeable and credible people on each topic we are interested in, by using advanced authors' reputation assessment and management techniques, and also focusing on the most relevant documents (e.g. web pages, blog posts, social media content, online comments, word/pdf documents, collected from various external sources) on each topic, by using documents' relevance assessment techniques. These assessments of authors' reputation and documents' relevance determine the order of content presentation in

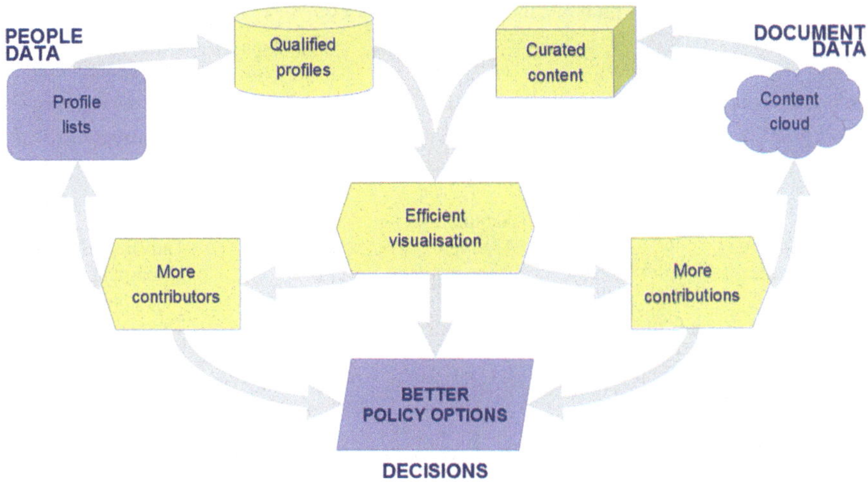

Fig. 4 Passive web 2.0 expert-sourcing—an overview

response to users' queries; therefore, experts' relevant content is given priority and higher visibility. This method is currently under development as part of the of the European research project 'EU-Community' (project. Eucommunity.eu/). An overview of this method is shown in Fig. 4. It consists of three processes: the first two of them crawl at regular time intervals external sources of profiles of individuals having extensive knowledge on various EU policies, and also of relevant documents, update the corresponding databases, and also assess the reputation of the former and the relevance of the latter. These databases are used by the third process, which processes users' queries and presents the results using visualization techniques. More information about this method is provided by Charalabidis et al. (2014b).

Theoretical Foundations

Based on the experience we gained from the development and some first pilot applications of the above methods, some theoretical foundations from previous political and management sciences research have been identified, which can be useful for the development of more ICT-based citizen-sourcing methods in the future, as well as for their evaluation. The development of an ICT-based citizen-sourcing method for supporting the formulation of public policy should enable the collection from citizens of useful information and knowledge concerning the main elements of the social problem that this public policy attempts to address; therefore, a theoretical foundation defining the main elements of a social problem is required. The 'Wicked Social Problems Theory' can be very useful for this purpose (see Section "Wicked Social Problems Theory"). Also, such a method should not be highly vulnerable to the inherent risks of crowd-sourcing (such as lack of representativeness and quality

Table 1 Summarization of proposed theoretical foundations

Wicked social problems theory	Crowd-sourcing risks	Absorptive capacity	Technology acceptance model	Innovation diffusion theory
Main problem elements:	Lack of sufficient, diverse, know-ledgeable crowd; digital divide; participation inequalities; crowd bias and manipulation	Components	Technology critical characteristics	Innovation critical characteristics
Topics; questions/issues; solutions/ideas; arguments		Acquisition Assimilation Transformation Application/ exploitation	Ease of use Usefulness	Relative advantage Compatibility Complexity Trialability Observability

of the content generated by the crowd, etc.); therefore, relevant theoretical foundations defining the inherent risks of crowd-sourcing are necessary (see Section "Crowd-Sourcing Risks"). Furthermore, an ICT-based citizen-sourcing method, in order to support substantially the formulation of a public policy, should support not only the identification and acquisition of relevant external information and knowledge from citizens, but also its dissemination within the competent government agency, and then its application and utilization for public policy making; therefore a theoretical foundation concerning the whole cycle of import, assimilation, and exploitation of external knowledge by firms is required. The 'Absorptive Capacity Theory' can be very useful for this purpose (see Section "Absorptive Capacity Theory"). Finally, the use of such an ICT-based citizen-sourcing method by a government agency for supporting the formulation of public policies constitutes a significant technological innovation in its policy formulation practices and processes; therefore, such a method should have the fundamental preconditions for a wide acceptance and diffusion. For this reason, a theoretical foundation is required providing the main characteristics that this technological innovation should have, in order to be widely adopted. The 'Technology Acceptance Model' and the 'Diffusion of Innovation Theory' Theory' can be very useful for this purpose (see Section "Technology Acceptance Model/Diffusion of Innovation Theory"). The above theoretical foundations are outlined next in Sections "Wicked Social Problems Theory" to "Technology Acceptance Model/Diffusion of Innovation Theory", and summarized in Table 1.

Wicked Social Problems Theory

Previous political sciences' research has revealed the increasing complexity and 'wickedness' of the problems of modern societies, which have to be addressed through appropriate public policies (Kunz and Rittel 1972, 1979; Rittel and Weber 1973; Head 2008). Societies have become more heterogeneous and pluralistic in terms of culture, values, concerns and lifestyles, and this made public policy problems 'wicked', i.e., lacking clear and widely agreed definition and objectives, and having many stakeholders with different and heterogeneous problem views, concerns and expectations. Rittel and Weber (1973) suggest that wicked social problems require 'second generation' methods, which include a first stage of consultation among problem stakeholders, aiming to formulate a shared definition of the problem and the objectives, and then a second stage of mathematical optimization analysis by experts of the well-defined at this stage problem. Subsequent research on this 'second generation' approach has revealed that its first stage can be substantially supported by the use of appropriate information systems, called 'issue-based information systems' (IBIS), which allow stakeholders to enter the main elements of the particular social problem, as perceived by them, which are: (a) the 'topics' (meant as broad discussion areas); (b) the 'questions/issues' (particular problems to be addressed within a discussion topic); (c) 'ideas' (possible alternative answers-solutions to questions/issues); (d) 'arguments' (positive or negative—evidence or viewpoints that support or object to ideas) (Kunz and Rittel 1979; Conklin and Begeman 1989; Conklin 2003). Therefore, the development of an ICT-based citizen-sourcing method for supporting the formulation of public policy should enable the collection from citizens of information and knowledge concerning the main elements of the social problem that this public policy attempts to address: the questions/issues, solutions/ideas and positive/negative arguments perceived by various problem stakeholder groups.

Crowd-Sourcing Risks

Another stream of crowd-sourcing research is dealing with the inherent risks and challenges of it; its main argument is that the outcomes of crowd-sourcing, mainly with respect to the quality and usefulness of the collected knowledge, might be uncertain, depending to a significant extent on the degree of managing some inherent risks of crowd-sourcing. This research stream identifies the most important of these risks, which are (Sharma 2010; Jain 2010; Agafonovas and Alonderiene 2013, Geiger et al. 2011): the lack of sufficient, diverse and knowledgeable, active crowd; digital divide related problems and the consequent participation inequalities (i.e. under-representation of some groups, and over-representation of some others); possible bias and manipulation of the crowd. Therefore the development of an ICT-based citizen-sourcing method for supporting the formulation of public policy should aim at overcoming or at least managing to a good extent the above risks.

Absorptive Capacity Theory

Previous management sciences' research has concluded that modern economy has become much more dynamic (as there are frequent and fast changes in most sectors), complex and 'knowledge intensive' than in the past, so organizations, in order to respond successfully, should become more innovative and flexible, and this necessitates the development of the capacity to identify and acquire useful external knowledge, assimilate it and then apply it towards achieving its organizational goals, which is termed 'absorptive capacity' (Cohen and Levinthal 1989, 1990; Zahra and George 2002; Camisón and Forés 2010). According to Zahra and George (2002) the four main components/dimensions of the absorptive capacity of an organization are: (1) acquisition capacity (= ability to locate, identify, value and acquire external knowledge that is critical to its operations or/and products and services); (2) assimilation capacity (= ability to absorb external knowledge; it can also be defined as the processes and routines that allow the new information or knowledge acquired to be analyzed, processed, interpreted, understood, internalized and classified); (3) transformation capacity (ability to develop and refine the internal routines that facilitate the combination of the previous knowledge base of the firm with the newly acquired or assimilated knowledge); (4) application or exploitation capacity (= ability, to incorporate acquired, assimilated and transformed knowledge into organizational operations and routines, not only in order to refine, perfect, expand and leverage existing routines, processes, competences and knowledge, but also to create new operations, competences, routines, products and services). Therefore, the development of an ICT-based citizen-sourcing method for supporting the formulation of public policy should support not only the acquisition of relevant external information and knowledge from citizens, but also its utilization: its dissemination within the competent government agency, and then its transformation, application and exploitation for public policy making.

Technology Acceptance Model/Diffusion of Innovation Theory

Extensive research has been conducted on the acceptance of technologies, and one of its main objectives is to identify characteristics of technologies that affect their level of acceptance. One of the most widely recognized and used outcomes of this research is the Technology Acceptance Model (TAM) (Davis 1989) and its extensions. According to the initial TAM the attitude towards using a technology, which finally determines the intention to use it and its actual use, is determined mainly by two critical characteristics of the technology: (a) its perceived 'ease of use' (= the degree to which potential users believe that using it would require minimal effort); (b) its perceived 'usefulness' (= the degree to which potential users believe that using it will enhance their job performance). Extensive research has been conducted based on the TAM for understanding better user acceptance of various types of ICT;

a comprehensive review of this research stream is provided by Hsiao and Yang (2011) and Marangunić and Granić (2015).

Furthermore, extensive research has been conducted also concerning the diffusion of innovations, in order to identify factors that favor it. One of the most widely recognized theories of innovation diffusion is the one proposed by Rogers (2003), which has been extensively employed for analyzing ICT-related innovations in both the public and the private sector (e.g. Raus et al. 2009; Loukis et al. 2011). The Diffusion of Innovation (DOI) theory of Rogers proposes a wider set of five critical characteristics of an innovation that determine the degree of its adoption: (1) Relative Advantage, defined as the degree to which an innovation is perceived as better than the idea, work practice or object it supersedes (it is similar with the usefulness of the TAM); (2) Compatibility, defined as the degree to which an innovation is perceived as being consistent with the existing values, past experiences, and needs of potential adopters; (3) Complexity, defined as the degree to which an innovation is perceived as difficult to understand, implement and use (it is similar with the ease of the TAM); (4) Trialability, defined as the degree to which an innovation may be experimented with on a limited scale basis; (5) Observability, defined as the degree to which the results of an innovation are visible to the external environment.

Therefore, an ICT-based citizen-sourcing method for supporting the formulation of public policy should have the above characteristics to a good extent so that it can be widely accepted and diffused in government agencies.

A Multi-Perspective Evaluation Framework

Based on the theoretical foundations presented in the previous Section "Theoretical Foundations" a multi-perspective framework has been developed for the evaluation of ICT-based methods of citizen-sourcing, which is shown in Table 2. It includes four evaluation perspectives:

1. a political perspective: it is based on wicked problems theory outlined in Section "Wicked Social Problems Theory" and aims to assess to what extent the particular method enables the collection from citizens of information and knowledge concerning the main elements of the social problem that a particular existing or under development public policy attempts to address;
2. a crowdsourcing perspective: it is based on previous theoretical work on the inherent risks of crowdsourcing outlined in Section "Crowd-Sourcing Risks" and aims to assess to what extent the particular method has the inherent risks of crowd-sourcing;
3. an absorptive capacity perspective: it is based on previous theoretical work on the absorptive capacity outlined in Section "Absorptive Capacity Theory" and aims to assess to what extent the particular method supports the whole cycle of import, assimilation and exploitation of external knowledge by government agencies;
4. a diffusion perspective: it is based on Rogers Diffusion of Innovation (DOI) theory outlined in Section "Technology Acceptance Model/Diffusion of Innovation

Table 2 A multi-perspective framework for the evaluation of ICT-based citizen-sourcing methods

Political perspective
To what extent the particular ICT-based citizen-sourcing method is useful/beneficial: – for assessing for a particular domain or an existing or under development policy the level of interest/discussion in the society? the attitudes/sentiments of the society (positive—neutral—negative)? the time-wise changes of the above (level of interest/discussion and attitudes/sentiments)? whether there is uniformity/homogeneity of the above among different groups? – for identifying relevant issues posed by citizens or needs of them? proposals for solving relevant problems or improving policies? arguments (positive or negative ones)? – and in particular for the early identification of new emerging relevant issues or needs in the society? new emerging proposals in the society for solving relevant problems or improving policies?
Crowdsourcing perspective
To what extent you agree with the following: the results provided (levels of interest, sentiments, issues, proposals, arguments, etc.) are representative (or at least indicative) of the ones prevailing in the society as a whole (and do not represent only some groups of citizens). the above are non-biased and non-manipulated. are reliable and of high quality. they can contribute positively to the development or improvement of public policies in the particular domain.
Absorptive capacity perspective
To what extent the particular ICT-based citizen-sourcing method is useful/beneficial for: recognizing/identifying and acquiring from its external environment useful knowledge for the development of public policies or improvement of existing ones. the internal dissemination of this new external knowledge within the government agency. the assimilation/integration of this new external knowledge in the existing knowledge base of the government agency. the exploitation of this new external knowledge, in combination with the existing knowledge possessed by the government agency, for the development of public policies or improvement of existing ones.
Diffusion perspective
To what extent you agree that the particular ICT-based method of citizen-sourcing, viewed as an innovation is better than other existing traditional or electronic methods used for similar purposes in the public policy development processes. is compatible with the public policy development processes, as they are applied in European Union countries, and can be integrated into these processes. is compatible with the needs, the mentalities and the values of the people designing and applying public policies. can be initially applied in a small scale in public policy making before proceeding to a large scale application of it. is in general easy to use. its application does not require extensive effort. its results are easy to understand.

Theory" (as it proposes a wider set of critical characteristics than the Technology Acceptance Model (TAM)) and aims to assess to what extent the particular method has the required characteristics for a wide adoption and diffusion.

Each of these four evaluation perspectives includes several evaluation dimensions, which can be used both for quantitative evaluation (i.e. for developing corresponding questions of an evaluation questionnaire) and for qualitative evaluation (i.e. as topics to be discussed in interviews).

Citizen-Sourcing Methods Evaluation and Comparison

In this section we provide an outline of the evaluations of the first three of the presented ICT-based citizen-sourcing methods in Sections "Active Web 1.0 Expert: Sourcing" to "Passive Web 2.0 Citizen: Sourcing" (as the fourth one presented in Section "Passive Web 2.0 Expert: Sourcing" has not yet been evaluated), which have been conducted as parts of the corresponding European projects, using parts of this evaluation framework presented in the previous Section "A Multi-Perspective Evaluation Framework" (as the whole evaluation framework had not yet been developed at the time of these evaluations), as well as a comparison of these methods. Furthermore, we provided references with more detailed descriptions of the methodologies and the findings of these evaluations.

Active Web 1.0 Expert: Sourcing

Two pilot applications of this method were organized concerning IBIS-based structured e-consultations on legislation under formation in the Greek and Austrian Parliaments. The Greek pilot e-consultation was about a bill on the 'Contracts of Voluntary Co-habitation', while the Austrian one was about a ministerial draft bill titled 'Child and Youth Welfare Law'. The evaluation of these two pilot applications was based on the TAM, so it focused on the ease of use by the citizens of this method and on its usefulness (with respect to improvements of discussion structure and quality). With respect to the former it has been concluded that this structured e-consultation is too difficult and demanding for less sophisticated users (e.g. in terms of education and general maturity). So it might not be appropriate for e-consultations with the general public, and would more suitable for e-consultations with experts. However, with respect to usefulness, it seems that the structured e-consultation is better than the normal unstructured one, especially for conducting important discussions, for both more and less sophisticated users, leading to more structured and higher quality discussions. More sophisticated users seem to perceive a higher usefulness of the e-structured forum tool than the less sophisticated users, since the former can much better use the complex discussion language and exploit to a larger extent the potential of these tools for structuring discussion. Our findings indicate that more discussion structure in citizen-sourcing can lead to higher quality of information, knowledge, ideas/proposals and arguments generation on a social problem of interest for a government agency; however, these can be less representative, reflecting the views experts, 'organized interest groups' (Gilens and Page 2014), or at least citizens having higher education, who have higher capacity for more substantial and influential participation in such structured discussions. More information on the methodology and the findings of this evaluation are provided in Loukis and Wimmer (2010, 2012).

Active Web 2.0 Citizen: Sourcing

A pilot application of this method was organized in co-operation with the Piedmont Regional Government, Italy, aiming to conduct a campaign and consultation in multiple social media concerning the implementation of a telemedicine initiative (initially applied in a limited area) at a large scale in the entire Piedmont region. An evaluation of this pilot application was conducted based on the political and diffusion perspectives of the evaluation framework presented in the previous Section "A Multi-Perspective Evaluation Framework". From a political perspective it was concluded that as this method is characterized by much less discussion structure than the previous ones, and also uses highly popular social media, it enables reaching wider and heterogeneous audiences in shorter time and at lower costs, conveying policy-related information to them, and also identifying a wide range of particular problems/issues they perceive with respect to a policy under discussion. The wealth of comments from different citizens' groups that such a multiple social media usage approach provides enables the identification of citizens' 'positive values' (things that citizens value) and also 'negative values' (things that citizens dislike) with respect to the particular policy or policy domain in general, which are very useful for the comprehensive design and evaluation of public policies that produce not only economic benefits, but also 'public value' (i.e. promoting collective values and preferences) (Cordella and Bonina 2012). However, this method seems to be less efficient in the generation of solutions and also in the facilitation of convergence among stakeholders' views. So, this ICT-based citizen-sourcing method allows collecting information, knowledge, opinions and ideas from wider groups of citizens (though they might be not representative of the whole society; also we cannot avoid the domination of some organized interest groups), but of lower quality than the previous method. From a diffusion perspective, it was concluded that this method, viewed as an innovation in policy formulation processes of government agencies, has the fundamental preconditions for a wide diffusion and adoption by government agencies: relative advantage, compatibility with existing values and processes, reasonable complexity, trialability, and observability. However, its compatibility, and in general the benefits that can be created by it, depend to some extent: (1) on the political tradition of the adopting government agency with respect to bi-directional communication with citizens in all phases of policy making, (2) on its familiarity with and experience in using social media for this purpose, and (3) on their positive general attitude towards innovation. More information on the methodology and the findings of this evaluation are provided in Ferro et al. (2013) and Charalabidis et al. (2014a).

Passive Web 2.0 Citizen: Sourcing

Two main pilot applications of this method were conducted. The first one was conducted by the Greek Parliament, and concerned the regulatory and legal framework of energy production and management. The second pilot application was conducted by the Austrian Parliament and aimed to monitor the ongoing public debate on open government data policies. An evaluation of these pilots was conducted based on the political, crowdsourcing and diffusion perspectives of the evaluation framework presented in the previous Section "A Multi-Perspective Evaluation Framework". From a political perspective it was concluded that this method of passive citizen-sourcing can provide considerable support for public policy making, enabling the low cost and fast assessment of citizens' feelings/attitudes concerning a prospective or existing policy, and also the identification of relevant issues/topics perceived by the society. By monitoring carefully selected high quality external sources (such as political blogs, fora, news websites, etc.) it is possible to collect high quality external information, knowledge, opinions, ideas and arguments, from both 'opinion leaders/influencers' (e.g. experienced journalists and experts, who create political content in the monitored sources), and 'average citizens' (who usually comment the above political content). However, this method poses some risks, concerning the possible intrusion into citizens' private sphere (so it is necessary to avoid monitoring sources in which contributors perceive their postings and discussions as private). From a crowdsourcing perspective, again there is some uncertainty concerning the representativeness of the content collected from the monitored sources (i.e. whether the results provided by this method reflect the general public opinion or not), and also about its reliability (i.e. whether they are non-biased, non-manipulated and of good quality). However, the important difference of this method from the previous ones is that it allows reducing these representativeness, quality and reliability risks by selecting an appropriate large and representative (e.g. with respect to political orientation) set of high reliability and quality sources to be monitored. From a diffusion perspective, it was concluded that this method of passive citizen-sourcing, viewed as an innovation, has most of the fundamental preconditions for a wide acceptance and diffusion. In particular, it offers strong relative advantage over the existing alternatives for the same purpose, and has high levels of trialability on a limited scale basis; also, it has a good level of compatibility with the policy formulation processes, and with the needs, mentalities and values of the people who design and apply public policies. However, this method does not seem to be easy to use, as it requires building complex models of the specific domain and also the particular policy we are interested in. More information on the methodology and the

Table 3 A comparison of the proposed ICT-based citizen-sourcing methods

	Active web 1.0 expert-sourcing	Active web 2.0 citizen-sourcing	Passive web 2.0 citizen-sourcing	Passive web 2.0 expert-sourcing
Active/passive citizen-sourcing	Active	Active	Passive	Passive
Target	Experts	General public	General public	Experts
Web paradigm	Web 1.0	Web 2.0	Web 2.0	Web 2.0
Use of external electronic spaces (not owned by government)	No	Yes	Yes	Yes
Content processing	No	Medium	Sophisticated	Sophisticated
Contributors' reputation assessment, management and use	No	No	No	Yes
Competitive contests—rewards	No	No	No	No
Support of the whole cycle of import, assimilation and exploita-tion of external knowledge	No	No	No	No
Content sources selection	No	No	Yes	Yes

findings of this evaluation are provided in by Loukis et al. (2015) and Androutsopoulou et al. (2015).

Comparison

A comparison of the four proposed ICT-based methods for citizen-sourcing is shown in the following Table 3, which reveals similarities and differences. An important similarity is that none of them includes competitive contests and rewards, which are central elements of private sector crowd-sourcing. Further research is required in order to examine to what extent we can incorporate in these methods some kind of competition and rewards (monetary or non-monetary). Another similarity is that none of these four methods supports the whole cycle of import, assimilation and exploitation of external knowledge, despite the strong arguments provided by the absorptive capacity research (see Section "Absorptive Capacity Theory') that it is necessary to increase the capacity of organizations not only for external knowledge exploration and acquisition, but also for internal assimilation and then exploitation of this external knowledge. All four proposed methods support only the first stages of this cycle, but do not support its highly important later stages. Therefore, it would be highly beneficial to examine how these ICT-based citizen-sourcing methods can be extended in order to support the internal dissemination and assimilation of the acquired external knowledge within the user government agency, and then its application and exploitation for public policy making. At the same time these four methods have important differences as to their main approach (active or passive), target (general public or experts), web paradigm used (web 1.0 or 2.0), type of electronic spaces used (government owned or external), level of processing of the collected external content and assessment/management/use of content contributors' reputation.

The evaluations of the first pilot applications of these ICT-based citizen-sourcing methods have shown that the active citizen-sourcing ones can provide to government agencies useful external content (e.g. concerning existing problems/issues, proposed solutions/ideas, arguments, comments, sentiments), which however might be not be representative of the perceptions, needs and values of the whole society. In these methods government agencies do not select the 'sources' of external content (though invitation of a closed and representative group of citizens might be an option). So it is possible that some organized interest groups dominate in these electronic discussions taking place in open government e-consultation spaces or social media accounts; previous literature has highlighted the strong and often disproportionate impact that some organized interest groups have on public policies (e.g. see Gilens and Page (2014)). Therefore, the content collected through such active ICT-based citizen-sourcing methods should be exploited for the formulation of public policies carefully, in combination with other relevant external content collected through other channels (e.g. through physical meetings with various stakeholder groups); this combination will enable distinguishing the parts of the former content that are broadly accepted and representative of the whole society from the parts of it that have lower acceptance and reflect views only of specific groups. On the contrary, the passive citizen-sourcing methods enable the selection of the sources to be monitored and this can reduce this representativeness uncertainty.

Furthermore, these first evaluations have shown that a critical success factor of the active citizen-sourcing methods is the engagement and participation of a large number of citizens. Therefore, a government agency before using it should to develop its network/community (e.g. followers, friends, subscribers, etc.) in the utilized social media. While in private sector crowd-sourcing the main motivation for the participating individuals/teams is to earn various types of rewards and/or professional recognition, in public sector crowd-sourcing the motivation for citizens to participate is quite different: very rarely there are some rewards (this happens in very few cases, such as the USA Challenge.gov platform (Mergel and Desouza 2013) reviewed in the Literature Review section)***; usually the main motivation of the participants in citizen-sourcing is to influence government agencies' policies and decisions towards directions believed as beneficial for the society, or promoting their own interests or/and values. So it is necessary government agencies adopting such active citizen-sourcing methods, as usually there is no monetary or non-monetary reward, to persuade citizens that their contributions (e.g. concerning existing problems/issues, proposed solutions/ideas, arguments, comments) are taken seriously into account in the formulation of public policies and decisions, and in general are exploited for the development of public policies aiming to generate 'public value' (promoting collective values and preferences) (Cordella and Bonina 2012). On the contrary, the passive citizen-sourcing methods do not rely on the direct engagement of large numbers of citizens, and their critical success factor is the selection of appropriate external sources (political blogs, fora, news websites, etc.) to be monitored.

Conclusions

The development of public policies for addressing the complex problems and needs of modern societies can greatly benefit from leveraging the 'wisdom of the crowds'. However, this necessitates the development of appropriate methods, supported by advanced ICT (policy informatics), which enable the automated collection of information, knowledge, ideas/proposals and opinions from citizens, and then the processing of them in order to calculate useful policy analytics. This chapter makes a contribution in this direction. It initially provides an overview of research conducted in this area by the research group of the author in the last decade, which has led to the development of four ICT-based citizen-sourcing methods. Then leveraging the experience gained from this research are proposed some theoretical foundations from previous political and management sciences research, which can be used for the future development of effective ICT-based citizen-sourcing methods for supporting public policy making, as well as for the evaluation of such methods. Based on them a multi-perspective evaluation framework has been developed. Finally, an outline of the evaluations of these ICT-based citizen-sourcing methods is provided, based on parts of this evaluation framework, as well as a comparison of them.

We believe that the research presented in this chapter has interesting implications for both research and practice in the area of policy informatics. With respect to research, it provides useful theoretical foundations from political and management sciences, which can be used for the required development of a wide range of advanced ICT-based citizen-sourcing methods in the future; useful for this can be also the multi-perspective evaluation framework we have developed. Furthermore, it proposes four fundamental typologies of methods, which can be used as a basis for further more advanced developments in this area. With respect to government policy formulation practice, our research provides four useful types of ICT-based methods for citizen-sourcing, which cover both 'active' and 'passive' citizen-sourcing, and can be very useful—as they are, or with appropriate adaptations, and possibly in combination—for collecting rapidly and at a low cost a wealth of external policy-relevant information, knowledge and ideas/proposals, both from the general public and the experts; these can significantly assist government agencies in becoming more 'extrovert', and developing effective socially rooted and acceptable public policies. Especially the proposed passive crowdsourcing methods allow overcoming two important problems that such citizen-sourcing initiatives currently face: the need of attracting and engaging large numbers of citizens (as they are using policy-related content that has already been developed in numerous external spaces), and the representativeness and quality uncertainties (which can be reduced by selecting a large and representative set of high quality sources).

Our main limitation is that only a few pilot applications of three out of the four proposed ICT-based citizen-sourcing methods have been conducted. Therefore, more application and evaluation of them is required, beyond the corresponding European projects, in order to understand better their value, strengths and weaknesses, and possibly to make required improvements and enrichments of them. Furthermore, more research is required in order to develop more advanced techniques for processing the content collected from citizen (e.g. based on previous research on data mining and text/opinion mining), in order to extract more knowledge from it and produce insightful policy analytics. Additional research is also required in order to gain a better understanding of the types of social problems, public policies, stages of the policy development cycle and solicited external information and knowledge each of them is more appropriate for, and also of how they can be combined. In general, it is necessary to conduct extensive research in the area of public sector citizen-sourcing, in order to reach a level of theoretical foundation, and also effectiveness and maturity, comparable to those of private sector crowd-sourcing.

References

Agafonovas A, Alonderiene R (2013) Value creation in innovations crowdsourcing - example of creative agencies. Organ Mark Emerg Econ 4(1):72–103

Androutsopoulou A, Charalabidis Y, Loukis E (2015) Using social media monitoring for public policy making – an evaluation. In: Proceedings of the 9th Mediterranean conference on information systems (MCIS 2015), Samos, Greece

Brabham DC (2008) Crowdsourcing as a model for problem solving: an introduction and cases. Convergence 14(1):75–90

Brabham DC (2012) Crowdsourcing: a model for leveraging online communities. In: Delwiche A, Henderson J (eds) The Routledge handbook of participative cultures. Routledge, Abingdon

Brabham DC (2013) Crowdsourcing. The MIT Press, Cambridge

Camisón C, Forés B (2010) Knowledge absorptive capacity: new insights for its conceptualization and measurement. J Bus Res 63(7):707–715

Charalabidis Y, Loukis E, Androutsopoulou A (2014a) Fostering social innovation through multiple social media. Inf Syst Manag 31:225–239

Charalabidis Y, Loukis E, Koulizakis Y, Mekkaoui D, Ramfos A (2014b) Leveraging European Union policy community through advanced exploitation of social media. In: Proceedings of IFIP sixth international conference on e-Participation - ePart 2014, Dublin, Ireland

Chesbrough H (2003a) Open innovation: the new imperative for creating and profiting from technology. Harvard Business School Press, Boston

Chesbrough H (2003b) The era of open innovation. Sloan Manag Rev 44(3):35–41

Chesbrough H (2006) Open innovation: a new paradigm for understanding industrial innovation. In: Chesbrough H, Vanhaverbeke W, West J (eds) Open innovation: researching a new paradigm. Oxford University Press, Oxford, pp 1–12

Cohen WM, Levinthal DA (1989) Innovation and learning: the two faces of R&D. Econ J 99(397):569–596

Cohen WM, Levinthal DA (1990) Absorptive capacity: a new perspective on learning and innovation. Adm Sci Q 35(1):128–152

Conklin J (2003) Dialog mapping: reflections on an industrial strength case study. In: Kirschner P, Buckingham Shum P, Carr C (eds) Visualizing argumentation: software tools for collaborative and educational sense-making. Springer, London

Conklin J, Begeman M (1989) gIBIS: a tool for all reasons. J Am Soc Inf Sci 40(3):200–213

Cordella A, Bonina CM (2012) A public value perspective for ICT enabled public sector reforms. Gov Inf Q 29:512–520

Davis FD (1989) Perceived usefulness, perceived ease of use, and user acceptance of information technology. MIS Q 13(3):319–339

Ferro E, Loukis E, Charalabidis Y, Osella M (2013) Policy making 2.0: from theory to practice. Gov Inf Q 30(4):359–368

Geiger D, Seedorf S, Schulze T, Nickerson RC, Schader M (2011) Managing the crowd: towards a taxonomy of crowdsourcing processes. In: Proceedings of American conference on information systems (AMCIS), 2011

Gilens M, Page BI (2014) Testing theories of American politics: elites, interest groups, and average citizens. Perspect Polit 12(3):564–581

Head BW (2008) Wicked problems in public policy. Public Policy 3(2):101–118

Hetmank L (2013) Components and functions of crowdsourcing systems – a systematic literature review. In: Wirtschaftsinformatik 2013 conference proceedings, Lipsia, Alemania

Hilgers D, Ihl C (2010) Citizensourcing: applying the concept of open innovation to the public sector. Int J Public Participation 4(1):67–88

Howe J (2006) The rise of crowdsourcing. Wired 14(6). Retrieved from http://www.wired.com/wired/archive/14.06/crowds.html

Howe J (2008) Crowdsourcing, why the power of the crowd is driving the future of business. Crown Business, New York

Hsiao CH, Yang C (2011) The intellectual development of the technology acceptance model: a co-citation analysis. Int J Inf Manag 31(2):128–136

Huizingh KRE (2011) Open innovation: state of the art and future perspectives. Technovation 31:2–9

Jain R (2010) Investigation of governance mechanisms for crowdsourcing initiatives. In: Proceedings of American conference on information systems (AMCIS), 2010

Kunz W, Rittel H (1972) Information science: on the structure of its problems. Inf Storage Retr 8(2):95–98

Kunz W, Rittel H (1979) Issues as elements of information systems. Working Paper No. 131, University of California, Berkley

Linders D (2012) From e-government to we-government: defining a typology for citizen coproduction in the age of social media. Gov Inf Q 29:446–454

Loukis E, Wimmer M (2010) Analyzing different models of structured electronic consultation on legislation under formation. In: 4th international conference on online deliberation (OD), Leeds, UK, June 2010

Loukis M, Wimmer A (2012) Multi-method evaluation of different models of structured electronic consultation on government policies. Inf Syst Manag 29(4):284–294

Loukis E, Spinellis D, Katsigiannis A (2011) Barriers to the adoption of B2B e-marketplaces by large enterprises: lessons learnt from the Hellenic Aerospace Industry. Inf Syst Manag 28(2):130–146

Loukis E Charalabidis Y, Androutsopoulou A (2015) Evaluating a passive social media citizensourcing innovation. In: Proceedings of IFIP eGovernment conference EGOV2015, Thessaloniki, Greece

Lukensmeyer CJ, Torres LH (2008) Citizensourcing: citizen participation in a networked nation. In: Yang K, Bergrud E (eds) Civic engagement in a network society. Information Age Publishing, Charlotte, pp 207–233

Majchrzak A, Malhotra A (2013) Towards an information systems perspective and research agenda on crowdsourcing for innovation. J Strateg Inf Syst 22(4):257–268

Marangunić N, Granić A (2015) Technology acceptance model: a literature review from 1986 to 2013. Univ Access Inf Soc 14(1):81–95
Mergel I, Desouza KC (2013) Implementing open innovation in the public sector: the case of Challenge.gov. Public Adm Rev 73(6):882–890
Nam T (2012) Suggesting frameworks of citizen-sourcing via Government 2.0. Gov Inf Q 29:12–20
Prpić J, Taeihagh A, Melton J (2015) The fundamentals of policy crowdsourcing. Policy Internet 7(3):340–361
Raus M, Flügge B, Boutellier R (2009) Electronic customs innovation: an improvement of governmental infrastructures. Gov Inf Q 26:246–256
Rechenberger T, Jung V, Schmidt N, Rosenkranz C (2015) Utilizing the crowd – a literature review on factors influencing crowdsourcing initiative success. In: Pacific conference on information systems (PACIS) 2015 proceedings. Paper 250
Rittel H, Weber M (1973) Dilemmas in a general theory of planning. Policy Sci 4:155–169
Rogers E (2003) Diffusion of innovations, 5th edn. The Free Press, New York
Sharma A (2010) Crowdsourcing critical success factor model strategies to harness the collective intelligence of the crowd. London School of Economics (LSE). Working-Paper 1–2010
Surowiecki J (2004) The wisdom of the crowds. Anchor Books, New York
West J, Salter A, Vanhaverbekec W, Chesbrough H (2014) Open innovation: the next decade. Res Policy 43:805–811
Zahra SA, George G (2002) Absorptive capacity: a review, reconceptualization, and extension. Acad Manag Rev 27(2):185–203

Euripidis N. Loukis is Associate Professor of Information Systems and Decision Support Systems at the Department of Information and Communication Systems Engineering, University of the Aegean, Greece. Previously he has taught at the Postgraduate Program 'Athens MBA' of the National Technical University of Athens and the Athens University of Economics and Business, at the postgraduate program of the National Academy of Public Administration, and at the University of Thessaly. Also, he has been Information Systems Advisor at the Ministry to the Presidency of the Government of Greece, and National Representative of Greece in the programs 'Telematics' and 'IDA' (Interchange of Data between Administrations) of the European Union. Dr. Euripidis Loukis has conducted extensive research in the areas of e-government, e-participation, ICT adoption, business value and impact, and decision support systems. He has been honoured with the International Award of the American Society of Mechanical Engineers (ASME)—Control and Diagnostics Committee, the Best Paper Award of the European and Mediterranean Conference on Information Systems (EMCIS) and the Most Innovative Research Award of the IFIP e-Government and e-Participation Conference.

Data Analytics for Policy Informatics: The Case of E-Petitioning

Loni Hagen, Teresa M. Harrison, and Catherine L. Dumas

Abstract To contribute to the development of policy informatics, we discuss the benefits of analyzing electronic petitions (e-petitions), a form of citizen-government discourse with deep historic roots that has recently transitioned into a technologically-enabled and novel form of political communication. We begin by presenting a rationale for the analysis of e-petitions as a type of e-participation that can contribute to the development of public policy, provided that it is possible to analyze the large volumes of data produced in petitioning processes. From there we consider two data analytic strategies that offer promising approaches to the analysis of e-petitions and that lend themselves to the future creation of policy informatics tools. We discuss the application of *topic modeling* to the analysis of e-petition textual data to identify emergent topics of substantial concern to the public. We further propose the application of *social network analysis* to data related to the dynamics of petitioning processes, such as the social connections between petition initiators and signers, and tweets that solicit petition signatures in petitioning campaigns; both may be useful in revealing patterns of collective action. The paper concludes by reflecting on issues that should be brought to bear on the construction of policy informatics tools that make use of e-petitioning data.

L. Hagen
School of Information, University of South Florida,
4202 E. Fowler Avenue, CIS 2031, Tampa, FL 33620-7800, USA
e-mail: lonihagen@usf.edu

T.M. Harrison (✉)
Department of Communication, University at Albany, SUNY,
BA 356, 1400 Washington Avenue, Albany, NY 12222, USA
e-mail: tharrison@albany.edu

C.L. Dumas
Informatics Doctoral Program, College of Engineering and Applied Sciences,
University at Albany, SUNY, BA 356, 1400 Washington Avenue, Albany, NY 12222, USA
e-mail: cdumas@albany.edu

J.R. Gil-Garcia et al. (eds.), *Policy Analytics, Modelling, and Informatics*, Public Administration and Information Technology 24,
DOI 10.1007/978-3-319-61762-6_9

Abbreviations

AP	Associated Press
EGRL	e-Gov Reference Library
ICT	Information and communication technology
LDA	Latent Dirichlet Allocation
NLP	Natural language processing
OECD	Organization for Economic Co-Operation and Development
SNA	Social network analysis
WtP	*We the People*

Introduction

As data and information play increasingly important roles in policy decision making, scholars have sought to apply new computational tools and techniques for a range of purposes related to policy analysis. The field of inquiry now emerging around these initiatives—policy informatics—is an interdisciplinary effort to leverage advances in computation and in information and communication technology (ICT) to address challenging problems in public policy and administration and to achieve useful innovations in governance and institutional processes (Chung and Zeng 2015; Dawes et al. 2014; Johnston 2015; PoliInformatics 2015). Scholars in public policy and management understand that data analysis is worth the investment of time and money (Crawford 2014) and that these practices will change the way that government works. One of the short-term goals of policy informatics is to use "technology to engage expertise wherever it exists, taking a broad view of what constitutes expertise" (Johnston 2015, p. 4). But the future envisioned by scholars of policy informatics will focus on the creation of policy analysis toolkits and applying these tools to specific circumstances so that policy analysts can match tools with context and provide time-sensitive advice for policy decision-makers (Howlett et al. 2009).

Achieving this goal thus embraces a challenging mix of intellectual tasks that involve creating and managing data, including new and hitherto unanalyzed datasets, as well as leveraging advances in computer and information science techniques, including machine learning. These tasks call upon scholars to re-align interests and relationships within the academic community and across research, education, and practice (Chung and Zeng 2015; Dawes et al. 2014; PoliInformatics 2015). Considering the rapidly changing information environment that has increased citizen access to political information and political actions such as boycotts, protest, and collective action groups (Dalton 2013), policy analysis will benefit from developing systematic approaches and new collaborative relationships that enable mining diverse information sources, such as open data and social media, and computational tools to produce meaningful analysis using these data.

To contribute to the development of policy informatics, in this chapter we discuss the benefits of analyzing electronic petitions (e-petitions) and the challenges of doing so. The rights of citizens to petition government can be traced from the British Bill of Rights promulgated in 1689, which asserts subjects' rights to petition the king for redress of grievances. Many other countries have since built on this British precedent, including the US where the right to petition is included in the First Amendment to the Constitution. However, with the creation of web-based platforms, e-petitions have become a contemporary and increasingly popular genre for the expression of opinion in the online public sphere. Such platforms make it easy and convenient for individuals to initiate petitions and gather support for them through signatures from supporters nationwide or globally. Indeed, e-petitioning is now a ubiquitous form of online collective action; few Internet users have neither seen nor signed an e-petition. Despite this, relatively little attention by social scientists has been devoted to this quintessentially twenty-first century form of political participation.

In what follows, we present a rationale for the analysis of e-petitions situated within the literature of public policy. In its contemporary form, we show that e-petitioning represents a novel and potentially consequential form of political action that policy makers should take seriously. From there we consider two general types of data and related data analytic techniques that offer promising information related to the analysis of e-petitions and that lend themselves to the future creation of policy informatics tools. The data likely to provide policy-related information includes textual data from the petition itself together with information that can be extracted from the textual data, such as estimated topics, sentiment, language choices, and names of locations, organizations and people featured. We show how topic modeling can be useful in extracting information from e-petition texts. A second type of data consists of information about the dynamics of petitioning activities including the numbers of petition initiators and signers, the social connections between them, and timing of petition initiation and events in the political and social environment. Included within this category is other information related to petitioning, such as tweets that refer potential signers to petitioning campaigns. We show how social networking analyses can be useful in extracting useful information from these dynamic processes. We discuss these two data analytic strategies, and illustrate them through examples drawn from our own research, focusing on the benefits and challenges of each. We conclude by reflecting on issues that should be brought to bear on the construction of policy informatics tools that make use of e-petitioning data.

What is new about our analysis is not so much the analytic techniques, as the application of them to this novel form of user-generated data. Analytic techniques such as machine learning, natural language processing, and social network analysis have been applied by computer and social scientists for some time to other types of data. However, the application of these techniques to the relatively novel data produced by electronic petitioning requires care in discerning appropriate uses and an appreciation for how to draw conclusions from this data.

In what follows, we argue that understanding the topics addressed by e-petitions has implications for the construction of public policy and policy analysis. Understanding the dynamics of e-petitioning, and its relationships with other forms of social media, can provide us with a greater appreciation of the strength and character of long term social movements and shorter mobilizations. E-petitions may inform policy analysts' considerations of policy goals and constraints by identifying social needs and oppositional concerns, as well as by estimating the strength of support for particular policy goals and options for achieving those goals.

E-Petitioning in the Policy Process

Public policy consists of the actions and decisions taken by government to solve the problems of citizens and improve their lives. The policy process is typically conceived as several steps through which policy is forged, beginning with problem identification and definition and proceeding subsequently through phases of agenda setting, formulation, choice making, and implementation (e.g., Downs 1972; Kingdon 1984; Birkland 2011). In theories of democratic governance, the public is expected to play a substantial role in public policy processes. However, what role citizens can or should play is an issue that has been the subject of long and considerable controversy. On the one hand, theories of democracy require some degree of active engagement of citizens in constructing, deliberating, and expressing choices regarding government action. On the other hand, the literature addressing policy analysis typically focuses on issues of efficiency and cost—benefit considerations. Citizen participation in policymaking is historically regarded by scholars with skepticism and opposition. Scholars have feared that citizen participation will overwhelm political systems with demands and that efforts to reform government will empower the wrong people and lead to regressive or ineffective politics. Recently, scholars have taken a comprehensive look at the most prominent public policy theories, concluding that the policy literature presents "a conceptual framework in which the public typically plays no active role. If they do play a role, theorists present a highly unfavorable view of them in enacting their role" (Muhlberger et al. 2011, p. 208). Within public administration contexts that consistently downplay the role of the public in policy formation, public managers may not be positioned to respond effectively to opportunities to incorporate public contributions to policy development.

Ironically, while theories of public policy generally ignore the public, it has been widely recognized that relatively new ICTs offer the potential to involve citizens in policy processes. As early as 2001, the Organization for Economic Co-Operation and Development (OECD) proclaimed that citizen involvement in policy-making is a "core element of good governance." The OECD endorses active participation by citizens in defining the process and content of policy making by "setting the agenda, proposing policy options, and shaping the policy dialogue—although the responsibility for the final decision or policy formulation rests with government" (OECD

2001). In subsequent years, numerous interventions to involve citizens in consultations and deliberations on topics provided by government practitioners or scholars have been designed and evaluated (see, e.g., Petrik 2009; Epstein et al. 2014; Towne and Herbsleb 2012). However, such systems rarely enable participants to *propose* policy topics for discussion; given existing designs, there are few opportunities for ascertaining what topics citizens would themselves suggest for policy making, if they had the chance.

In contrast, a petition is a form of discourse in which individuals express opinions about actual and potential policy decisions, and sometimes propose their own ideas for policy actions. Petition texts are composed by individuals, independent of traditional political party, interest group, and polling processes (although no doubt potentially complementary to such processes). Petitions thus convey input from voices that can contribute to policy development processes. The right to petition government is part of the historical legal frameworks of many countries but since 2001 and using new ICTs, numerous national experiments with electronic petitioning ("e-petitioning") have emerged in OECD and other countries, specifically Scotland, Wales, Great Britain, Germany, South Korea, and Australia. More recently in the United States, the Obama Administration launched *We the People* (WtP) in 2011 as a part of its open government policy; this platform explicitly requests that petition initiators take a position on an issue, propose or change an administration policy, or call on Congress to act on an issue (https://petitions.whitehouse.gov/petition/create). Thus, the historic right of citizens to petition their governments has transitioned into a technologically-enabled process, unmediated in principle by pollsters, political party, news media, or researchers, that empowers citizens to use their own words to make a suggestion to government about a policy topic of their own choosing.

The ability to e-petition governments may be viewed as acts of participation in the initial phases of the policy making process, i.e., problem identification and agenda setting. Supplementing petitioning systems with other Web 2.0 capabilities, as many e-petition platforms do, enables petition initiators to use social media applications such as Twitter and Facebook to circulate information about petitions to their social networks and encourage signatures from others in support of their ideas. It is apparent that petition initiators mobilize their social networks to support their petitions, in hopes that substantial signature support will bring visibility to their causes and perhaps a government response. Some government e-petitioning systems are also enhanced by related discussion forums, contributing to their participatory character.

E-petitioning systems have been enormously popular. In the US alone, as of April 2016, over 23 million users have used We the People (WtP) to generate over 446,000 petitions that have accumulated nearly 33 million signatures; the platform registers an average of over 20,000 signatures a day (Goldman 2016). To be sure, the vast majority of petitions do not attract substantial support. These numbers do indicate, however, that the public is quite motivated to make policy suggestions to the US federal government. By March 2016, Change.org, a commercially owned petitioning platform, boasted over 140 million users from all over the world targeting decision makers on local, national, and global levels seeking some type of action, depending on the issue. The very popularity of petitioning would seem to present an argument against its use in policymaking, since the act of cataloging and making sense of so many policy suggestions sounds overwhelming, thus creating the problem feared earlier of citizen participation producing too many demands.

At the same time, petitioning systems may also be seen as providing potentially fruitful views of the kinds of topics citizens are thinking about, their suggestions for addressing such subjects, and the social networks that are mobilized in support of particular policy suggestions. Furthermore, the number of signatures for each petition provides a natural way to assess the level of interest and support for the topics and courses of actions suggested by petitioners. Thus, petitioning systems may present a natural laboratory for considering citizen inputs to policy making, provided it is possible to make use of data generated in the petitioning process.

At this time, there has been little analysis of e-petitions in general and very little use made of the textual content or the mobilizing activities associated with such petitions from a policy analysis perspective. Given that the current volume of e-petitions is both enormous and increasing, the major obstacle to studying their contents is the lack of efficient data analytic methods and strategies.

A Novel Form of Political Participation: Functions of E-Petitioning

Since early in the history of its diffusion, scholars have speculated about if, and to what extent, individuals would use the Internet for political action. Relatively early it became clear that political uses of ICTs reinforced traditional forms of political engagement. For example, the results of the 2009 PEW Internet survey result showed that the "well-off and well-educated" are more likely to participate in online activities that "mirror offline forms of engagement" (Smith et al. 2009, p. 1). However, the report also called attention to "the development of new forms of communication on the internet—like blogs and social networking sites" that "potentially expands the opportunities for civic engagement" (Smith et al. 2009, p. 49).

More recent studies of the impact of the Internet on political activism now point to some of these effects. For example, Nam (2012) measured online political activity using three binary variables: political participation online, forwarding political messages, and visiting political websites. As in past research, he found that the internet *reinforces* existing patterns in offline participation, but he also found that the Internet *mobilizes* political participation among people with less interest in politics (Nam 2012). Similarly, Bimber et al. (2015) found that when those with low interest in politics use the Internet for political information the likelihood of their engaging in political talk *increases* (Bimber et al. 2015, p. 36). The transition from apolitical talk to political talk is easy in private conversation, and digital media makes this especially easy since it "blur[s] boundaries between public and private life as well as between news and discussion" (Bimber et al. 2015, p. 28). Thus, engaging in political talk on the Internet may begin to take place even in the absence of political interest.

These studies suggest that the changing nature of political participation using ICTs requires a research strategy that takes these new forms of participation seriously. The study of e-petitioning is relevant in this respect because it reflects important ICT-induced structural changes in participation and collective action. Scholars have begun to argue that e-petitions can fulfill substantial political functions in a society depending on the historical and institutional background of the society (Lindner and Riehm 2011). While direct and visible changes are not easily attained, e-petitions provide individuals with the ability to contribute opinions to the deliberative process. Lindner and Riehm (2011) consider the German Federal Parliament e-petitioning system to be successful based on increasing e-petition submissions and newly registered users as well as users' perceptions that the system was a useful and important tool for influencing. E-petitioning systems thus may provide citizens an "opportunity to be included in what was viewed as more democratic interaction" thereby supporting deliberative processes (Macintosh et al. 2002, p. 10). This form of participation can be more inclusive because users access the system at their convenience and reach a wider range of social groups (Macintosh et al. 2002).

As Bochel (2012) has observed, e-petitioning is the most successful e-participation system in terms of providing a mechanism for large numbers of people to express their opinions to governments. At the same time, she also notes that e-petitions merely allow people "to air their views, with little or no 'real' participation or empowerment" (Bochel 2012, p. 7), which may give the majority of petitioners and signers the feeling that their contributions are not valued, causing disillusionment among e-petition system users. While actual policy change as a result of e-petitioning happens infrequently, the Obama Administration promised a response within 60 days to any petition that meets a threshold signature level, now set at 100,000. At the time we write, it is not clear how the Trump Administration will manage the We the People petitioning website.

However, e-petitions can function as powerful indicators of public opinion regarding government policy, potentially contributing to the responsiveness of government (Lindner and Riehm 2011, p. 5). There have been notable examples of governments' decisions to change policy as the result of e-petitioning campaigns. The Brown government of the UK withdrew a plan for road pricing after an online petition attracted over 1.8 million signatures against road pricing (Van Dijk 2012, p. 112). Similarly, a WtP petition, "Make Unlocking Cell Phones Legal," endorsed by the Obama Administration in part on the basis of having achieved its signature threshold, was used to pressure the U.S. Congress into passing a bill, "Unlocking Consumer Choice and Wireless Competition Act."

Beyond conveying citizens' concerns and opinions to governments, e-petitions also convey such information to other citizens. Before e-petitioning was available, media reports or opinion polling was the best way to know peer citizens' interests (Coleman 2003, p. 97). Now journalists use e-petition websites as a news source, reporting on selected e-petitions (Chasmar 2014; Jaya 2015; Liebelson 2013; Wan

2013). Moreover, since e-petitioning platforms are available to everybody, individuals can visit e-petition websites themselves and garner unfiltered information about peer citizens' interests and opinions, facilitating the ability to mobilize supporters in signature drives to promote issue-based campaigns (Bimber et al. 2008; Lindner and Riehm 2011; Lynch 2012; Stempeck 2012).

Thus, by providing citizens with a formal channel to submit requests to the government, e-petitions can potentially function to facilitate citizens' integration into the political system (Jungherr and Jürgens 2010; Lindner and Riehm 2011). When appropriately designed, e-petitions can complement existing democratic institutions (Bochel 2012, p. 2). Indeed, government e-petition platforms may be viewed as *government-sponsored* efforts to facilitate and integrate citizens' participation into the process of policymaking. E-petition systems are popular and potentially effective tools for public participation in policymaking. Analysis of the topics addressed in petition texts and the dynamics of signing behavior can make it possible for policy makers and government leaders to identify and respond to issues the public is concerned about. The challenge is finding ways to make sense of and use the vast amounts of data that are generated.

Data Analytic Strategies for Studying E-Petitioning

E-petitions are a unique information source in that they represent petitioners' expressions of opinion using their own frames and choice of words; such information is not generally available in traditional information sources such as news outlets, survey results, direct discussion (Shipley and Utz 2012) and statistics. The specific data created through e-petitioning includes petition texts and their linguistic characteristics, signatures and their rates of accumulation, demographic and social characteristics of petitioners and signers where available, and social media traces related to discussion about e-petitions and signature solicitation (e.g., Twitter streams, hashtag campaigns, Facebook posts, etc.). The process of soliciting signatures via harnessing one's social network through Twitter, Facebook, and other mechanisms, which is endorsed and facilitated by many e-petitioning platforms, also provides indicators of organizational mobilization through social networking that may be key to understanding the significance of an incipient or blossoming movement, as well as appreciating what kinds of individuals are involved.

The use of e-petitioning data for policy informatics requires substantial thought about appropriate strategies for data analysis. The few existing research studies have addressed characteristics of e-petition systems (Bochel 2012), dynamic e-petition behaviors and characteristics of users (Jungherr and Jürgens 2010), technical and procedural features of e-petition systems (Lindner and Riehm 2009), and the factors influencing signature accumulation (Hale et al. 2013). Below we describe two major strategies for analyzing e-petitioning through two computational techniques—natural language processing (NLP) and social network analysis (SNA)—to analyze new forms of textual and network data.

Strategy One: Topic Modeling for Analyzing Textual Data

Traditionally, human coding has been used for content analysis of textual data. The major benefit of human coding is that humans can map words in documents to topic categories developed by domain experts. The major disadvantages to human coding include the high labor costs involved with topic category development and the manual coding of text, as well as reliability issues stemming from the somewhat subjective nature of human coding (Quinn et al. 2010). Developing category schemes together with manually reading and sorting text are challenging tasks for even expert coders (Grimmer and King 2009). Further, human coding cannot keep pace with the large volumes of texts from internet archives such as e-petition databases. This is why automatic methods such as those described below are appealing for categorizing e-petition text.

The use of automated content analysis—statistical natural language processing (NLP) and text mining—in the study domain of e-governance, while not yet common, is at least represented. We searched version 11.0 of the e-Gov Reference Library (EGRL), which contains "7,553 references of predominantly English-language, peer-reviewed work" in electronic government and governance" (Scholl 2015) for examples of text-based analyses. Using 5 queries, we retrieved a total of 56 articles: 15 articles with "text mining (or text-mining)," 12 articles with "categorization," 9 articles with "text analysis," 6 articles with "machine learning" and 14 articles with "natural language processing." Of these, NLP is the most sophisticated approach; we discuss it further below.

Supervised NLP The best-known use of NLP for government administration has focused on "e-Rule Making" (Cardie et al. 2008; Kwon et al. 2006; Purpura et al. 2008; Shulman et al. 2008). The major NLP tasks of this project address processing and organizing large volumes of public comments so that policy analysts could efficiently and systematically understand and summarize them. These studies compared classification results produced by human annotators and by supervised statistical NLP models, and reported that such "supervised" models could outperform human annotators. This means that a large number of texts can be coded and categorized quickly because the process of categorizing can be automated (Quinn et al. 2010). However, the major disadvantage of supervised NLP is its high execution cost since the creation of the initial category scheme involves human annotators.

Unsupervised NLP and Topic Modeling A very limited number of studies have used unsupervised NLP to explore the potential of this new technology to automatically find patterns from big textual data. Methods applied by Evangelopoulos and Visinescu (2012), Ekstrom and Lau (2008), and Lin et al. (2015) used unsupervised learning methods. Topic models are examples of unsupervised natural language processing models that *discover* topics and themes by observing the distribution of words in large volumes of text documents. Topic models have quickly become one of the most popular probabilistic text modeling techniques in machine learning (Aggarwal and Zhai 2012, p. 107; Wei and Croft 2006) since they enable "soft clustering of terms and documents into a fixed number of topics" (Boyd-Graber et al. 2014, p. 226). Topic models refer to Latent Dirichlet Allocation (LDA), the original generative topic model (Blei et al. 2003), and its extensions (Blei et al. 2004; Chang et al. 2009; Grimmer 2010; Mcauliffe and Blei 2008). LDA and "topic models" are terms often used synonymously, but LDA is a special case created by Blei et al. (2003) (Graham et al. 2012). Numerous extensions of LDA have been developed to improve or refine the resulting output (Blei et al. 2004; Blei and Lafferty 2006; Teh et al. 2004), but Chang et al. (2009) found that LDA outperformed newer algorithms in producing more semantically coherent topics.

Only a small number of scholars in public policy and political science have used topic modeling for automatic content analysis of textual big data to discover themes from political texts (Grimmer and Stewart 2013; Quinn et al. 2010; Roberts et al. 2013). In their political agenda study of the U.S. Senate, Quinn et al. (2010) extracted topics from legislative speech data (a total of 118,065 speeches) to infer the relative amount of legislative attention paid to topics (Quinn et al. 2010, p. 224). Quinn et al. (2010) stated that the topic model is "a valuable first step toward using" text to estimate the ideological preferences of actors and, further, to discover insights from text regarding the underlying political landscape (Quinn et al. 2010, p. 226).

Grimmer (2010) used topic modeling to measure the priorities emphasized by US senators in their press releases. Using 24,236 press releases, Grimmer (2010) found that senators in the same state put out semantically similar press releases compared to senators representing different states (Grimmer 2010). The finding contradicted a previous study by Schiller (2000), who argued that senators representing the same state tend to express dissimilar priorities in order to compete for more media and public attention (Schiller 2000). Grimmer (2010) underscored the limitations of Schiller's study since it used a limited number of newspaper statements, and further stressed that his new discovery was made possible by the ability to use topic modeling with large volumes of data.

Roberts et al. (2013) developed a structural topic model to investigate the differences between the way topics are represented in Chinese news sources compared to other countries' news sources (Roberts et al. 2013, 2014). This study demonstrated that their structural topic model could identify different framing by the Associated Press (AP) and the Chinese state-owned Xinhua news source in reporting the same Taiwanese election. While AP—an outside news source—viewed the event as an "electoral competition," the Xinhua presented the event from their own interest of "unifying with Taiwan" (Roberts et al. 2013, p. 3). The study showed that topic modeling is a useful tool for social science researchers in extracting emerging topics from texts independent of potentially biased perspectives.

Topic model outputs are "fully data-driven" and therefore are not dependent on users' particular perspectives or experiences (Mimno 2012, p. 3:10). In contrast, *interpretation* of the topic models outputs depends on the goals of the analysis as well as the levels of domain knowledge and the perspectives of the interpreters (Croft et al. 2010; Yau et al. 2014).

Application to E-Petitioning So far, little attention has been paid to indexing or describing the kinds of petitions that garner enough support to appear on WtP. The only categorization scheme for WtP petitions has been 39 conventional subject categories provided by the White House that mirror to some extent the organizational division of the federal bureaucracy (e.g., Foreign Policy, Health Care). In prior research, we found these categories accounted for less than 2% of variance in signature accumulation (Harrison et al. 2014) (since April 2016, these have been reduced to 18 similar categories). However, our research group (see, e.g., Hagen et al. 2015a) has used topic modeling successfully to extract naturally emerging topics from e-petitions. In our most recent use of this approach, we found 15 emergent topics (e.g., veteran, religion_gay, children, investigation, marijuana, sentence, cancer_research, secession, china, awareness [of special days and specific people] school_visa, military, national park, white_genocide, and gun), illustrating a more finely grained and nuanced set of subjects discussed by petition initiators and their signers between 2011 and 2014, that explained a statistically significant 8% of the variation in signature accumulation (Hagen et al. 2016). It is particularly interesting that one of these topics—white genocide—while lost in the WtP categorization scheme and registering a low number of signatures, has thematically unified a set of petitions submitted over time that appears to reflect middle class resentment of immigrants and racial minorities that is now galvanizing electoral discourse in the 2016 election season. The array of topics discovered through this type of modeling was able to explain greater variance in signature accumulation compared to that explained by other more traditional linguistic and semantic information derived from the e-petitions, such as sentiment, extremity, repetition, urgency, and informativeness, in our regression models. Our analysis thus exposed topics in citizen discourse that are both more specific and more highly related to signature accumulation than those used by the Obama Administration. We suggest that such emergent topics provide a better view of citizens' policy interests than conventional a priori issue categories.

Challenges in Using Topic Modeling We have found topic modeling useful for the analysis of e-petition data with relatively low human cost. But one of the major challenges of using topic modeling has been the lack of reliable and generalizable processes, a problem that grows in importance when using ephemeral social media data (Zamith and Lewis 2015). E-petitions are likely to reflect social concerns within a dynamic political environment; indeed, they are a means by which analysts may take the "pulse" of political attitudes. Topic modeling algorithms used for petition analysis must be flexible enough to conform to new language within changing datasets. At the same time, frameworks for structured analysis should present recommendations related to data collection, pre-processing choices, training the data, topic generation, evaluation, and analysis, and must be created to minimize the chances of capitalizing on sample specific variation. Only by following careful procedures and making judicious decisions at data analytic choice points will we be able to generate reliable and valid analyses of e-petitions and create policy informatics tools that make such analyses more feasible. Our work (Hagen et al. 2016) provides one vision of how such a framework might work.

Strategy Two: Analysis of Networks, Signatures, and Other Petition-Related Data

E-petitioning also lends itself to forms of data analyses that are more familiar to e-government and policy informatics scholars. A promising approach focuses on the application of social network analysis and related procedures to the study of signature accumulation patterns on specific petitions. Most of the current e-petitioning platforms call upon initiators to solicit support from their social networks and provide them with direct access to the social media tools (e.g., Twitter and Facebook) for doing so. Understanding how petition signatures are solicited and accumulate provides substantial understanding of the social dynamics in collective action for particular policy goals. We describe two major tools below and illustrate how we have used them to illuminate the dynamics of collective action.

Social Network Analysis Large-scale online networking applications, such as the Internet, Facebook, LinkedIn, and Twitter have increasingly become the focus of social network analysis (SNA) to understand the roles of participants and the nature of interactions among them in connected social environments. SNA examines how individuals engage in collective action by observing ties between individual actors and the transmission of information and behavior among individuals. These methods have been applied to behaviors in various networked systems (Newman 2010) and, as we discuss below, they can be used to study the diffusion of e-petitions together with signature accumulation. The notion of centrality is commonly used to characterize the level and type of importance of a participant in a social network (Easley and Kleinberg 2010; Newman 2010). The notion of community (or cluster) is used to identify a group of nodes with similar behavior in a social network. There are several ways to define similarity in behavior and algorithms are available for partitioning the nodes of a social network into communities according to those definitions (Newman 2010).

Market Basket Analysis The primary goal of market basket analysis is to identify patterns of *co-occurrences* of objects. The basic idea can be easily understood by applying it to the example of a supermarket transaction (or market basket), which consists of a set of items bought by a customer. By analyzing transactions that occur over a period of time, managers can identify sets of items that are frequently bought together by customers. The supermarket is interested in knowing which items to place in adjacent shelves for customer convenience. We have used market basket analysis to understand which petitions are signed by the same people.

Application to E-Petitioning Using these related techniques, our work has produced evidence of mobilization to attract signatures for similar petitions. In two studies using WtP petition data, we focused on 33 petitions involving gun control that were initiated during the week following the Sandy Hook tragedy on December 14, 2012 (see Dumas et al. 2015a, b). Within a few hours of the shootings, a national conversation about gun control commenced through petitioning on WtP. Numerous petitions were initiated, some supporting stricter gun control laws, others favoring increased funding for mental health care, some advocating armed guards in schools, and still others in support of maintaining the status quo of current gun control policy. On Friday, December 21, 2012, President Obama issued a response to all 33 petitions, which had attracted over 503,000 signatures.

We divided the petitions into two groups differentiating between those that expressed a clear preference in favor of greater gun control and those that expressed other preferences; this produced a cluster of 12 "pro gun control" petitions, and a remaining group of 21. The group of 12 "pro gun control" petitions were sorted further into three thematic clusters: a group in support of gun control laws, which we labeled "establish gun laws"; a group advocating banning the sale and use of assault weapons which we labeled "assault weapons"; and a final group consisting of only one petition calling for Congress to repeal the second amendment, labeled "repeal the 2nd amendment". We also sorted the remaining group of 21 petitions into three thematic clusters: a group in support of law-abiding gun owners ("support law abiding gun owners"); a group advocating investment in the improvement of mental health care ("invest in mental health care") and a final group advocating using firearms to guard our schools ("guard our schools").

In one study (Dumas et al. 2015b), we focused on the 12 "pro" gun control petitions and, in the second (Dumas et al. 2015a), we looked at the 21 petitions that argued against gun control, advocated improvements in mental health care, or proposed arming protectors within the school system. We used market basket analysis to explore questions about whether individuals who signed one "pro" gun control petition also signed other "pro" gun control petitions, which turned out to be the case. We also found that individuals who signed one petition favoring existing gun control policies or favoring alternatives to stricter gun control also signed other petitions within the same cluster. We interpreted this as evidence that, during this time of national crisis, individuals were strategically signing several petitions that represented a similar, and apparently preferred, policy option.

We also used social network analysis to determine if there were groups of individuals who signed similar "pro" gun control petitions, thus suggesting the creation of "communities" of individuals whose actions were similarly aligned in support of gun control policy proposals. We did the same analysis for the other 21 petitions. We found evidence of the centrality of certain signers of multiple petitions indicating their importance in the mobilization of discrete online communities in favor of or opposed to gun control to address the Sandy Hook tragedy. It is useful to see how these linkages among policy proposals are structured into communities that support particular policy proposals that conform to a theme. Behavior such as this suggests that e-petitioning was used strategically by individuals to express their opinions and influence the future of gun control policy.

More graphic evidence of strategic collective action can be seen in the use of tweeting to solicit signatures in support of e-petitioning campaigns. In current research, we examine a popular petition initiated on Change.org that galvanized world attention for a period of time in May 2014 (Dumas et al. 2016). A Change.org petition entitled "All World Leaders Bring Back Nigeria's 200 Missing School Girls

#BringBackOurGirls" was created on May 1, 2014, generating over 4500 tweets (in English) requesting petition signatures, and covered by numerous news venues. The petition attracted over 1.1 million signatures by May 2015 with most accumulating between May 1 and June 1, 2014. In an effort to describe the course of this social movement, we are using publicly available signature data from Change.org and Twitter data to re-construct the social networks of petition signers who then tweeted to their Twitter followers. We wish to determine how many of the followers subsequently signed this petition and how specific actors contributed to the diffusion and success of this campaign. Since the signature data and tweet data are aligned chronologically, it will also be possible to quantitatively assess the relationship between the two time series, which gives rise to the more general possibility of developing forecasting models of petition signature growth in the future.

Challenges in Using Social Network Analysis The principle challenge in using social network techniques to analyze petition dynamics lies in being able to trace connections between individuals who sign more than one specific petition as well as between an individual who signs a petition and then subsequently tweets to their followers to solicit signatures. Of course, we are not interested in specific identities (and are sensitive to privacy considerations), but this type of analysis is not possible without being able to create linkages between related behaviors. This is feasible, to different extents, using the data available from petition platforms and Twitter APIs.

E-Petitions and Policy Informatics

Policy informatics toolkits may take many forms and address numerous subjects, but we have argued that one of the most promising type of toolkit will focus on the significant amount of unexplored data that is available through e-petitioning. Taken together, our work on US e-petitions provides two kinds of information not previously available when considering how the public might be involved in policy development. First, we demonstrate that the enormous body of discourse represented in e-petitions can, through topic modeling, provide important insights about the kinds of policy topics that are of interest to the American public. The emergent topics we have found are related to petitions that are signed more (or less) frequently and thus represent themes and subjects more likely (or less likely) to be endorsed by individuals politically active enough to initiate and sign petitions. Second, our social networking analysis suggests that individuals use the US petitioning platform to engage in systematic collective action in support of their policy preferences, well beyond the one-shot "liking" on Facebook pages and single petition signatures often criticized by those advancing "slactivist" hypotheses (e.g., Morozov 2009). Indeed, the thematic market baskets of petitions co-signed by individuals represent policy options currently under active discussion in the public sphere.

However, at this point, it is also worth noting that e-petitions can rightfully be seen as one form of "big data" and e-petitioning datasets are thus subject to the now familiar caveats that are becoming characteristic of data science. Datasets must be scrupulously assembled and cleaned; data structures and data points must be thoroughly understood. Marcus and Davis (Marcus and Davis 2014) remind us that big data is susceptible to many types of flawed analysis, and thus may produce multitudes of meaningless correlations. Big data may also be tainted by changes in social patterns that affect the way that data accumulates and is produced or technical complexities in the way that data is collected or presented.

The approaches we think can lead to the development of a policy informatics toolkit are computationally intensive and require the collaboration of computer and social scientists, the former to create data sets and execute data mining methods that enable the discovery of patterned behavior and the latter to provide domain expertise that bears on what questions are worth pursuing and what outcomes are meaningful. The use of big data in policy informatics cannot be wholly empirically driven, as it frequently is in business contexts. We advocate instead the creation of informatics toolkits that are based firmly in theory driven research, and our work takes this perspective. A policy informatics toolkit is an outgrowth of such theory driven work, and not a substitute for it.

We have argued that e-petitions are a valuable source of information about what issues and policy options the public cares about and that topic modeling can provide an emergent and refined view of topics that are latent within these textual datasets. The analysis provides a snapshot of topics that are the subjects of authors' petitions. Once this snapshot is provided, policy analysts can delve further into interpreting the topics or exploring the social forces behind these topics. For example, an analyst responsible for producing immigration policy may benefit from knowing the issues raised by highly-educated and skilled foreign workers complaining about policies regarding acquiring visas by the U.S. government. The e-petition categorization analysis of Hagen et al. (2015b) provides a quantitative and qualitative basis for making statements about discontent by skilled foreign workers who call attention to problems regarding immigration policies.

But topic modeling by itself does not provide other important information such as the formation and role of issue-oriented online mobilization related to certain social actions. For this, social network analysis, and the analysis of signature accumulation in relation to other social media, complemented by more traditional forms of data analysis such as, for example, time series regression, can be used to explore the dynamics of e-petitioning behavior that may reflect the social organization of policy movements. We already know that signature growth is most likely to happen early in the life span of a petition (Hale et al. 2013; Hagen et al. 2016). And it seems likely that signatures may accumulate in response to social media campaigns, such as Twitter (Dumas et al. 2016). These basic observations suggest that forecasting models of petition growth and potential viral spread can be created and that, with further research, such models might be extended to predict more direct forms of political action.

Beyond these instrumental considerations, e-petition systems are popular and potentially effective tools for public participation in policymaking. E-petitions can be used as indicators of public opinion regarding government policy and as a reservoir of unfiltered information about peer citizens' interests and opinions. Opinions regarding issues expressed in petition texts and the level of support for them may inform policy options by identifying areas of public discord, controversy, and opposition. Thus, analysis of these texts and the dynamics of signing behavior can provide policy makers and government leaders with the ability to appreciate issues of public concern, perhaps even before such issues become more prominent and problematic.

References

Aggarwal CC, Zhai C (2012) A survey of text clustering algorithms. In Aggarwal CC, Zhai C (eds) Mining text data. Springer, pp 77–128. Retrieved from http://link.springer.com/chapter/10.1007/978-1-4614-3223-4_4

Bimber B, Stohl C, Flanagin AJ (2008) Technological change and the shifting nature of political organization. In: Chadwick A, Howard PN (eds) Routledge handbook of internet politics. Routledge, New York, pp 72–85

Bimber B, Cunill MC, Copeland L, Gibson R (2015) Digital media and political participation: the moderating role of political interest across acts and over time. Soc Sci Comput Rev 33(1):21–42

Birkland TA (2011) An introduction to the policy process: theories, concepts, and models of public policy making, 3rd edn. M. E. Sharpe, Armonk

Blei DM, Lafferty JD (2006) Dynamic topic models. In: Proceedings of the 23rd international conference on machine learning. ACM, pp 113–120. Retrieved from http://dl.acm.org/citation.cfm?id=1143859

Blei DM, Ng AY, Jordan MI (2003) Latent Dirichlet allocation. J Mach Learn Res 3(4/5):993–1022

Blei DM, Griffiths D, Jordan MI, Tenenbaum M (2004) Hierarchical topic models and the nested Chinese restaurant process. Adv Neural Inf Proces Syst 16:17

Bochel C (2012) Petitions systems: contributing to representative democracy? Parliam Aff:1–18

Boyd-Graber J, Mimno D, Newman D, Airoldi EM, Blei D, Erosheva EA, Fienberg SE (2014) Care and feeding of topic models: Problems, diagnostics, and improvements. In: Handbook of mixed membership models and their applications, pp 225–255

Cardie C, Farina C, Aijaz A, Rawding M, Purpura S (2008) A study in rule-specific issue categorization for e-rulemaking. In: Proceedings of the 2008 international conference on digital government research. Digital Government Society of North America, Montreal, Canada, pp 244–253. Retrieved from http://dl.acm.org/citation.cfm?id=1367832.1367874

Chang J, Boyd-Graber J, Gerrish S, Wang C, Blei DM (2009) Reading tea leaves: how humans interpret topic models. In: Neural information processing systems. British Columbia, Vancouver

Chasmar J (2014) White House responds to Tahmooressi petition: "We respect the rule of law." Retrieved September 1, 2014, from http://www.washingtontimes.com/news/2014/aug/31/white-house-responds-tahmooressi-petition-we-respe/

Chung W, Zeng D (2015) Social-media-based public policy informatics: sentiment and network analyses of U.S. immigration and border security. J Assoc Inf Sci Technol. doi:10.1002/asi.23449

Coleman S (2003) Making parliamentary democracy visible: speaking to, with, and for the public in the age of interactive technology. In: Chadwick A (ed) Handbook of internet politics. Cambridge University Press, Cambridge/New York, pp 86–98

Crawford S (2014) Chicago is your big (friendly) brother. Retrieved June 15, 2015, from http://www.bloombergview.com/articles/2014-06-19/chicago-is-your-big-friendly-brother

Croft WB, Metzler D, Strohman T (2010) Search engines: information retrieval in practice. Addison-Wesley, Boston

Dalton RJ (2013) Citizen politics: public opinion and political parties in advanced industrial democracies. CQ Press, Los Angeles

Dawes SS, Helbig N, Nampoothiri S (2014) Workshop report: exploring the integration of data-intensive analytical skills in public affairs education. Retrieved from https://www.ctg.albany.edu/publications/reports/egovpolinet_workshopreport?chapter=&PrintVersion=2

Downs A (1972) Up and down with ecology – the "issue-attention cycle". Public Interest 28:38–50

Dumas C, LaManna D, Harrison T, Ravi SS, Hagen L, Kotfila C, Chen F (2015a) E-petitioning as collective political action in *We the People*. In: IConference 2015, Newport Beach, CA, 24–27 March. https://www.ideals.illinois.edu/bitstream/handle/2142/73665/215_ready.pdf?sequence=2

Dumas C, LaManna D, Harrison TM, Ravi SS, Kotfila C, Gervais N, Hagen L, Chen F (2015b) Examining political mobilization of online communities through e-petitioning behavior in *We the People*. Big Data Soc 2(2).doi:10.1177/2053951715598170

Dumas C, Atrey A, Lee J, Harrison T, Fake T, Zhao X, Ravi SS (2016) E-petition information diffusion in online social networks. In: Proceedings of the 17th annual international conference on digital government research. doi:10.1145/2912160.2912227

Easley D, Kleinberg J (2010) Networks, crowds and markets: reasoning about a highly connected world. Cambridge University Press, New York

Ekstrom JA, Lau GT (2008) Exploratory text mining of ocean law to measure overlapping agency and jurisdictional authority. In: 9th annual international conference on digital government research (dg.o 2008), pp 53–62

Epstein D, Newhart M, Vernon R (2014) Not by technology alone: the 'analog' aspects of online public engagement in policymaking. Gov Inf Q 31(2):337–344

Evangelopoulos N, Visinescu L (2012) Text-mining the voice of the people. Commun ACM 55(2):55–62. doi:10.1145/2076450.2076467

Goldman J (2016) Redesigning We the People: improving the way you engage with the White House. https://medium.com/@Goldman44/redesigning-we-the-people-c8ce93f4280#.7y8dl69u7. Accessed 18 May 2016

Graham S, Weingart S, Milligan I (2012) Getting started with topic modeling and MALLET. Retrieved April 25, 2015, from http://programminghistorian.org/lessons/topic-modeling-and-mallet.html

Grimmer J (2010) A bayesian hierarchical topic model for political texts: measuring expressed agendas in Senate press releases. Polit Anal 18(1):1–35

Grimmer J, King G (2009) Quantitative discovery from qualitative information: a general-purpose document clustering methodology. In: APSA 2009 Toronto Meeting Paper. Social Science Research Network, Toronto, Canada. Retrieved from http://ssrn.com/abstract=1450070

Grimmer J, Stewart BM (2013) Text as data: the promise and pitfalls of automatic content analysis methods for political texts. Polit Anal 21(3):267–297. doi:10.1093/pan/mps028

Hagen L, Harrison T, Uzuner O, Fake T, Lamanna D, Kotfila C (2015a) Introducing textual analysis tools for policy informatics: a case study of e-petitions. In: Proceedings of the 16th annual international conference on digital government research, 10–19, ACM, NY. doi:10.1145/2757401.2757421

Hagen L, Uzuner O, Kotfila C, Harrison T, LaManna D (2015b) Understanding citizens - direct policy suggestions to the federal government: a natural language processing and topic modeling approach. In: Proceedings of the 48th Hawaii international conference on system sciences (HICSS-48). doi:10.1109/HICSS.2015.257

Hagen L, Harrison TM, Uzuner Ö, May W, Fake T, Katragadda S (2016) E-Petition popularity: do linguistic and semantic factors matter? Gov Inf Q. doi:10.1016/j.giq.2016.07.006

Hale SA, Margetts H, Yasseri T (2013) Petition growth and success rates on the UK No. 10 Downing Street website. In: Proceedings of the 5th annual ACM web science conference. ACM, New York, pp 132–138. doi:10.1145/2464464.2464518

Harrison TM, Dumas C, Kotfila C, Lamanna D, Ravi SS (2014). *We the People:* U.S. e-petitioning as technology-mediated social action. Accepted for presentation at the annual conference of the international communication association, Seattle, WA

Howlett M, Ramesh M, Perl A (2009) Studying public policy: policy cycles and policy subsystems, 3rd edn. Oxford University Press, New York
Jaya P (2015) White House responds to Anwar petition - Nation | The Star Online. Retrieved May 18, 2015, from http://www.thestar.com.my/News/Nation/2015/03/13/white-house-reponds-anwar-petition/
Johnston EW (ed) (2015) Governance in the information era: Theory and practice of policy informatics, 1st edn. Routledge, New York
Jungherr A, Jürgens P (2010) The political click: political participation through e-petitions in Germany. Policy Internet 2(4):131–165. doi:10.2202/1944-2866.1084
Kingdon J (1984) Agendas, alternatives, and public policies. Little, Brown and Co., Boston
Kwon N, Shulman SW, Hovy E (2006) Multidimensional text analysis for eRulemaking. In: Proceedings of the 2006 international conference on digital government research. Digital Government Society of North America, San Diego, pp 157–166.
Liebelson D (2013) The 7 craziest Chinese petitions submitted to the White House. Retrieved April 9, 2015, from http://www.motherjones.com/politics/2013/05/chinese-censorship-white-house-petition
Lin F-R, Chou S-Y, Liao D, Hao D (2015) Automatic content analysis of legislative documents by text mining techniques. In: Proceedings of the 48th Hawaii international conference on system sciences (HICSS-48), pp 2199–2208
Lindner R, Riehm U (2011) Broadening participation through e-petitions? An empirical study of petitions to the German parliament. Policy Internet 3(1):1–23. doi:10.2202/1944-2866.1083
Lindner R, Riehm U (2009) Electronic petitions and institutional modernization. International parliamentary e-petition systems in comparative perspective. JeDEM 1(1):1–11
Lynch R (2012, March 26) Trayvon Martin case gives Change.org a boost – and vice versa. Los Angeles Times. Retrieved from http://articles.latimes.com/2012/mar/26/nation/la-na-nntrayvon-martin-case-poll-petition-20120326
Macintosh A, Malina A, Farrell S (2002) Digital democracy through electronic petitioning. In: McIver WJ, Elmagarmid AK (eds) Advances in digital government. Springer US, pp 137–148. Retrieved from http://link.springer.com/chapter/10.1007/0-306-47374-7_8
Marcus G, Davis E (2014) Eight (no, nine!) problems with big data. New York Times, http://nyti.ms/1kgErs2. Accessed 17 May 2016
Mcauliffe JD, Blei DM (2008) Supervised topic models. In: Advances in neural information processing systems, pp 121–128
Mimno D (2012) Computational historiography: data mining in a century of classics journals. J Comput Cult Herit 5(1):3
Morozov Y (2009) The brave new world of slacktivism. http://neteffect.foreignpolicy.com/posts/2009/05/19/the_brave_new_world_of_slacktivism
Muhlberger P, Stromer-Galley J, Webb N (2011) Public policy and obstacles to the virtual agora: insights from the deliberative e-Rulemaking project. Inf Polity 16(3):197–214. doi:10.3233/IP-2011-0235
Nam T (2012) Dual effects of the internet on political activism: reinforcing and mobilizing. Gov Inf Q 29:S90–S97. doi:10.1016/j.giq.2011.08.010
Newman MEJ (2010) Networks: an introduction. Oxford University Press, New York
OECD (2001) Citizens as partners: information, consultation, and public partnerships in policy making. Retrieved from http://www.ecnl.org/dindocuments/214_OECD_Engaging%20Citizens%20in%20Policy-Making.pdf
Petrik K (2009) Participation and e-democracy how to utilize Web 2.0 for policy decision-making. In: Proceedings of the 10th annual international conference on digital government research: social networks: making connections between citizens, data and government, Puebla, Mexico, 2009, pp 254–263
PoliInformatics (2015) About PoliInformatics. Retrieved June 15, 2015, from http://poliinformatics.org/

Purpura S, Cardie C, Simons J (2008) Active learning for e-rulemaking: public comment categorization. In: Proceedings of the 2008 international conference on digital government research. Digital Government Society of North America, Montreal, pp 234–243
Quinn KM, Monroe BL, Colaresi M, Crespin MH, Radev DR (2010) How to analyze political attention with minimal assumptions and costs. Am J Polit Sci 54(1):209–228. doi:10.1111/j.1540-5907.2009.00427.x
Roberts M, Stewart B, Tingley D, Airoldi E (2013) The structural topic model and applied social science. In: Advances in neural information processing systems workshop on topic models: computation, application, and evaluation
Roberts, M. E., Stewart, B. M., Tingley, D., Lucas, C., Leder-Luis, J., Gadarian, S. K. et al (2014). Structural topic models for open-ended survey responses. Am J Polit Sci, 58(4), 1064–1082.
Scholl HJ (2015) EGRL. Retrieved from http://faculty.washington.edu/jscholl/egrl/
Shipley R, Utz S (2012) Making it count: a review of the value and techniques for public consultation. J Plann Lit 27(1):22–42
Shulman SW, Hovy E, Kwon N, Huisman E (2008) Tools for rules: technology transfer and electronic rulemaking. In: Hawaii international conference on system sciences, proceedings of the 41st annual, p 200
Smith A, Schlozman KL, Verba S, Brady H (2009) The demographics of online and offline political participation. Retrieved from http://www.pewinternet.org/2009/09/01/the-demographics-of-online-and-offline-political-participation/
Stempeck M (2012) The internet didn't make Trayvon national news, but it did sustain the story. Retrieved from https://civic.mit.edu/blog/mstem/the-internet-didnt-make-trayvon-national-news-but-it-did-sustain-the-story
Teh YW, Jordan MI, Beal MJ, Blei DM (2004) Hierarchical Dirichlet processes. J Am Stat Assoc 101:1566
Towne WB, Herbsleb JD (2012) Design considerations for online deliberation systems. J Inf Technol Polit 9(1):97–115
Van Dijk J (2012) The network society. Sage, London/Thousand Oaks
Wan W (2013) Frustrated Chinese complain to Obama. The Washington Post. Retrieved from http://www.washingtonpost.com/world/asia_pacific/frustrated-chinese-complain-to-obama/2013/05/09/131e04a8-b8ad-11e2-92f3-f291801936b8_story.html
Wei X, Croft WB (2006) LDA-based document models for ad-hoc retrieval. In: Proceedings of the 29th Annual International ACM SIGIR Conference on Research and Development in Information Retrieval. ACM, New York, pp 178–185. http://doi.org/10.1145/1148170.1148204
Yau C-K, Porter A, Newman N, Suominen A (2014) Clustering scientific documents with topic modeling. Scientometrics 100(3):767–786. doi:10.1007/s11192-014-1321-8
Zamith R, Lewis SC (2015) Content analysis and the algorithmic coder: what computational social science means for traditional modes of media analysis. Ann Am Acad Pol Soc Sci 659(1):307–318. doi:10.1177/0002716215570576

Loni Hagen is an assistant professor at University of South Florida. She uses natural language processing and data mining techniques to automatically extract meaningful information from large volumes of text data. Her current research interests are in use of computational techniques for policy decision making in the domains of e-participation, privacy, cybersecurity, and health emergency communication.

Teresa M. Harrison is Professor of Communication in the Department of Communication and an affiliated faculty member in the interdisciplinary Information Science Doctoral Program, both at the University at Albany, SUNY. She is also a Faculty Fellow at the Center for Technology in Government at the University at Albany. She studies the use of communication media in government contexts and the relationships between communication and democratic practices and processes.

Catherine L. Dumas is a PhD Candidate in Informatics in the College of Engineering & Applied Sciences (CEAS) at the University of Albany, State University of New York. Her research interests include: analyzing e-petitioning systems and behaviors using data mining techniques and social network analysis, online collective action, information diffusion in social networks, HCI, usability testing, and open source.

Visualization Practice and Government: Strategic Investments for More Democratic Governance

Evert A. Lindquist

Abstract Practitioners and advocates see great potential to lever visual tools of all kinds to illuminate complex challenges, share information, create collective insight, and more efficiently show the results of meetings, data-dives, scans of social media, etc. Advocates of using visualization in public sector contexts typically focus on its potential for: describing and analyzing complex challenges; tracking social media and other streams of big data; and using infographics and other techniques to communicate information, insights, and messages to elected leaders and various publics. However, with the exception of digitally-driven services, little interest has focused on how visualization can better show how government works, its fabric, shifting contours, and complexity. This paper reviews different visual practice domains and their underlying craft logics and motivations. It makes a distinction between the application of visual tools in support of the 'instrumental' functions of government (associated with competing for, securing, and wielding the power) as opposed to 'democratic functions' (advising elected leaders, engaging citizens, and furthering accountability). It argues that the greatest inroads for visual tools has been in the more 'instrumental areas'—supporting political parties and sitting governments for monitoring and service delivery—as opposed to furthering the larger aspirations of 'democratic governance'. It concludes that for governments to invest more in visual tools for democratic governance will require political leadership and improved capacity inside government, and that sustained, systematic research will is required to monitor the take-up and influence of digital tools in these areas.

E.A. Lindquist (✉)
School of Public Administration, University of Victoria,
HSD Building, 3rd Floor, Victoria, BC, Canada
e-mail: evert@uvic.ca

J.R. Gil-Garcia et al. (eds.), *Policy Analytics, Modelling, and Informatics*, Public Administration and Information Technology 24,
DOI 10.1007/978-3-319-61762-6_10

Introduction

This is an exciting time to be learning about or applying one or more visualization techniques to deal with complex design and strategic challenges. Fueled by a variety digital possibilities for creating or sharing visual products, there is no shortage of interest in the various disciplines associated with visualizated (Lindquist 2015). Practitioners and proponents see even greater potential to lever visual tools of all kinds to illuminate complex challenges, share information, create collective insight, and more efficiently show the results of meetings, data-dives, scans of social media, etc. Advocates of using visualization in public sector contexts typically focus on its potential for: describing and analyzing complex challenges; tracking social media and other streams of big data; and using infographics and other techniques to communicate information, insights, and messages to elected leaders and various publics. However, with the exception of digitally-driven services, which rely on dashboards and other monitoring repertoires, little interest has focused on how visualization can better show how government works, its fabric, shifting contours, and complexity.

This paper suggests that, while there are great hopes for how visualization tools and techniques can improve public sector governance, and many practitioners are indeed employing different visual methods for their clients, the greatest inroads have occurred in the more 'instrumental areas' of governance associated with competing for, securing, and wielding the power and authorities we associate with 'government'—supporting political parties, sitting governments, and the critical tasks of monitoring and service delivery—as opposed to furthering the larger aspirations of 'democratic governance' by better advising elected representatives, engaging citizens, and ensuring accountability. Using visualization to further 'democratic governance' refers to how governments might find new ways to share information with the public to further understanding, participation, deliberation, and accountability—it is a critical complement to the more instrumental side of governance as practiced by governments. Since systematic evidence on the use of visualization tools in government is scant, this paper considers how visualization techniques could be used to further democratic governance and sets out an agenda for more systematic research.

This chapter has five sections. Section 1 on "Visualization: Diverse, Overlapping Practice Domains and Aspirations" reviews different visual practice domains—information visualization, graphics and information display, and visual facilitation for thinking and strategy—and provides an overview of these overlapping approaches and considers their underlying logics and motivations. Section 2 on "Beyond Performance: Ways to Better Inform and Monitor Government" considers how digital tools might further the instrumental versus democratic functions of government, recognizing they are inter-related in democratic systems. Section 3 on "Where Have Governments Invested in Visualization?" explores where governments have made make strategic investments in visualization, while Section 4 on "Next-Generation Visual Investments: Furthering Democratic Aspirations" identifies the three areas (advising elected leaders, engaging citizens and stakeholders, and fostering accountability) which could benefit from considerably more investment and a strategic approach for sharing a larger portfolio of richer, more detailed information from across the diverse visual practice domains. Section 5 "Conclusion: Supporting Visual Tools to Further Democratic Governance" suggests that for governments to invest more in visual tools for democratic governance will require a leap of faith on the part of political leaders, and that monitoring the take-up and influence of digital tools for these purposes requires a sustained, systematic research.

Visualization: Diverse, Overlapping Practice Domains and Aspirations

Before exploring the potential of visualization to better inform public administration and democratic governance, we must acknowledge the many visual practice traditions, which can add distinct value in different governance contexts. What follows provides a brief overview the three domains of visualization practice—information visualization and data analytics, graphics and information display, and visual facilitation for thinking and strategy—recognizing that there are overlaps among them. We do not have the space here to show a sufficiently diverse range of visualizations form different domains, but readers are encouraged to visit web sites and the books and articles below to see a stunning array of examples.[1]

Information Visualization and Data Analytics

The field of information visualization is well-established, driven by the latest developments in digital technologies, tracing its origins to early mapping and graphing techniques (Friendly 2008; Tufte 1990, 1997). It has developed at the intersection of the fields of computing, engineering, graph analysis, data management, cognitive psychology, software development, human-computer interface, etc. (Ware 2004) Contributors come from the full range of scientific, social science and humanities disciplines. The field of information visualization and day analytics has been motivated by the need to visually represent increasingly large data-sets generated or accessed in every field that engages in the collection of data (including disciplines across the sciences, social sciences, humanities, etc.), corporate and government data on clients and operations, and various forms of social media, to enhance how humans can analyze and learn from this information (Simon 2014). This field has been institutionalizing with an expanding number of conferences, journals (e.g.

[1] See, for example, Steele and Ilinsky (2010), Lima (2011), and McCandless (2014), or web sites like http://d3js.org; http://www.visualcomplexity.com; https://public.tableau.com; http://flowingdata.com; http://infosthetics.com; and http://www.smallmeans.com/new-york-times-infographics/.

Information Visualization) and university research centers, courses, and programs. Many boutique and top consulting firms have been levering these techniques, along with data mining and visual analytics, to improve their market share.

The field of information visualization is breathtaking in diversity in terms of techniques and applications to diverse "practice domains" (e.g., geosciences, biology, medical imaging, physics, health, monitoring social media, etc.), but collectively all these practitioners seem motivated by a fundamental and shared premise: data—when found, accurately transformed and well-represented, and properly matched with other data—will help improve awareness of issues and analysis, and improve decision-making. Data is as varied as scientific measurements, information packets, representations of concepts and variables (such as social or economic data), images, texts, and documents. While information and scientific visualization outputs can be stimulating and aesthetically pleasing, they have common repertoires for accessing and cleaning up data—along with several basic ways to represent that data (Chen 2006; Ward et al. 2010). Chen (2006) identified different ways to display and visualize data: graphs, trees, and cones; proximity and connectivity techniques (such as semantic distance and word search, multi-dimensional scaling, and network analysis); clustering and classification (e.g., dividing data into sub-sets and taxonomies, cluster-seeds); use of glyphs (e.g., using symbols on charts to convey additional information); creating virtual structures (e.g., WordNet, Wordle, etc.); representing data from networks (scale, small or large, topological, nodes, etc.); and data and information dashboards (Few 2013). After years of calls for more research and training in information analytics and visualization (Chen 1999; Gramwell et al. 2010), university programs, degrees and certificates have proliferated.

Graphics and Information Display

While the possibilities for 'big data' and information visualization have taken up much of the marketing bandwidth about visualization, there have long been efforts to creatively and better display concepts, ideas, complexity, and data with graphs, graphic design and other forms of visual renderings (e.g. Tufte 1990). This domain can be divided into: (1) the broad field of graphics, which has long explored and celebrated innovative ways to convey information for scientific, professional and advertising; and (2) the increasing number of magazines, web site capabilities, and newspapers investing in visual renderings or 'infographics' of issues and stories, referred to as 'infographics' (Meirelles 2013; Landow and Ritchie 2014).

Much of this is not new: think of the use of graphics and renderings in architectural, planning, engineering, and other scientific publications over many decades—which relied on drawings and printed displays of one kind or another. However, all of this been dramatically enhanced by digital platforms and software for production and display. Web sites and books proliferate showcasing the most intriguing efforts (e.g. Baer 2010; McCandless 2009, 2014). Baer (2010) and Steele and Ilinsky (2010) include representations as varied as: social and market network analysis;

legislative voting patterns; aviation flight patterns and subway maps; text-related applications such as Wordle, searching *New York Times* data-bases, and monitoring Wikipedia.

These visual traditions are broad and diverse, ranging from exploring new programs and algorithms for producing visualizations, to showcasing the remarkable and beautiful examples of visualization, to exploring applications in an ever-increasing array of fields, to developing theoretical constructs, and to exploring the cognitive dimensions of processing and interpreting visualizations. Baer (2010) defines the field as "the translating [of] complex, unorganized, or unstructured data into valuable, meaningful information" (p. 12). Practitioners include graphic designers, information architects, interaction designers, user experience designers, usability and human-factors specialists, human-computer interaction specialists, and plain language experts (pp. 14–15). An important focus of this realm of visualization has been to engage and persuade audiences with presentations, animations, and narratives, the latter either animating the creation of the visualization or necessary for sharing its meaning and relevance (e.g. Atkinson 2008; Heer and Robertson 2007; Duarte 2010; Segel and Heer 2010).

Visual Facilitation for Thinking and Strategy

Another rapidly growing community are visual and graphics assisting clients in grappling with complexity with sketching, often involving elaborate renderings of challenges and strategies. Often known as graphic recorders, graphic facilitators, and visual practitioners, essentially they engagingly sketch the evolution and key conclusions of meetings and conferences over a day and more, often in substantial and dynamic diagrams attempting to capture the movement, enthusiasm and vision of participants (Sibbet 2010, 2011, 2013; Roam 2009).[2] However, there are other professionals who have long used diverse visual techniques for different purposes:

- *systems thinking*. Some facilitators work with decision-makers and stakeholders to better understand in the context of problems and interventions the issues, surrounding complexity, diverse interests and perspectives, the task and institutional factors at play, and, through dialogue, identify pragmatic ways for improving the situation. Participants are encouraged to commit perspectives, perceptions, and even emotions to paper. Such 'rich pictures' and other kinds of sketches are shared and debated (Senge 1990; Checkland 1999; Checkland and Poulter 2010; Chapman 2004; Chapman et al. 2009).
- *education and therapy*. Many drawing and sketching techniques that are associated with systems thinking have long been used in education and therapy. Educators and facilitators sought new ways to encourage children, youth and adults to evince

[2] A key vector for the visual facilitation community is the International Forum of Visual Practitioners (IVFP), founded in 1995 (see www.ifvp.org) with an annual conference.

feelings, develop broader organizing frameworks, and appeal to different cognitive styles for communication (i.e. Hyerle 2009; Winkel and Junge 2012). However, many practitioners see relevance for other challenges, including organizational development.

- *foresight and scenario-building*. Some practitioners have long worked with clients to imagine different contingencies, futures, and scenarios. They seek to get participants to think broadly and creatively and to better appreciate dynamic, complex environments. Such work is often highly visual, with participants encouraged to share their ideas on walls or whiteboards, explore connections among key variables, and develop narratives and images of future states, often summarized with visual imagery (e.g. Rosell et al. 1995; Ringland 1998; Müller and Shwarz 2016).
- *simulations*. These include models of how systems work and evolve (markets, social, organizations, natural), which can alter key variables to understand system properties. The results can be conveyed visually to engage analysts and audiences (Johnston 2015). Other examples include airplane cockpit training devices, climate-change models, or multi-actor game simulations. Increasingly 'slider' technologies are used to not only share data sets and encompassing models, but also to let users play with them, changing assumptions, variables and filters so that users (sometime working in teams) can better appreciate linkages among variables and how they affect outcomes.
- *performance models*. Evaluators of programs and performance build 'logic models' linking inputs and activities of programs to outputs and desired outcomes (McDavid et al. 2013). Although the resulting diagrams are linear, delineating logic models is visual and iterative, leading to parsimonious representations of more complex realities.[3]

Most practitioners working in these areas would not consider themselves 'visualists' per se, but see the value of visual tools for facilitating groups in sense-making and strategic dialogue about complex challenges, and for capturing complexity at levels of analysis, often over time. Having moved through discover and analysis with clients, these practitioners use visual tools to convey final insights to them.

In general, visualization practice and writing often taps into approaches that many associate with 'systems thinking' (but see Sibbet 2010) and reflective practice (Schon 1993), which seeks to surface tacit knowledge and diverse perspectives, and bring more holistic perspectives to individuals, groups, and communities to address complexity and wicked problems. Many visual practitioners have different styles and specializations, but there is convergence in approaches and techniques (see Margulies and Valenza 2005; Hyerle 2009; Sibbet 2010; Blackwell et al. 2008). These include Venn diagrams, concept mapping, bubble maps, mind maps, thinking maps, systems

[3] These diagrams usually leave out details on the state of organizational capabilities and culture, political dynamics and commitment, resource allocation, client perspectives, and environmental change.

feedback loops, mind-scaping, thinking hats, visual journeys, assumption trees, icebergs, influence circles, etc.[4]

Looking Across Visual Domains: Comparative Advantage and Underlying Motivations

Figure 1 (next page) summarizes this overview by identifying three broad domains, recognizing distinct kinds of practice within each domain and that often the domains overlap. Strategic visualization practitioners are often the most focused on addressing specific challenges of decision-makers: they seek to assist clients in capturing the nature of problems and developing strategies to address them. This differs from supplying policy-makers from afar with data or perspectives driven by data, or by raising the profile of issues, trends or anomalies by creating intriguing visual representations and releasing them into the public domain. Conversely, the strategic visualization approaches usually do *not* rely on computer-mediated visualizations of findings from data-sets (larger or small), instead relying on hand-sketched renderings to move conversations along. However, strategic visualization practitioners and their clients can be informed by data and rendering from the other visualization practitioners. Some graphics recorders and facilitators use digital technology to assist with drawing and to store and transmit them to clients and others.

There is no doubting the enthusiasm of all visual practitioners about how their tools and skills can inform strategic thinking, the design and delivery of services, and public decision-making. We know that good visual representations can seize and arrest attention. Elsewhere I argue that a client orientation and craft logic permeate the work of visual practitioners, whether their representations draw on data, focus more on graphics and information displays, or assemble the insights of people and organizations (Lindquist 2017). Regardless of the type of visual practice, arriving at good representations is as much about 'fit' with client needs and contexts, as the inherent quality of the representation and what informs it (Vande Moere and Purchase 2011; Bederson and Shneiderman 2003). All visual practitioners wrestle with showing context and providing focus for the purpose at hand. Moreover, there is a general view that more visualization, more graphics and displays, and more visual facilitation will lead to better understanding, decisions, and service delivery outcomes. Even 'independent' actors—such as scholars or public intellectuals undertaking visual representation—see their clients as the 'public' and that they serve the public interest by contributing to the marketplace of ideas and information. They presume that more information is better than less.

Many visual practitioners would argue that the client orientation described above does not do justice to their craft and what motivates them. Many–particularly work-

[4] Horn's (1998) classic *Visual Language: Global Communications for the 21st Century* used sketches to assist policy-makers and citizens to comprehend and think address complex policy challenges and wicked problems.

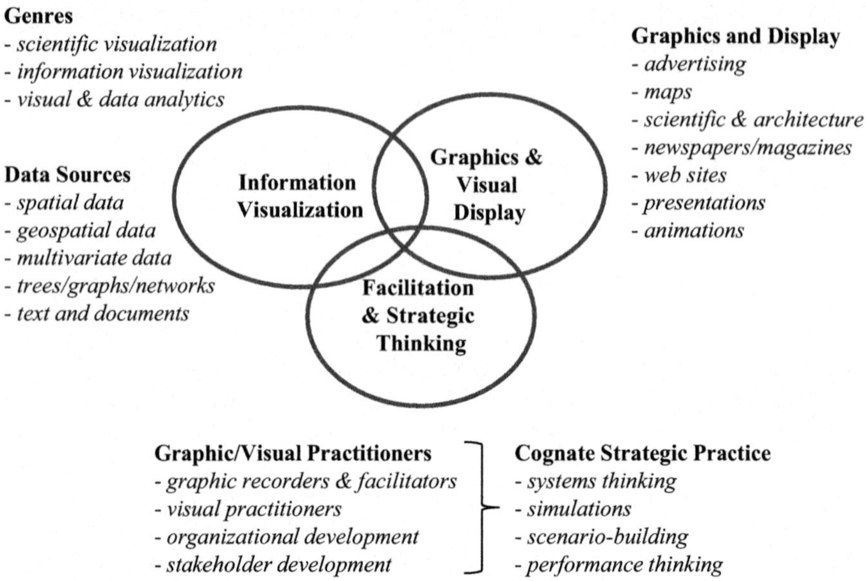

Fig. 1 Three visualization domains

ing in the visual facilitation domain–would strongly argue that their goal is to bring people together, to facilitate expression and recognition of all voices in an organization or community, to foster constructive and informed dialogue, and to arrive at final representations and reports which can influence a larger circle of stakeholders. Another way to put this is that many practitioners and advocates of visualization are motivated by the values of inclusiveness and open deliberation, which might be called collaborative and democratic aspirations. In some cases we can see these as distinct but complementary values to a client orientation.[5]

However, this is not the same thing as thinking about the overall effect of having more actors relying on visualizations for analysis, communication, and influence in market, governance or community contexts. Client logic does not explore the intersection of interests, agency and power in broader organizational and governance contexts. Indeed, many clients are purchasing visualization services not simply to better understand situations and move forward, but to gain advantage.

This leads to a more sobering perspective on how visualization will intersect with and add value to the normal dynamics of markets, governance and society. In a public governance context, actors will seek visual support as part of efforts to persuade and achieve desired outcomes. There is the inevitable prospect of multiple

[5] For others, the aesthetic or creative impulse looms larger, particularly those aggregating administrative and other kinds of data (think of social media, value evidence-based or evidence-informed decision-making) might be driven either more narrowly by fidelity to data and its representation—which, for many practitioners, remains paramount regardless of the clients served and broader context in which that creativity and work proceeds.

actors in many different forums and stages of the agenda-setting and policy-making process sharing visualizations which—even if not intended by their progenitors—are competing with other streams of information and with each other to inform various stakeholders. We turn to these topics in the next two section of this chapter.

Beyond Performance: Ways to Better Inform and Monitor Government

Governments stand as important sites for application of visual tools for two reasons. First, even the largest consulting firms, ostensibly serving private sector clients, have as their biggest clients governments of all kinds (often comprising well over half of their revenues). Second, for visual practitioners and clients seeking to influence policy and societal outcomes, governments are natural target audiences. As is the case with policy analysis and many professional services, governments are not only the recipients of visual products, but invest in them. Anyone who steadily monitors news media and advertising, as well as the public administration literature, knows that much is being made of digital government, the open data movement, improving analysis and engagement with social media and other online tools, the application of visual analytics to data sets gathered by governments, and improvement of service delivery by means of innovation labs (which rely on qualitative methods and often visual tools to elicit, assess, and share insights with and about clients, often in co-production mode).

These movements and initiatives, however, are proceeding in fits and starts, and should not be seen as representing the universe or even the most important possibilities for visual methods. Here we explore the context in which many of these initiatives proceed, why some have been taken up and others less so. We go on to identify several kinds of visualization initiatives which are more promising from a democratic governance perspective (in the sense of being transformative and briefing life into other visual and digital initiatives), but will likely be contested because of the stakes involved in policy-making and democratic oversight.

From Instrumental and Ideological Considerations to Furthering the Public Interest

Governments may be obvious targets for and consumers of visualization practice and products, but they and the contexts in which they operate have their own cultures, constraints, and logics. These should be acknowledged before suggesting where to make contemporary investments in visualization. Governments and public service institutions are instrumental and goal-oriented entities: duly elected governments seek to implement their agenda and maintain power. Professional public servants and the organizations they serve in have a duty to advise governments and assist with implementing policy and administrative decisions, while serving the broader public interest. Balancing the views of governments and the broader public interest has always been difficult, and public servants—executives, analysts and front-line staff—have always had to carefully navigate the cross-pressures in those domains, particularly in Westminster governance systems where loyalty to the government, maintaining cabinet confidentiality, and other requirements for administrative secrecy, make it difficult for governments to 'open up' in the way that many advocates for more deliberation and sharing of information would like.

As instrumental entities governments and public servants are more likely to gravitate to new techniques or more effective ways to secure better or more accurate data and information on issues, citizens, and societal sectors. No matter the ideology of a government, such investments give ministers and the public servants advising them an informational advantage in dealing with other actors, much like private-sector communications and branding efforts (Marland 2016), on top of their positional advantage. Meetings the demands needs of duly elected governments, though, should be balanced with other democratic aspirations: fostering accountability, sharing information with citizens and taxpayers, and working in the broader public interest. However, the enthusiasm for these goals should be tempered by realism about the instrumental and ideological tides to navigate.

The Challenge: Surmounting Performance Parsimony with Richer Pictures

Despite the burgeoning of ways to display information over the last 20 years, it has been remarkable how little take up there has been by governments. Indeed, where public reporting is concerned, the trend has been in the opposite direction, towards more opaque reporting, notwithstanding the open government movement. The 'New Public Management' movement of the late 1980s and 1990s brought along a collection of ideas emerging about how to better manage government in a more business-like and innovative manner (Aucoin 1995). These ideas included creating new incentives and structures to deliver services, relying more on public markets and competition, treating recipients of services as 'clients' and 'customers', and new ways to account and report on government performance, including accrual budgeting and accounting (Kelly and Wanna 2002). All of these initiatives were connected to improving performance reporting, placing more emphasis on outputs and outcomes, as opposed to the traditional focus on inputs and activities for budgeting and accounting (McDavid et al. 2013).

The result was that governments began to provide less detail on inputs and activities, and set out goals to be achieved, larger budget categories (to allow flexibility in inputs and activity mixes), and parsimonious performance indicators. One result was the loss of readily available detailed administrative data—at least for the public—on inputs and the costs of specific activity streams (Lindquist 1998). Moreover, *without* this information it has not been easy to assess outputs and outcomes achieved. This was further complicated by what evaluation specialists term the 'attribution challenge', where outcomes are usually influenced by a host of factors beyond the 'outputs' generated by any one public organization (McDavid et al. 2013). From a visualization perspective, one can see that 'logic modelling'—long an underpinning of program evaluation and performance measurement—is a linear visualizing repertoire which not only identifies government aspirations for addressing certain problems or gaps, but also how they intervene. It narrowly focuses on instruments, targets and indicators without much description.

Over the last many years, then, despite the oft-repeated mantra of open data along with more open government—many governments have become more opaque and the open government movement has not materially changed this reality. Even though some observers provocatively declared that "the New Public Management is dead" as a result of the arrival of digital tools (Dunleavy and Margetts 2006), very little NPM thinking and ways of doing things have fallen off the governance and management table. Indeed, the proliferation of 'co-production' and other collaborative and network models in the era of distributed governance fueled by the digital era, often called the New Public Governance (Osborne 2006). This is creating a huge need for more sophisticated ways to depict and monitor the performance of new ways of delivering public services. This will require 'dynamic', 'interactive' and 'citizen-initiated' accountability models which will work on a continuous as opposed to an episodic and relatively predictable basis (Schillemans et al. 2013), which will create demands for not only the collection and analysis of data, but better ways to display it and show complexity and linkages (Lindquist 2017). We can see this as governments moving towards the 'performance governance' posture envisioned by Bouckaert and Halligan (2008) close to decade ago. The abiding challenge of better describing how government works remains very much on the table, and visualization practitioners have much to contribute here.

Increasing the quality of reporting on government, however, is only one facet of an agenda for improving governance in a rapidly evolving world. In Canada, for example, many observers have worried about concern about whether Westminster government and our federal structures can handle the current and emerging governance challenges, many of which transcend levels of government—involving national, provincial, and local governments as well as First Nations and aboriginal organizations—and working across sectors (Lindquist and Eichbaum 2016). Making sense of all of these actors—along their roles, needs, and authorities—let alone communicating this to the public, and facilitating negotiations and decision-making, suggests a role for visual tools, but we see little use of them. This raises larger questions about whether governments are interested and capable of taking up the digital tools to open-up government (Lindquist and Eichbaum 2016; Lindquist and Huse 2016). The sensibilities and inclinations of the progenitors, advocates and users of visual tools are often wrought up in digital-era thinking, even if many visual traditions can trace their inspirations and roots well before modern digital tools became available. Accordingly, we now briefly consider where the take-up in visualization and digital tools has been greatest in recent years.

Where Have Governments Invested in Visualization?

Over 25 years ago, Stinchcombe (1990) suggested that organizations 'grow' towards the information they need to deal with crucial contingencies. It is a useful point of departure for considering the take-up in visualization methods by government. What follows just below considers the domains in which the greatest strides have been made. Section 4 on "Next-Generation Visual Investments: Furthering Democratic Aspirations" outlines areas were progress has been uneven or tentative.

There has not been a systematic comparative study on the take-up of visualization techniques in government, simply because of the diversity of tools and the sheer number of policy and administrative domains in which they can be applied makes that impossible for a single study (and to draw conclusions by tracking one set of tools or a single domain would not have external validity). However, casual observation suggests that the largest investments in visualization, particularly where data and visual analytics are concerned, have been in the following areas:

- *Security activities*. The internet arose from US government-sponsored research, and much of the innovation for early information and visual analytics was funded by the Department of Homeland Security (Thomas and Cook 2005). Early funding underwrote the design of better software and hardware used widely in the field of information visualization, including many well-known commercial products. These capabilities are used to monitor and sift through vast amounts of data from social media, telephone and mobile traffic, satellite imagery, travel data, etc., and to generating great visual representations of that data and to facilitate analysis and inferences.
- *Political parties: fundraising and elections*. In many jurisdictions the political class have been early and significant investors in privately held data-bases of donors and voting intentions on a riding-by-riding basis. This, along with regular polling and digital-calling capabilities, have provided party strategists not only with block-by-block and often finer grained intelligence on contributions and issues. Complementing this are increased capabilities to monitor and analyze social and other media. Such data easily lends itself to different visualizations and different levels of aggregation. Access to such intelligence also explains why elected leaders often feel more confident when dealing with public service policy advisors on priorities and the public interest.
- *Service delivery operations*. Many jurisdictions have significant digitally-enabled service delivery platforms, particularly in countries which have centralized provision of many services from across several departments in single administrative entities (i.e. Service Canada, Australia's CentreLink). Digitally-enabled service delivery leads to ongoing collection of administrative data and opportunities for

analysis across service channels and points of delivery, including tracking, identifying patterns, and finding more efficient and effective ways to deliver services. Such tracking continues if service is distributed to non-government providers, and is often protected for privacy or proprietary reasons.

- *Data-rich agencies*. Many government entities sit on massive amounts of data, whether in the areas such as taxation, natural resources, weather, a wide variety of science-based research capabilities, and general-purpose national and sub-national statistical agencies. While such entities may not have been originally seen as investments in visualization capabilities, they usually have either incentive or public mandates to demonstrate their relevance as institutions (if not bound by confidentiality requirements), show how public and administrative data can speak to contemporary issues, and sometimes to generate revenue. Whatever the motivation, such entities have not only joined the 'open data' movement but also have invested in innovative ways—such as slider and filtering software—to allow users to peruse and manipulate data, often in a multi-dimensional way. Providing open data sets encourages outside entrepreneurs and researchers to invest in visualization by essentially sharing the information at no cost, and if they can get outsiders to use and review data, this supplements internal analytic capabilities.
- *Design labs*. These emulate the Danish MindLab approach and create space for innovation—often tapping digital and visual methods—to better understand the experience of clients (dealing with a program, multiple interactions with different parts and levels of government, or broader life phases and pathways) as well as to identify new approaches to service delivery. Researching and then creating visual images of client challenges and experiences, can lead to design gaps and pilots of new approaches. The jury is out as to whether such units are strongly supported by departments or, if they are adornments for governments ostensibly advancing innovation (Meyer and Rowan 1977).
- *Foresight capabilities*. Several national governments have invested heavily in developing 'foresight' capabilities (Schmidt 2015). These entities are more likely to invest in visual tool capabilities because they have proven useful for recognizing, aggregating, and integrating disparate insights and variables into imagined futures. Visual tools can also encapsulate key themes and scenarios emerging from foresight work (Müller and Shwarz 2016).

What is striking about these practice domains is that, with exception of data-rich organizations that supply open data (e.g. transportation, weather, natural resource, and economic agencies) most visualization activity is rarely public, making it difficult to track and monitor. Even the literature which reviews open-data portals and open-government initiatives seems to have concluded that governments release minimal amounts of information, does not array data and information which could really be used to hold government to account, has not delved into whether accountability and transparency has improved, and has largely concluded that 'digital' tools have reinforced previous patterns of power in government and society (Lindquist and Huse 2017). These domains of government activity tend to be critical and not surprisingly the visualizations used for analysis, decision-making, and monitoring are more likely to be confidential or secret. It raises questions about 'open data' initiatives and whether important or simply safe data has been released—not as important for evaluating or strategizing about governance.

Moving away from information visualization, governments continue to be significant consumers *and* producers of graphics and visual displays, although it is not clear that these are liberally produced as part of ongoing public education and engagement strategies. As digital capabilities proliferated across all types of visual practice, and with governments typically procuring such services, most innovation has occurred within firms and the broader communications industry. Such procurement tends revolve around 'information-out' visual displays, which nevertheless are not inconsistent with rich offerings of information for engagement. While the take-up of visual recording and facilitation has been gaining recognition as more qualified facilitators come onto the market, this occurs in a bottom-up manner, relying on word-of-mouth and a client interest in a facilitation-plus experience. Visual facilitation is more expensive than regular facilitation, and therefore only larger corporations and public agencies can afford the rates, particularly when multiple recorders and facilitators are needed. Moreover, one can ask whether one-off visual facilitation and recording engagements are really that productive, when many visual facilitators believe that their diagrams need to be debated and negotiated over longer periods to fully develop a shared vision of how to move forward. However, such longer processes require significantly more resources on the part of sponsors.

Next-Generation Visual Investments: Furthering Democratic Aspirations

Casual observation suggests that the use of visual tools have tended to focus on gaining intelligence on or anticipating the challenges coming to government or as internal tools for managing and monitoring the implementation of policy and service plans. Far less attention has been paid to explaining how government works as a basis for informing elected leaders, citizens, and key stakeholders, and the argument of this paper is that visual methods can make an important contribution in this regard. What follows identifies three domains in the policy and governance process which would benefit from strategic investments in visualization capabilities as part of broader commitments to better inform our democracies. Likewise, these domains are ripe for nuanced, systematic empirical research.

Better Advising Elected Leaders and Legislatures

No recent studies have been undertaken on the use of visualization tools by elected leaders, whether ministers or elected representatives in legislatures. There has been far more interest on the use of research by government officials and leaders, flowing from the evidence-based decision-making movement (Nutley et al. 2007; Sa and Hamlin 2015).

A key challenge will be ascertaining what kind of visualizations and information elected leaders would prefer to have, since each one will have their own cognitive style and preference for how to consume information (which creates a further challenge as leaders come and go). They are not likely to trust or accept a representation developed for someone else, and in a format they are not familiar with, in terms of origins and underpinnings. To have confidence in the advice and information shared with them, elected leaders need to be aware of the people and process producing the insights; for visual tools this requires building trust and confidence in those responsible for producing visualization and not just those responsible for the data and policy analysis. Conversely, if leaders are comfortable with and engaged by visual methods, an upside risk emerges for officials: leaders will demand more visual representations and expect quick turn-around as they seek to tweak and modify them. This implies that governments will need sufficient internal 'visual' capacity to quickly make necessary reformulations and adjustments.

As leaders learn and preferences evolve, and as public servants must anticipate and respond to new leaders with different preferences, it is possible to see how visual communication is just one more communicating and advising modality (others include face-to-face discussions, briefing notes, meetings with stakeholders, tours and site visits, etc.). Advisors and policy analysis units, regardless of their preferred modes for professional communication, must be communicatively multifaceted and expand their portfolio with new visual ways to convey similar points, not only to handle new issues but also the preferences of different leaders. Building trust, so that powerful visual representations can be used to advise on critical issues, requires credibility, familiarity, and a broader base of knowledge from previous interactions: briefing on other issues and providing background materials for visits, consultations, and strategic planning, which later allows for the honing of visualizations conveying information and advice on particular issues as they arise.[6]

Better Informing Citizen and Stakeholder Engagement

Digital tools are enabling new ways for governments to canvass citizens' views, aspirations and preferences: through surveys, reaction polls, and experiments on service delivery platforms. The data can be more efficiently collected, aggregated, analyzed, and shared by governments. However, this is not the same thing as fostering dialogue and capturing the subtleties and richness of those discussions and evolving points of view.

[6] This presumes that advisors and leaders have time to develop a knowledge base and work out the particulars of the advising relationship, with visualization as part of that mix—however, often events are thrust on all concerned, making the choice of mode higher risk.

A key issue concerns providing sufficient information to citizens so that elicited views are better informed and more useful to those ultimately using them: governments often provide rather parsimonious information, insufficiently detailed for those engaged to feel confident about the advice they are giving. Such participants are typically not interested in sharing thinly informed views—rather, they want to know more about the problem, the context and history, constraints and trade-offs, and so on, and then share a more considered view. Many citizens want to more fully understand linkages and dynamics of key variables associated with the issues they are asked about, and how different kinds and levels of interventions might result in different outputs or outcomes. This suggests that providing visual tools, such as simulations and more tactile, interactive displays, could be important.

More governments are experimenting with more open processes, such as citizen juries and open budget processes, although these methods have long existed. However, such engagement requires widespread confidence in the representativeness of citizens selected to participate in such processes and in the quality of the information at hand. But it is a difficult for sponsors and facilitators to convey and represent the nature of deliberation in such a way that it informs those *not* directly engaged with the dialogue. Not all engagement need rely on new digital engagement tools, and many traditional approaches (conferences, hearing, workshops, town halls, surveys, workbooks, etc.) can be part of the mix depending on the issue and available time and resources. However, the big question is how to capture such proceedings in richer ways than sound-bites and linear reports: there are many possibilities for designers to capture proceedings, sources of information, and insights with mixes of visual tools, but sufficient financial resources would be required.

Furthering External Accountability

One the great frustrations of citizens, the media, watchdogs and other observers concerns the inability of 'getting to the bottom' of breakdowns or issues. This often reflects a disjuncture between the public's need for simple answers and seemingly inevitably more complicated contexts, which more linear accounts and reporting do not easily capture. As noted above, performance reports are concise by design, and usually too parsimonious to provide sufficient information on the flow of activities, what resources were mobilized or deployed to deliver programs and services, and the intersecting responsibilities and flows of resources and information. Accountability, of course, does not have to only be about monitoring failure, and here we consider what kind of information and visualization would be useful to ensure ongoing monitoring of diverse programs. There are many examples of governments erecting portals dedicated to sharing diverse indicators of 'performance' for their jurisdictions[7] (but usually this information is not connected to what levels

[7] For example, see: Finland's dynamic 'Findicator' at www.findikaattori.fi; Australia's Progress at www.abs.gov.au/ausstats/abs@.nsf/Lookup/by%20Subject/1370.0~2013~Main%20

of government and programs are implicated in enabling or militating against success), and likewise information on the progress of programs or service undertakings could be shared.

From an information vantage-point, a well-supported accountability regime would supply year-over-year data on key indicators and variables, concise histories of the policy domain under consideration, and good spatial and multi-level governance snapshots of pertinent agencies and their respective roles and responsibilities (within and across levels of government) as well as contributions from citizens, firms or non-profit organizations. It would be important to provide a good sense of relevant operations and capacity in terms of central capabilities and various delivery points (and not simply relying on a GIS map with points of interest—rather, imaginative ways to show local capacities and pertinent linkages). They would be similar to the bundle of material which some central agencies keep on tap to monitor departments and agencies (e.g. see Lindquist 2009 on Canada's Management Accountability Framework), but more effort could be made to organize and transmit such information in accessible visuals and user-friendly platforms. Finally, to deal with the attribution challenge concerning the influence of policies on desired outcomes, there would be a real opportunity to develop rich pictures or info-graphics. Such images would identify not only stakeholders and key variables influencing those outcomes, but also attempt to depict estimates of the strength of those connections.

Furthering Democratic Governance: Prospects for Moving From Promise to Progress

Finding new ways to advise elected leaders, engage citizens and other stakeholders outside our governments, and improving accountability are abiding challenges and making progress would be no small achievement. An optimistic perspective suggests that moving forward on these three fronts simply involves more transparency about the basics of how government works: for whom and what it works with, where it works from, and how it works. There is nothing top-secret or sensitive about such background information and, for democratically-inclined observers, governments *should* show such context. Indeed, providing richer information on instrumental activities will make it more likely to anticipate where uncertainties and issues might arise (which could be a government program, contracted services, or even a networked or collaborative delivery system across governments or other sector providers). This would not remove robust debate, but such debates can be informed or not, and surely the former is more desirable.

Features~Homepage~1; Scotland Performs at www.gov.scot/About/Performance/scotPerforms; and Measuring Ireland's Progress at www.cso.ie/en/statistics/measuringirelandsprogress/ and http://www.irelandstat.gov.ie/. Similar government performance dashboards and portals can be found for Edmonton and the states of Washington and Virginia.

A more sober and pessimistic perspective notes that richer renderings of how government works might well be seen by elected governments and top officials as arming certain actors or interests over others. As noted earlier, while much has been made about 'open data' and many other 'open government' initiatives and undertakings, it is not clear these have made modern governments any more transparent or accountable, except at the local level where there can be stronger connections between governments and citizens, and the services are less controversial and there is less need for secrecy. It suggests that political leaders might have to exercise more leadership, that new practices and forums (deliberative spaces) might have be opened up in support of these functions, and that citizens would also have to embrace and see value in them.

Conclusion: Supporting Visual Tools to Further Democratic Governance

Governments have considerable work ahead to expand the application of visualization tools beyond the instrumental needs of governments to more democratic activities such as advising elected representatives, citizen and stakeholder engagement, and improving accountability. Moving in these directions will require shrewdly building the right mix of visualization capacities in public service institutions, which will involve non-trivial investments, enlightened support from top elected leaders of government, and sufficiently engaged citizens.

Public service institutions need to continue to build their capacity to experiment, nurture and support movement in these directions. However, given the diversity in issues, information needs, clients across government—pointing to intriguing mixes and 'just-in-time' demand for visual tools—it will be unlikely that every government department or agency could create its own central capacity. More promising would be developing an inventory of visual capabilities required across departments and agencies. Such expertise could be shared across boundaries, and selective investments made to deal with priority issues and housed in departments and agencies associated with them. Focusing initial efforts on priority issues would help build the repertoires for developing visual products, requisite supporting expertise across programs and organizations, and the new sensibilities of ministers and public service executives. Developing 'quick-response' capabilities would be crucial, particularly when advising ministers, elected representatives and legislative committees. Over time, other departments and agencies should be able to support governments and central agencies in meeting higher expectations about sharing data and information with legislatures and citizens; moreover, producing it for one democratic purpose (e.g. accountability) would better prepare government other democratic purposes (e.g. advising elected representatives and engaging citizens). Such an emergent and evolutionary strategy could expand information-sharing and use of visualization tools.

Supporting visualization in government for advising, engagement, and accountability does not involve just supplying financial resources and administrative capacity: more fundamentally, it will require a commitment by successive political leaders of governments to more fully share information outside the core of government with more stakeholders. In this sense it needs be understood that the ambition of using visual tools for information-sharing is transformative, moving democratic and professional practice onto a new plane, but it will not remove politics and vigorous debate about the issues and the facts—indeed, there will bound to be challenges of visual renderings and information supplied, and the prospect of contending visualizations. Moving to this new plane of governance and democratic practice will not be for the faint of heart: not only will governments need to think about sharing rich information, citizens and outside experts will need to become more visually 'literate' and more critical consumers, so as not to be dazzled by visual virtuosity or attractive but specious renderings.

Looking forward, all of this suggests a rich research agenda. This chapter has provided a high-level survey of the diverse domains of visualization and their promise for improving democratic governance. It has concluded that the use of visual tools by government to further democratic governance had lagged behind their use for acquiring and exercising power. Nevertheless, there is an expanding number of visual tools and literature on the promise of visualization, increasing familiarity with those tools, and arguably more kinds of data which can be corralled, analyzed and displayed with them. A new generation of systematic research studies are needed to track and assess whether governments take-up these tools and if elected leaders in executive or legislative branches feel better informed, if citizens and stakeholders are more meaningfully engaged and informed, and if accountability and transparency is improved, or if the adoption of digital tools are more like donning a fig leaf, serving to amuse and look modern even as the traditional patterns of power are preserved.

Acknowledgements The first version of this paper was presented to the 4th Annual Visual Methods Conference, University of Brighton, England, 16–18 September 2015, and written in honour of Glen Milne and David Sibbet, respectively Ottawa-based and San Francisco-based consultants, educators, and designers. Decades ago they saw the potential of visual tools for engagement and conveying complexities to further democratic governance, knowing that appreciation of the 'whole' is critical for furthering strategic change and collaboration. The second version benefitted from succinct and useful comments from three anonymous reviewers.

References

Aucoin P (1995) The new public management: Canada in comparative perspective. Institute for Research in Public Policy, Ottawa

Atkinson C (2008) Beyond bullet points. Microsoft, Redmond

Bardach E (2012) A practical guide for policy analysis: the eightfold path to more effective problem solving, 4th edn. Sage, Thousand Oaks, CA

Baer K (2010) Information design workbook: graphic approaches, solutions, and inspiration + 30 Case studies. Rockport, Beverly
Bederson BB, Shneiderman B (2003) The craft of information visualization: readings and reflections. Morgan Kaufman, San Franciso
Blackwell AF et al (2008) Strategy roadmaps: new forms, new practices. In Stapledon G, Howse J, Lee J (eds) Diagrams 2008. Springer, pp 127–140
Bouckaert G, Halligan J (2008) Managing performance: international comparisons. Routledge, Abingdon
Chapman J (2004) System failure: why governments must learn to think differently, 2nd edn. Demos, London
Chapman J, Edwards C, Hampson S (2009) Connecting the dots. Demos, London
Checkland P (1999) Systems thinking, systems practice. Wiley, New York
Checkland P, Poulter J (2010) Soft systems methodology. In: Reynolds M, Howell S (eds) Systems approaches to managing change. Open University, Springer-Verland, London, pp 191–242
Chen C (1999) Information visualization and virtual environments. Springer, London
Chen C (2006) Information visualization: beyond the Horizon, 2nd edn. Springer-Verlag, London
Duarte N (2010) Resonate: present visual stories that transform audiences. Wiley, Hoboken, NJ
Dunleavy P, Margetts H (2006) New public management is dead—long live digital-era governance. J Public Adm Res Theory 16(3):467–494
Few S (2013) Information dashboard design: displaying data for at-a-glance monitoring. Analytics Press
Grammel L, Tory M, Storey M-A (2010) How information visualization novices construct visualizations. IEEE Trans Vis Comput Graph 16(6):943–952
Heer J, Robertson GG (2007) Animated transitions in statistical data graphics. IEEE Trans Vis Comput Graph 13(6):1240–1247
Horn RE (1998) Visual language: global communication for the 21st century. MacroVU, Bainbridge Island, WA
Hyerle D (2009) Visual tools for transforming information into knowledge. Corwin Press, Thousand Oaks
Johnston EW (ed) (2015) Governance in the information era: theory and practice of policy informatics. Routledge, New York
Kelly J, Wanna J (2002) Twin reforms: crashing-through with accrual-output price budgeting in Australia: technical adjustment or a new way of doing business? Int Rev Public Adm 34(1):94–111
Landow J, Ritchie J (2014) Infographics: the power of visual storytelling. Wiley, Hoboken
Lima M (2011) Visual complexity: mapping patterns of information. Princeton Architectural Press, New York
Lindquist EA (1998) Getting results right: reforming Ottawa's estimates. In: Pal LA (ed) How Ottawa spends 1998–1999: balancing act: the post-deficit mandate. Oxford University Press, Toronto
Lindquist EA (2009) How Ottawa assesses department/agency performance: treasury board's management accountability framework. In: Maslove AM (ed) How Ottawa spends 2009–2010: economic upheaval and political dysfunction. McGill-Queen's University Press, Montreal, pp 47–88
Lindquist EA (2011) No reform left behind: multiplicity, integrating frameworks and implications for New Zealand's Centre-of-Government and Public Sector Improvement, Ch. 2. In: Gill D, Ryan B (eds) Future state. Victoria University Press, Wellington, pp 46–84
Lindquist EA (2015) Visualization meets policy making: visual traditions, complexity, strategic investments, Ch. 4. In: Johnston EW (ed) Governance in the information era: theory and practice of policy informatics. Routledge, New York, pp 62–82
Lindquist EA (2016) performance monitoring and the management accountability framework: recent developments, perspective, and insights from the literature. Paper prepared for the Treasury Board of Canada Secretariat

Lindquist EA (2017) Visual methods and beyond: exploring the logics and influence of competing domains of visual practice. Vis Methodol 5(1):59–66
Lindquist EA, Eichbaum C (2016) Remaking government in Canada: dares, resilience and civility in westminster systems. Governance 29(4):553–571
Lindquist EA, Huse I (2017 forthcoming) Accountability and monitoring government in the digital era: promise, realism and research for digital-era governance in Canada. Paper prepared for a special issue of Canadian Public Administration on 'Governance in the Digital Era'
Margulies N, Valenza C (2005) Visual thinking: tools for mapping your ideas. Crown House, Bethel
Marland A (2016) Brand command: Canadian politics and democracy in the age of message control. University of British Columbia Press, Vancouver
McCandless D (2009) Information is Beautiful (London: Collins, 2009). Also published as The Visual Miscellaneum: A Colorful Guide to the World's Most Consequential Trivia
McCandless D (2014) Knowledge is beautiful: impossible ideas, invisible patterns, hidden connections—visualized. Harper Design
McDavid JC, Huse I, Hawthorn LRL (2013) Program evaluation and performance measurement: an introduction to practice, 2nd edn. Sage, London
Meirelles I (2013) Design for information: an introduction to the histories, theories, and best practices behind effective information visualizations. Beverley, Rockport
Meyer JW, Rowan B (1977) Institutionalized organizations: formal structure as myth and ceremony. Am J Sociol 83(2):340–363
Müller AW, Shwarz JW (2016) Assessing the functions and dimensions of visualizations in foresight. Foresight 18(1):76–90
Nutley SM, Walter I, Davies HTO (2007) Using evidence: how research can inform public services. Policy Press, Bristol
Osborne SP (2006) The new public governance? Public Manag Rev 8(3):377–387
Ringland G (1998) Scenario planning: managing for the future. Wiley, New York
Roam D (2009) The back of the Napkin: solving problems and selling ideas with pictures, Expanded Edition. Portfolio, New York
Rosell S et al (1995) Changing maps: governing in a world of rapid change. Carleton University Press, Ottawa
Sa C, Hamlin D (2015) Research use capacity in provincial governments. Can Public Adm 58(3): 468–486
Schmidt JM (2015) Policy, planning, intelligence and foresight in government organizations. Foresight 17(5):489–511
Schon DA (1993) The reflective practitioner: how professionals think in action. Basic Books, New York
Schillemans T, Twist MV, Vanhommerig I (2013) Innovations in accountability. Public Perform Manag Rev 36(3):407–435
Segel E, Heer J (2010) Narrative visualization: telling stories with data. IEEE Trans Vis Comput Graph 16(6):1139–1148
Senge PM (1990) The fifth discipline: the art and practice of the learning organization. Doubleday, New York
Sibbet D (2010) Visual meetings: how graphics, sticky notes & idea mapping can transform group productivity. Wiley, Hoboken
Sibbet D (2011) Visual teams: graphic tools for commitment, innovation, and high performance. Wiley, New York
Sibbet D (2013) Visual leaders: new tools for visioning, management, and organizational change. Wiley, New York
Simon P (2014) The visual organization: data visualization, big data, and the quest for better decisions. Wiley, New York
Steele J, Ilinsky N (2010) Beautiful visualization: looking at data through the eyes of experts. O'Reilly, Cambridge
Stinchcombe AL (1990) Information and organizations. University of California Press, Oakland

Thomas JJ, Cook KA (2005) Illuminating the path: the research and development Agenda for visual analytics. Department of Homeland Security, Washington

Tufte ER (1990) Envisioning information. Graphics Press, Chesire

Tufte ER (1997) Visual explanation: images and quantities, evidence and narrative. Graphics Press, Chesire

Tufte ER (2001) The visual display of quantitative information, 2nd edn. Graphics Press, Chesire

Vande MA, Purchase H (2011) On the role of design in information visualization. In Kerren A, Plaisant C, Stasko JT (eds) State of the field and new research directions. Inf Vis J 10:4, October pp 356–371

Ward M, Grinstein G, Keim D (2010) Interactive data visualization: foundations, techniques, and applications. AK Peters, Natick

Ware C (2004) Information visualization: perception for design. Morgan Kaufman, San Francisco

Winkel M, Junge MB (2012) Graphic facilitation and art therapy: imagery and Metaphor in organizational development. Charles C. Thomas Publishing Ltd, Springfield

Wilson JQ (1989) Bureaucracy. Basic Books, New York

Evert A. Lindquist is Professor of Public Administration at the University of Victoria and Editor, *Canadian Public Administration*, the flagship journal of the Institute of Public Administration of Canada. He served as Director of the School of Public Administration (1998–2015), professor at the University of Toronto (1988–1998), and the first Treasury Board of Canada Secretariat Visiting Fellow (1992–1994), and the ANZSOG-ANU Chair in Public Management (2010-2011). He has been principal investigator for a SSHRC partnership development grant on 'Digital Governance: Transforming Government for the Digital Era'.

Can the French Republic Be Digital? Lessons from the Last Participatory Experience on the Law-Making Process

Christophe Premat

Abstract A new law on the Digital Republic was voted in France at the end of 2016. It focused on the definition of digital rights, Net neutrality and the idea of a public service of data. The law was coproduced by the citizens as it was possible to amend the bill proposal on the Net. In fact, it was the first time in France that a bill was digitally amended by citizens. The chapter aims at analyzing the participatory method as a law-making option in France. The question is to understand whether or not this participatory method challenges the traditional way of governing in France. Is it the implementation of a citizen control of the law-making process or a single communicative tool?

The first part of the chapter describes the relations between the European and the national contexts of this law. The second part of the chapter explains the participatory tool, introduces the actors and analyzes the impact of the method on the law-making process. The third part of the chapter studies the emergence of an open government data in France. This digital strategy is promoted in order to win the competition of digital nations (Premat, Smarter as the new Urban Agenda: a comprehensive view of the 21st century city, 207–224, 2015). Moreover, the last part of the chapter confronts this new open data strategy with the surveillance laws voted in 2015 that control the Internet to trace suspicious behaviors.

Introduction

The digital republic was presented in the early Autumn 2015 as a new form of governance. Prime Minister Manuel Valls presented it as a specific law but also as a law-making model. The inclusion of citizens at an early stage was clearly a step forward for a more participatory process. If the citizen inclusion gives a strong touch of modernity to this law, the content should be analyzed as there was a tension between the open access procedures and the surveillance context that was implemented in

C. Premat (✉)
Department of Romance Studies, Classics of Stockholm University, Stockholm, Sweden
e-mail: christophe.premat@su.se

J.R. Gil-Garcia et al. (eds.), *Policy Analytics, Modelling, and Informatics*, Public Administration and Information Technology 24,
DOI 10.1007/978-3-319-61762-6_11

other laws. The aim of this law is to enshrine new digital rights and make French administrations and citizens move forward in the digital era. Is it a sharing governance based on an interaction between citizens, lobbies and administrations? Is the digital Republic an adaptation to the growing ideology of open government or a deep transformation of the law-making process? In other words, is there a transformation of the political framework due to the use of digital mechanisms?

Is there a captation of resources from some organizations (Freeman and Langbein 2000) that can have more legitimacy in this new sharing governance? I would like to analyze the law-making process to see whether it is a new form of governing or a technical adjustment linked to political goals. Analyzing the law-making process means describing the whole legislative process and referring to the important declarations or discussions on this law. In France, the legislative process is characterized by a parliamentary discussion in both chambers, the National Assembly and the Senate. This case-study is worth studying as it can be applied to other bill proposals in the future. In other terms, is it a communicative toy linked to a form of total lobbying (Nownes 2006, p. 38) or a real participatory implementation?

In order to analyze the implementation of the digital experiment, it is important to refer to the central concept of Zygmunt Nauman, the liquidity (Bauman 2005). The societies are more and more liquid and the digitization reinforces this feeling. The liquid society is linked to the ideas of uncertainty, permanent change and evolution. The idea of mobility is highly promoted in France and Europe and the digitization is often seen as a synonym for modernity, innovation and transformation. The digital modernity is associated with the simplification of administrative tasks, often perceived as long and bureaucratic. In 29 March 2013, the governement announced a phase of modernization with the acceleration of digital facilities. The goal was to avoid long procedures and simplify too many administrative procedures. In 17 July 2013, 200 measures were adopted to facilitate some administrative procedures such as applying for a new Identity Card. A bill was passed on 22 July 2014 to facilitate the norms for companies and citizens. Another package of 170 measures was adopted on 3 February 2016 which was completed by 30 other measures on 26 October 2016.[1] The idea of simplifying overwhelming norms for people was first made true before being integrated in a new law on Digital Republic that redefined the act of citizenship in France.

In this context, the digital commitment of citizens and the possibilities of affecting the laws can be seen as a deep characteristic of the liquid modernity. The laws which are classified as solid social and juridical elements do not escape this tendency. This is why the political attempt of codifying digital matters is necessary to protect citizens and allow them to control the tool. Citizens can act as consumers to modify the content of public policies that can be beneficial for them. Digital matters are not only an expression of pure market, they can also help to reshape the public sphere in the contemporary societies. In this perspective, the notion of digital

[1] http://www.modernisation.gouv.fr/sites/default/files/fichiers-attaches/enquete_complexite_administrative_2014_synthese.pdf (Retrieved 20 November 2016).

republic is presented in France as a new reenchantment of citizen participation (Bauman 2003, p. VII).

The debate surrounding the digital rights was initiated at the European level as some ideas such as net neutrality, digital death were already dealt with in the European Parliament. As the French parliamentary agenda was postponed because of the terror attacks and a series of surveillance laws, this parliamentary discussion could finally begin in January 2016 with a discussion in the National Assembly. In the French system, a bill can eventually be adopted after two readings in both chambers, the National Assembly and the Senate. The second part of the chapter focuses on the relation between the participatory inclusion of citizens and the parliamentary debate in France about digital Republic. If this law was already announced in 2014, it took a longer time to prepare the text that would be discussed in the parliament.

A European Context of Digitization

The European Union put the digitization on the top priorities of its agenda for the next years. Furthermore, the idea of digitization makes cultural resources available for a broad public. The European Commission encouraged the setting of a framework for digitization of cultural resources accessible to the public. The Commission´s recommendation of 27 October 2011 clearly assigned the digitization of resources as a goal in order to create jobs, growth, opportunities and a common heritage (*Official Journal of the European Union* 2011). At the same time, the idea is to protect citizens from different risks emerging with the use of new technologies. For instance, internet neutrality has first been discussed. It means that the new technologies are not a property but a common good that should be preserved from any propaganda or marketing (Sennett 2006, p. 153). There is a neutrality to be observed for consumers regarding the platform which is used.

Net neutrality means in that sense that all data on the Internet should be considered as equivalent by governments and Internet service providers. In other words, there can be no discrimination regarding user, site and platform. Tim Wu was the first researcher pointing out this debate in 2003 (Wu 2003). Broadband policies became as important as other public services. In the European context of digitization, there is the idea of protecting the access of consumers to the Internet supplies. On 11 September 2013, the European Commission adopted a few legislative procedures entitled "Connected Continent: Building a Telecoms Single Market" which aimed at creating sustainable digital jobs. The debate did not only concern digital economy but also the equality of opportunites for citizens. This is why the European Parliament voted in October 2015 for the first European Union-wide net neutrality rules. The European rules prevent from an intern competition among State members concerning internet providers. In other words, the goal is to create a common market where the platforms do not discriminate the contents for consumers. For instance, it will not be possible to restrict the access to the internet and different services such as Facetime or Skype. Mobile providers have to share a common policy, there can be no prioritisation of traffic in the internet access service. These rules were based on the recommendations of the Body of European Regulators for Electronic Communications (BEREC) in 2011 that highlighted the necessity of Net neutrality in the European Union. These rules are new and need to be transcripted into national legislations in order to be valid. For the first time, there was a difference made between digital economy and minimal digital goods. Citizens cannot be discriminated by Internet providers, their open access to specific legal contents cannot be slowed down. Furthermore, National Regulatory Authorities have to monitor and control market developments by respecting these rules.

The rules are clear to avoid a strong technological dependency, this is the reason why the regulatory process is so important for the protection of citizens´ consumption. At the same time, the goal is to sustain a digital growth where business possibilities can be expanded thanks to the access to the Internet. All national legislations are supposed to transcript this new frame for European countries.

The Parliamentary Process in France

The final law is a product of a dialogue between the Senate and the National Assembly. In other words, lobbyists have more opportunities to encourage MPs to amend the bill thanks to the system of two readings. A mixed committee is built after the first reading in the National Assembly and in the Senate (Carcassonne 1984). If both chambers happen to agree on a draft, then the law can be adopted. Otherwise, a second reading in both chambers is initiated. If there is still a disagreement, the National Assembly has the final word as the MPs are directly elected. The government can help its own majority in the National Assembly to adopt a final draft. At the same time, the MPs cannot introduce new amendments in the final reading, they only discuss amendments that were already dealt with in the previous readings.

For each reading, the procedure is the following one: a text is first announced in the Council of Ministers. The Council transfers to the Assembly a bill that is discussed in the first parliamentary Chamber. In each assembly, a MP from the governmental majority is elected in the law committee to make a report on the bill. She/he is appointed as the rapporteur for the bill. She/he has a special position as she/he will defend the draft and organize the discussions with the government. The rapporteur receives a technical assistance from the committee's staff in order to analyze and deposit some amendments to the text. The writing of a legislative text is a collective process and the rapporteur is like an arrangor in music, she/he has to compose the final draft by integrating different proposals.

The text is examined and modified in the law committee, other proposals can be made by some other committees. After the first technical and political discussion, the text is presented in the plenary session of the National Assembly. Some amendments that were initially rejected by the committee can be reintroduced during that process. As a matter of fact, the parliamentary discussion is long and complex with

Table 1 The parliamentary filter

The arena where the text is examined	The key-persons	Time-process/action
National Assembly	The State Secretary for digital affairs/The rapporteur	The rapporteur is appointed by the public bill committee one month before the examination of the bill
National Assembly	The rapporteur/the committee staff	Hearings of the associations/trade-unions and groups that have an expertise on the question
National Assembly	The rapporteur/the president of the public bill committee/the position of the majoritarian group inside the committee/the State Secretary	Presentation of the report to the members of the public bill committee. A new draft with a vote on the different amendments
National Assembly	The State Secretary/the rapporteur	Discussion of the text issued from the public bill committee
Senate	The rapporteur	Nomination of a rapporteur for the text
Senate	The rapporteur/the public bill committee	Hearings of the associations/trade-unions and groups that have an expertise on the question
Senate	The rapporteur/the president of the public bill committee/the position of the majoritarian group inside the public bill committee/the State Secretary	Presentation of the report to the members of the public bill committee. A new draft with a vote on the different amendments

Source: own analysis of the parliamentary process in France

two different steps, a technical step within the committees and a discussion in plenary session. The agenda-setting is a key-word as the readings of the draft have to be scheduled on time. Table 1 illustrates the parliamentary process for the examination of a bill.

For the State Secretary for digital affairs, it was really important to initiate a dialogue with both rapporteurs and build a consensus on different parts of the text. If the Senate chooses to rewrite the text of the National Assembly, a public bill committee composed of 7 senators and 7 MPs from the National Assembly is built. If both committees agree on a version, the text can be adopted by both assemblies. If not, there is a second reading of the text in both chambers. The text is finally adopted when both assemblies accept the same modifications, otherwise a new public bill committee composed of senators and MPs is appointed. In case of disagreement, the National Assembly has the final word. The problem is that in each assembly, the text has to be put on the parliamentary agenda. The control of the time-process is all the more important as it affects the results of the bill. Introducing a participatory procedure gives more complexity as the State Secretary for digital affairs has to control both processes, the participatory and parliamentary agendas. To analyze the procedure, it is important to come back to the participatory procedure itself that prepared the parliamentary discussion. This bill allows the State secretary of Digital affairs to create a consensus around a participatory norm which seems modern and innovative in terms of political roles (Lagroye 1994, p. 15). Citizens and MPs interact, their role changes as the MPs do not have anymore the privilege of creating the norm, they have to control and complete what the government and the citizens prepared. In terms of public policies, it is a cultural shift (Jobert 1985). The citizens used to be associated with the evaluation of public policies, now they are required to give some guidelines in the law-making process. By just describing the time-process, the inversion of roles is noticeable. The research in political science has widely focused on the agenda-setting within the participatory model (Schneider and Janning 2006). In this perspective, the citizens could formulate public policies through forums (Boussaguet and Muller 2005). Here, the government already proposed a draft that the citizens were supposed to amend through a first digital phase. The agenda was modified, the government controls the initiative and includes the citizens in the bill conception (Boutinet 2004).

Does Digital Rime with Participation?

The digital procedure was presented by the State Secretary of digital affairs, Axelle Lemaire, as a political innovation. The idea is to use new technologies in order to encourage the citizens´ participation in the law-making process. This is a top-down procedure where the government opens a participatory process. Figure 1 indicates the different steps of this citizen inclusion.

There was a period start with a draft proposed by the government. The citizens were invited to react by making remarks and/or amending the draft. During the process, some contributions were selected and the contributors were invited to meet the State Secretary for digital affairs and discuss the content of the proposals. The criteria that led to the selection of some of the proposals were not that clear, it reflects the governmental will of communicating on the inclusive process as well as the indetermine aspect of an experimental tool. No explanation was found on the profile of selected citizen groups. Why were some associations met when the deadline was even not reached? This fact shows that this procedure was biassed as the Cabinet used the proposals that confirmed the expectations. The State Secretary had expectations around the important topics (net neutrality, loyalty of platforms, possibility of erasing data, digital death) that should be discussed in the law. The Cabinet for digital affairs had an important role as it organized the procedure and prepared all the meetings with associations and digital lobbies.

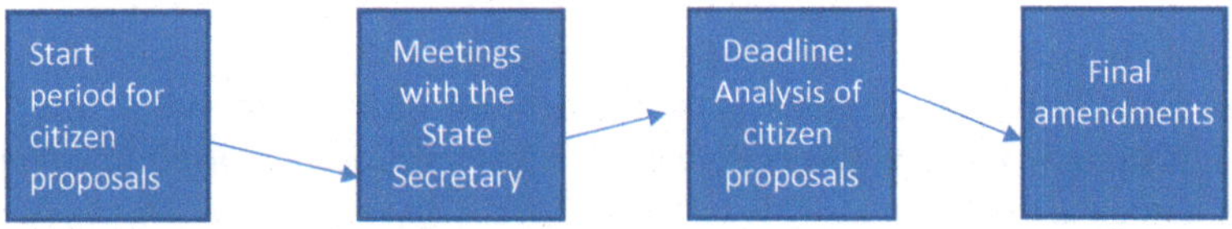

Fig. 1 The different steps of the law on Digital Republic

As Nownes wrote, "lobbyists meet personally with legislators and/or their aides. Lobbyists meet with legislators in many different places, including their personal offices, meeting rooms in government buildings, and informal venues such as bars and restaurants [...] When lobbyists meet with legislators or legislative aides, they often emphasize how a proposed bill will affect a legislator´s constituents" (Nownes 2006, p. 17). In this case, the digital platform was built to select the proposals that matched the initial intention. The proposals were conceived by single citizens, associations or sometimes digital entrepreneurs that have an interest in the regulation of digital economy. Claiming that digital republic simply encourages new democratic tools would be too early. The idea of democracy is deeply associated with confrontation of ideas and face-to-face relations (Fishkin 1995, p. 4). As Richard Sennett wrote, "the very idea of democracy requires mediation and face-to-face relations; it requires deliberation rather than packaging" (Sennett 2006, p. 135). According to Sennett, cultural capitalism is characterized by the fact that citizens consume politics in a similar way to marketing. Citizens can have the illusion of taking part in the law-making process but at the same time the government controls the participatory agenda by selecting the contributions. Sennett deals with the sense of democracy developed by Hannah Arendt and compares the democratic innovation (creating new rules even though the former rules work) with the technological progress. The democratic forum and *Media Lab*[2] share a common strategy: the idea is to produce rules in a changing world and anticipate what can happen. This is a characteristic of what Bauman called liquid modernity (Bauman 2005).

The French Secretary State of digital affairs was aware of the necessity of improving the digital culture in France. This participatory procedure follows a strong engagement of the Secretary State for a digital culture that implies a deeper knowledge of those technological tools. The objective is to democratize the use of new technologies to avoid a dependency that would have social consequences. As a matter of fact, it is of paramount importance to learn pupils how to encode at a very early age.[3] Educating people to encode helps to understand the possibilities of new technologies. Citizens can use new technologies thanks to an appropriate digital culture. Rewriting the bill by modifying the amendments suppose that citizens and associations have time to understand the technical process. The cooperation requires a deep understanding of the technological process. The real need for expertise is here prevailing. In his works, Richard Sennett compared the "citizen-as-consumer" with the "craftsman": "In labor, the crafstman is more than a mechanical technician. He or she wants to understand why a piece of wood or computer code doesn't work; the problem becomes engaging and thereby generates objective attachment" (Sennett 2006, p. 170). Richard Sennett worked on the cities and their social relations. He always refers to the craftsman as there is a concrete relation between the

---

[2]The *Media Lab* is a transdisciplinary research centre of MIT that aims to improve humanity thanks to a radical understanding of what technologies are.

[3]"Apprendre à coder à l´école, pourquoi ce n´est pas encore gagné", *Le Nouvel Obs,* 7 November 2014, http://tempsreel.nouvelobs.com/education/20141107.OBS4417/apprendre-a-coder-a-l-ecole-pourquoi-ce-n-est-pas-encore-gagne.html (Retrieved 5 May 2016).

Table 2 The presentation of the law on Digital Republic in France

Date/Period	Steps	Arena	Actors
4 October 2014–4 February 2015	National concertation (citizen consultation)	Digital platform with 4000 contributions (report on "digital ambition")	Government
18 June 2015	Presentation of a report/ Announcement of the digital law-making process		Digital National Council
26 September 2015–18 October 2015	Procedure of cocreation of the law	Digital platform	Government/lobbies/ associations/citizens/ NGOs
18 October 2015	Results of the procedure (20,000 participants/140,000 votes, 8000 arguments)	30 articles for the draft	Government
6 November 2015	Bill draft transmitted to the State Council		State Council
9 December 2015	Presentation of the bill draft	Council of Ministers	Government

Source: http://www.gouvernement.fr/action/pour-une-republique-numerique (Retrieved 5 May 2016)

craftsman and his/her work. In another book, Richard Sennett explains the promotion of the words "cooperation" and "solidarity" at the beginning of the nineteenth century when Paris had a universal exhibition (Sennett 2012, p. 38). It could be said, following that argument, that in digital affairs, the role of engineers is highlighted as they have to find out solutions and repair. The objective of the State Secretary on digital affairs is to democratize this engineering process to avoid depending on external technological capibilities. By introducing a digital participation of citizens, it will help to create a deeper digital culture in the country and a specific interest for these new technologies that affect and transform everyday life (Rodotà 1999).

Digital Republic was seen as an arena to experiment a new mode of participation. The goal is to propose a text that the citizens can amend and discuss before the parliamentary discussion. In fact, the participatory mode aims at breaking the existing gap between citizens and politicians in a representative system. Normally, in France, the first draft of the bill is presented to the Council of Ministers before entering the parliamentary arena with a double-reading process through the National Assembly and the Senate. Table 2 reminds the selected steps of the law on Digital Republic in France.

The final text was finally adopted on 29 September 2016. The Cabinet of the State Secretary helped to organize meetings with citizens and associations that took part in the digital construction of the bill. This is important to analyze the communicatory tools from the Ministry of Digital Affairs to see if the so-called selected citizens came with new ideas. The comparison between the main communicative tools with the different drafts are the empirical data that have to be scoped to be able to evaluate the emergence of new ideas. The government controlled the procedures and opened a space for citizens to amend the law.

The digital platform http://www.republique-numerique.fr was created to receive all the contributions and select them in a transparent manner. During the reception of contributions, the internautes were invited to vote on the different contributions. By analyzing the profile of the respondents on the platform, it can be concluded that associations working on digital matters were prevailing. Their proposals got most of the votes. One of the associations is "La Quadrature du Net"[4] that defends the citizen rights on the net. On the ten contributions that obtained most of the votes, six contributions were from "La Quadrature du Net". This association is supported by different NGOs such as Electronic Frontier Foundation,[5] Open Society Institute[6] and Privacy International.[7] Other institutions took part in the participatory process such as the University consortium of research Couperin.org,[8] the digital platform of scientific publication Cairn.[9] The first contribution that was selected was an amendment on the article 11 of the first draft concerning net neutrality. A short paragraph explained the amendment:

> Operators play a key role in access to information and digital development. They include controlling internet access of each subscriber and implement the technologies needed to ensure the availability, quality and speed of flow. The use of these technologies should not lead to favor certain services or restrict access to subscribers, by clamping or blocking access to certain content, services or applications online (protocols, websites, etc.), apart from their legal obligations. Now, the practices of ISPs will be precisely framed: it is especially forbidden to deteriorate the quality of dissemination of a site for the benefit of another, and more broadly to limit consumers' access unduly to open internet. ARCEP (Electronic Communications Regulatory Authority and Post) will be the competent authority to ensure compliance with the principle of net neutrality.[10]

The European principle of net neutrality is here affirmed, there were 1949 votes which concern this matter. Only the citizens registered on the platform were allowed take part in the evaluation process: they commented the content of the contributions and voted for the best ones. The objective of the contribution is to enshrine the European principle in the national legislation: an operator cannot reduce the bandwidth to free websites and increase it to other websites that have a fee. Net neutrality means that service providers have to give the same speed to the internet users. The cable companies have to respect this principle which was afterwards proposed in the

[4] https://www.laquadrature.net/fr/qui-sommes-nous (Retrieved 6 May 2016).

[5] https://www.eff.org (Retrieved 6 May 2016).

[6] https://www.opensocietyfoundations.org (Retrieved 6 May 2016).

[7] https://www.privacyinternational.org (Retrieved 6 May 2016).

[8] http://www.couperin.org (Retrieved 6 May 2016).

[9] http://www.cairn.info (Retrieved 6 May 2016).

[10] http://www.republique-numerique.fr/projects/projet-de-loi-numerique/consultation/consultation/opinions/section-1-neutralite-de-l-internet/article-11-neutralite-de-l-internet/versions/preciser-le-perimetre-d-application-de-la-neutralite-du-net (Retrieved 6 May 2016).

draft of December 2015 (the article 19).[11] The code of postal and electronic communications is modified in order to include net neutrality and the regulation of the European Parliament on the free roaming adopted on the 25th of November 2015. This article was adopted without changes at the National Assembly during the first parliamentary discussion. The law on "Digital Republic" included other important measures such as the disqualification of anti-concurrency practices from digital platforms. Digital platforms have to give fiscal information to the French administration.

The title of the bill "Digital Republic" aims to promote a new method in order to include citizens to the law-making process. In his statement, the State Council recommended to change the title of the bill and replace it by "Bill on citizen rights in a digital society".[12] The State Secretary maintained this title to underline the political empowerment that can be used to renew the image of the French republic. This law is an example of a new political mentality. Other MPs such as Patrice Martin-Lalande proposed a systematic implication of citizens in the law conception.[13] Patrice Martin-Lalande is a MP from the Conservative Party which proves that there is a strong consensus around the digital implication of citizens. In France, the political leaders do not want to deprive the institutions from the control of the agenda, they want to associate the citizens to the improvement of the law. This explains why the law on Digital Republic was mainly approved by all the political parties represented in the Parliament. The coconstruction of the law implies a definition of the different roles. The creation of a digital instrument also influences the definition of the process. In the bill of digital Republic, the government created the whole platform and proposed the citizens to amend the law. This is an experiment on the way of using digital tools in the production of juridical norms (Unger 1998).

The participatory process revealed the profile of the main actors that chose to take part in the discussion. This helped to create a repertoire of single-issued organizations that have an expertise in digital affairs. This method is all the more interesting as there is a discussion in France on how to register all existing lobbies as it is done in the European Parliament. The National Assembly has specific rules for the MPs. According to the article 79 of the rules of the National Assembly, a MP cannot be member of a group interest that can interfere with the law-making process.[14] The mandate is representative which means that every MP should vote without being dependent on any exterior pressures. She/he has to deliver an objective statement by hearing the maximum amount of points of views. In addition to this, the articles 433-1 and 433-2 of the Penal Code punish bribes and traffic of influences. A parliamentary report was written in 2013 to propose a few recommandations for the creation of a public register of lobbying organizations.[15] A comparison

[11] http://www.assemblee-nationale.fr/14/projets/pl3318.asp (Retrieved 6 May 2016).

[12] http://www.republiquenumerique.fr/media/default/0001/02/30232b6f96673c5c91483841fa4866a5f9ad54d2.pdf (Retrieved 7 May 2016).

[13] http://www.assemblee-nationale.fr/14/propositions/pion3686.asp (Retrieved 20 November 2016).

[14] http://www.assemblee-nationale.fr/connaissance/reglement.asp (Retrieved 8 May 2016).

[15] http://www.assemblee-nationale.fr/representants-interets/rapport_bureau_2013.pdf (Retrieved 8 May 2016).

showed that lobbies had different strategies in the European countries. Sometimes, they present themselves as consultants and advisers and some national legislations consider them as a part of the political life. Most of lobbies have more power than MPs as they use resources to promote a single-issued perspective. They have an expertise, they pay lobbyists to influence the modifications of the laws (Nownes 2006). The National Assembly appointed a person in charge of deontology to be able to identify the possible interest conflicts between the MPs and the external influences. The MPs can obtain help and assistance when they are unsure about the validity or the objectivity of a decision. This Code of Conduct shows that the ethical aspect is important to renew the image of the political system.

The participatory process could be a real occasion to identify and reveal the organizations that wish to interact in the law-making process. The influence is neutralized as the government opens the first steps of the law-making process before having a discussion inside the Parliament. The pros and cons, the main points of the discussion are thus clearly identified before the parliamentary discussion. In other words, the digital tool explicitly points out the slowness of the parliamentary process. The circulation of a text between the two parliamentary assemblies (the Senate and the National Assembly) makes it difficult to reach a consensus after a few readings, especially when the political majority differs from one assembly to the other. The description of this participatory tool shows that it contributes to the simplification of the representative system. The inclusiveness is limited to the mere improvement of the law. Contrary to what has been announced, the law on Digital Republic used the participatory tool to rebuild the format of the law-making process. In the field of political theory, the debate focused on the opposition between the participatory norm and the hierarchical structure of the political process (Freund 1988, p. 82). With the implementation of digital amendments, the debate relates to the technical arrangement of the law.

The control of the technical staff is also important as many assistants in the Parliament and in the government cabinets have strong links with lobbies. Their penetration should be controlled and the Code of Conduct should be reinforced to avoid this phenomenon. In the National Assembly, a Code of Conduct was adopted by the board of the National Assembly the 26th June of 2013.[16] There is an existing repertoire but the associations and companies are not really classified.[17] No organization working on digital affairs is registered, the list should be updated. The National Assembly has the possibility to create study groups that are different from lobbies but there can be an influence of lobbies on specific topics. The study groups replace the former intergroups that existed a few years ago (Pontier 1982, p. 820).

The bill on digital republic can bring more transparency in the political work by neutralizing the political role of lobbies. If lobbies can be traced by this new mode, then it is easier to create the conditions for a well-balanced parliamentary discussion. To sum up, the participatory process could give more legitimacy to the politi-

---

[16] http://www2.assemblee-nationale.fr/14/representant-d-interets/repre_interet (Retrieved 8 May 2016).

[17] http://www.assemblee-nationale.fr/representants-interets/liste.asp (Retrieved 8 May 2016).

cians by showing which organizations want to influence the law-making process. In a way, it could re-politize the political sphere which is more and more dominated by the influence of lobbies. Without this phase, it is almost impossible to trace all the influences of lobbies during the parliamentary process. Some of the amendments are easily recognizable as they were directly proposed to the MPs whereas other amendments do not reflect the influence of lobbies. The participatory process clarifies the parliamentary discussion as the positions of the single-issued organization are identified. The law of digital republic reveals a will of including citizens to the elaboration of the law, but it was affected by the legislative atmosphere dominated by security and surveillance questions. If it is a filter to improve the parliamentary discussion, the openess of digital tools can be contradicted by other sectors. The Home Office could also prevent those tools from being used for security reasons.

The Paradox Trend: Open Data and Surveillance

In the discussion of the law, most of the technical contributions came from digital platforms, NGOs, universities and lobbies. These organizations are afraid of the surveillance of mass data as they need them in order to expand projects. Entrepreneurs and politicians cooperate with each other to define a proper use of big data (Boullier 2015). The State Secretary for digital affairs has been promoting for two years the idea of the "French high tech"[18] by encouraging the digital hubs and the new start-up companies. She has been working on the idea of sharing economy[19] (Cheng 2016) where the flexibility of work create an added value. The digital strategy is seen as a way of creating a new growth. Digital platforms aim at including these new tools and defining digital rights for citizens and internet users.

A distinction can be made between digital entrepreneurs and digital influencers. Digital entrepreneurs refer to people who work in digital economy whereas digital influencers promote actively the idea of sharing economy, which means new work organizations mainly based on the use of digital tools. This flexibility can create conflicts as some companies invest the market by using this kind of economy. This was the case in France with the conflict between Uber company and the taxis. This conflict began 2 years ago when taxi-drivers complained about an illegal concurrency as workers from Uber do not need to pay a specific licence to transport people (Cannon and Summers 2014). A parliamentary report was published in June 2014 to propose a few legislative recommendations that can prevent these conflicts from happening.[20] The idea was to guarantee a fair concurrency in terms of licence, access to open data and tarification. The legislation aimed at fighting against fraudulent

[18] *Le Monde,* 12 November 2014, "L´offensive "French Tech" d´Axelle Lemaire".

[19] "La France dévoilera sa stratégie nationale pour l´économie collaborative à l´Automne", http://www.usine-digitale.fr/editorial/la-france-devoilera-sa-strategie-nationale-pour-l-economie-collaborative-a-l-automne-annonce-axelle-lemaire.N330731, *L´Usine digitale,* 20 May 2015.

[20] http://www.assemblee-nationale.fr/14/rapports/r2063.asp (Retrieved 8 May 2016).

licences. The concurrency should concern services offered by different taxi-drivers companies. It is an example showing how the digital localization of services transformed the economy and the facilities. The regulation is thus important to avoid the apparition of non qualified taxi-drivers. Many strikes were organized even though the government appointed a mediator to resolve the conflict between this company and the taxi-drivers. The National Assembly made some proposals to avoid this conflict in order to protect the licences of taxi-drivers. The new technologies illustrate that digital companies transform the organization of markets. The i-phone applications modify the consumers´ attitudes and create a specific relation between firms and citizens. The discussion on the bill proposal of Digital Republic dealt with the access to open data. Giving more access to big data would be a decisive step to create digital growth (Chakravorti et al. 2015).

According to James G. March, "given that we wish to describe a system as satisfying the postulates of conflict, and given that we wish to describe this system as "acting", "behaving", "choosing", or "deciding", we are required to introduce some mode of conflict resolution. By saying that the system does in fact "act", we accept the proposition that the system *in some sense* prefers some state of the world to other possible states of the world" (March 1962, p. 666). It is natural to say that everybody should adapt to the use of new technologies as it can create more opportunities and more exchanges in a quicker and efficient way. But at the same time, the system of opportunities is a part of the existing economical system where business firms are more empowered than citizens. The collection of data encourages the game of digital entrepreneurs that understood the benefits of such a tendency. James C. March showed in the sixties that the firms defined the political rules (March 1962, p. 673). If it is natural in political science to deal with political coalition, the firms share a same ideological goal, a "superordinate goal" (March 1962, p. 665) that creates the conditions of a strong social control. The idea of a growth of the digital sector is shared by all digital entrepreneurs. The companies and the public authorities can define common priorities in order to accompany a new economical tendency.

James G. March was one of the first scientists studying the behavioral aspect of firms and their political aspect. During the first reading of the bill draft in the National Assembly, the MPs discussed this role as well as the digital platforms' central position with the GAFA (Google-Apple-Facebook-Amazon). The collection of big data is a real problem as it creates a strong dependency. The loyalty of digital platforms emerged during the participatory process and was reinforced in the parliamentary discussion. It is an example of how the participatory discussion can prepare and improve the discussion inside the Parliament. At the same time, the lobbying was important to avoid the development of common licenses. Some MPs proposed to introduce the definition of the "common goods", which was also supported by some of the associations and NGOs included in the participatory process.[21] The

[21] "Loi numérique: l´assemblée rejette les amendements sur les communs", *L'Humanité*, 21 January 2016 http://www.humanite.fr/loi-numerique-lassemblee-rejette-les-amendements-sur-les-communs-596527.

dependency to the digital corporations was seen as a strong threat against the sovereignty of the countries. The use of data should be done carefully to avoid the collect of information. During the parliamentary discussion, an amendment creating a Committee of Digital Sovereignty was adopted.[22] The idea was to provide the State with the possibility of having its own digital platforms to avoid the transfer of data. In addition to this, the European Court of Justice invalidated in 2015 the transfer of personal data of European citizens to American companies.[23] The bill on Digital Republic questioned this principle as many data were collected by companies. The parliamentary phase insisted more on the aspects of economical intelligence and defence.

At first sight, it is obvious that open data and surveillance contradict each other. But everything depends on the way open data governance is seen. If one of the goals of the digital republic is to encourage a digital growth (Premat 2015), then the use of data by companies can be a source of benefits. The release of big data creates new marketing targets (Mellet 2011) and certainly contributes to the evolution of digital capitalism (Schiller 2009). The law of digital republic seemed to contradict other laws that reinforced the surveillance of data because of terror attacks. In June 2015, the Parliament finally adopted a bill on the modernization of the French Intelligent Services. The role of the government is reinforced with the creation of an independent body whose task is to follow up the new techniques of collecting and sharing information.[24] The bill gave a legal base for new techniques that include the interception of electronic data in case of suspicion of terror attacks. The Constitutional Council censured the proposal of extended international surveillance in its decision 2015-713 DC.[25] The topic was reintroduced in a bill written by MPs in September 2015,[26] that allowed the interception of international communications for security reasons. The adoption of a series of surveillance laws seem to contradict the spirit of the digital Republic. In fact, there is a strong concurrency on the digital strategies of different States. The use and collection of big data is a challenge as it can serve the interests of transnational corporations that can be more powerful than national economies. It is of no surprise that in France like in other European countries, technological surveillance and information are gathered and processed using sophisticated technologies that watch, trace and detect coded behaviours (Raab 2012, p. 17).

---

[22] "Le contresens de la souveraineté numérique", 29 January 2016, http://www.lesechos.fr/idees-debats/cercle/cercle-148468-le-contresens-de-la-souverainete-numerique-1196326.php.

[23] "Les conséquences de l´invalidation de l´accord "Safe Harbor" sur les données personnelles", *Le Monde,* 6 October 2015.

[24] http://www.vie-publique.fr/actualite/panorama/texte-discussion/projet-loi-relatif-au-renseignement.html (Retrieved 8 May 2016).

[25] http://www.conseil-constitutionnel.fr/conseil-constitutionnel/francais/les-decisions/acces-par-date/decisions-depuis-1959/2015/2015-713-dc/communique-de-presse.144139.html (Retrieved 8 May 2016).

[26] http://www.assemblee-nationale.fr/14/propositions/pion3042.asp (Retrieved 10 May 2016).

Conclusion

The bill of Digital Republic revolved around principles regarding the access and the use of new technological tools. The parliamentary discussion revealed a new method to create the law in France. Since this discussion, every bill has to include an aspect concerning digital affairs. Furthermore, this law enshrined the principles of net neutrality and digital oblivion. The participatory process was a first step to prepare a parliamentary discussion. This experience was all the more interesting as the government used it as a new method to associate citizens to the law-making process. The analysis of the profile of contributors reveal that the associations, the NGOs and the digital platforms were the main actors of that process. In other words, this participatory project is useful for the government as it aims to select the appropriate experts that improve the quality of the law. It tends to replace the parliamentary hearings that collected the arguments and the point of view of experts in a short period of time before the parliamentary discussion in committees and in plenary sessions. As a matter of fact, this participatory process looks like a pre-parliamentary discussion without MPs to anticipate the agenda and the main points of the discussion around these questions. It gives some legitimacy for the government to avoid some conflicts in the Parliament even though the parliamentary arena is shortcut to some extent. As a matter of fact, it is possible to see this method as a reform of the representative system. The government controls the agenda and the main questions and uses citizen groups and lobbies to amend and improve the bill draft.

The participatory process is in fact necessary to structure the relation between the active forces (lobbies, NGOs, organizations, associations) and the political power. For the government, the process is mainly used as a capture of expertise and an early connection to the main actors of the digital network. This phase has to be improved to register all the active actors of the sector and see how the law-making process can be shared. Some technical points need to be more visible about the platform and the collection of contributions and the link between votes and contributions. How were the participatory amendments selected? Is it possible to structure the participatory process by implementing a classical participatory tool to select the contributions? A citizen jury, a focus group or a consensus conference can also be set to control the procedure and avoid the governmental communication on it. It is all the more necessary to create a face-to-face tool as the procedure is digital. There is some space to improve the process by establishing a participatory tool to control the application of the law when it is finally adopted. The top-down process can be suspected of legitimizing strong political decisions whereas the main idea is to limit and control the influence of lobbies and organizations. In the meantime, the National Assembly initiated digital consultations of citizens on some questions. These consultations are not quite detailed as contrary to the bill on digital republic, they were not an invitation to amend a legislative text. They were more like a digital survey to communicate on the openness of the National Assembly. It could be more useful to collect some ideas about the possible legislative proposals that the MPs could make. In a nutschell, the possibility of digital citizen amendments is an efficient filter for the work of the representatives and for the discussion between the government and the parliament. At the same time, it should not be a pure communicative tool, this is why the process should be controlled by independent citizen groups.

Recommendations

1. Create a repertoire of all associations, NGOs and lobbies that took part in the process to identify the actors of the sector and the degree of expertise.
2. Do organize thematic meetings with contributors during the participatory process.
3. Create a citizen jury that controls the process and discusses the selection of contributions. Face-to-face interactions are necessary to check the different steps of the participatory process. The capture of expertise should be enlarged so that new points of view can be included.
4. Do not multiply the expertise reports that do not improve the process. The capture of expertise should help to structure the field in an interactive way.
5. Create a parliamentary group work that interacts with the citizen jury to have stronger links between the citizen debate and the parliamentary discussions.
6. Modify the French Constitution to create a right for citizen amendments. As the lobbies and organizations have the possibilities to use parliamentary amendments in order to influence the law, a form of open Parliament should allow citizens to depose some amendments that can be examined by the MPs.
7. Suppress all digital consultations and replace them by a form of law evaluation made by citizens. That could help the MPs to control the application of the legislative process.

Key Terms and Definitions

- Capture of expertise: successful interaction between citizens and groups that give an accurate value to the law-making process.
- Citizen amendments: modifications of the law implemented by citizens and organizations outside the parliament.
- Digital entrepreneurs: people working in digital companies such as start-up companies.
- Digital influencers: people actively promoting the idea of sharing economy.
- Digital oblivion: possibility to erase some data concerning a person or an organization after a certain period of time. This right is granted in order to protect the privacy of citizens.
- Digital republic: definition of a new method to reinforce the digital relations of citizens to republic. It is the title of the bill in France that defined for the first time the idea of net neutrality.
- Digital rights: new rights granted to citizens in terms of access to internet.

- Electronic surveillance: interception of internet communication by political powers.
- Net neutrality: principle that defines the equal access of citizens to internet without any restriction from the service providers.
- Sharing economy: economy promoting a new work organization mainly based on the use of digital tools.

References

Bauman Z (2003) Intimations of postmodernity. Routledge, London

Bauman Z (2005) Liquid life. Polity, Cambridge

Boullier D (2015) Les sciences sociales face aux traces du big data: société, opinion ou vibrations? Revue française de science politique 65(5):805–828

Boussaguet L, Muller P (2005) L´impact du forum politique sur la formulation des politiques publiques. Politiques et management Public 23(3):41–59

Boutinet J-P (2004) Vers une société des agendas, une mutation des temporalités. PUF, Paris

Cannon S, Summers LH (2014) How Uber and the sharing economy can win over regulators. Harvard Business Review, [https://hbr.org/2014/10/how-uber-and-the-sharing-economy-can-win-over-regulators]

Carcassonne G (1984) La résistance de l´Assemblée Nationale à l´affaiblissement de son rôle. Revue française de science politique 34:910–921

Chakravorti B, Tunnard C, Chaturvedi RS (2015) Where the digital economy is moving the fastest. Harvard Review Business, [https://hbr.org/2015/02/where-the-digital-economy-is-moving-the-fastest]

Cheng M (2016) Sharing economy: a review and agenda for future research. Int J Hosp Manag 57:60–70

Fishkin J (1995) The voice of the people, public opinion and democracy. Yale University Press, New Haven

Freeman J, Langbein LI (2000) Regulatory negotiation and the legitimacy benefit. N Y Univ Environ Law J 9:60–151

Freund J (1988) Politique et impolitique. Sirey, Paris

Jobert B (1985) L´État en action. L´apport des politiques publique. Revue française de science politique 35(4):654–682

Lagroye J (1994) Être du métier. Politix 7(28):5–15

March JG (1962) The business firm as a political coalition. J Polit 24(4):662–678

Mellet K (2011) Marketing en ligne. Communications 88:103–111

Nownes AJ (2006) Total lobbying. Cambridge University Press, New York

Official Journal of the European Union, 27 Oct 2011

Pontier J-M (1982) Le rôle des intergroupes au Parlement français. Revue française de science politique 32(4):810–836

Premat C (2015) Smart cities in a digital nation: are Swedish cities enough innovative? In: Ramon Gil-Garcia J, Pardo TA, Nam T (eds) Smarter as the new Urban Agenda: a comprehensive view of the 21st century city. Springer, Cham, pp 207–224

Raab CD (2012) Regulating surveillance: the importance of principles. In: Ball K, Haggerty KD, Lyon D (eds) Routledge handbook of surveillance studies. Routledge, London, pp 377–385

Rodotà S (1999) La démocratie électronique: de nouveaux concepts et expériences politiques. Apogée, Rennes

Schiller D (2009) Digital capitalism: stagnation or contention? (Conference) https://www.youtube.com/watch?v=oQuzo7NfjE0

Schneider V, Janning F (2006) Politikfeldanalyse, Akteure, Diskurse und Netzwerke in der öffentlichen Politik. Verlag für Sozialwissenschaften, Wiesbaden
Sennett R (2006) Culture of the new capitalism. Yale University Press, Yale
Sennett R (2012) Together: the Rituals, pleasures and politics of co-operation. Yale University Press, Yale
Unger RM (1998) Democracy realized, the progressive alternative. Verso, New York
Wu T (2003) Network neutrality, broadband discrimination. J Telecom High Technol Law 2:141–176

Christophe Premat is an Associate Professor in the Department of Romance Studies and Classics at Stockholm University and an Associate Researcher in Political Science at the Institute of Political Studies of Bordeaux (Centre Émile Durkheim). Since 2014, he has served as deputy for the Third constituency for French residents overseas in the Assemblée Nationale in Paris. He is a member of the editorial board of the review *Sens Public*, an international online journal of social sciences. In 2015, he co-edited a handbook on French-German relations, entitled *Handwörterbuch der deutsch-französischen Beziehungen* (Nomos) and in 2014 he published a book entitled *L´idée d´une nouvelle représentation politique* (Edilivre).

Part IV
Tools and Applications

Policy Makers' Perceptions About Social Media Platforms for Civic Engagement in Public Services. An Empirical Research in Spain

Manuel Pedro Rodríguez Bolívar

Abstract Policy informatics and Social Network Analysis are new set of theories seeking to explain how governments are embracing new technologies for improving policy making and relational links with citizens. Under this framework, local governments are now moving to the creation of networks for making public decisions and monitoring public actions. This chapter seeks to identify the perceptions of policy makers in Spanish municipalities about the influence of Web 2.0 technologies on: (a) the relational benefits among members of the networks; (b) the capacities of actors to lead these networks and their influence on the design, construction, and management of public sector services; and (c) the capacities of social networks for improving government legitimacy and accountability. Findings indicate that policy makers do not take advantage of the capabilities that have Web 2.0 technologies to enhance the participation of citizens in public sector management, and they only use social media tools as new channels for unidirectional communication and representation of the local government, which limits the capacity of these technologies for citizen engagement in public policies.

List of Abbreviations

IAB	IAB Spain Research
ICT	Information and communication technologies
IDA	International Development Association
SNSI	Spanish National Statistics Institute

M.P. Rodríguez Bolívar (✉)
Department of Accounting and Finance, Faculty of Business Studies, University of Granada, C/ Campus Universitario de Cartuja s/n, P.O. Box 18071, Granada, Spain
e-mail: manuelp@ugr.es

J.R. Gil-Garcia et al. (eds.), *Policy Analytics, Modelling, and Informatics*, Public Administration and Information Technology 24,
DOI 10.1007/978-3-319-61762-6_12

Introduction

Local governments are now moving to the creation of networks to taking public decisions and monitoring public actions. These networks require new models of governance in which public participation is their central attribute (Munro et al. 2008). These participatory environments are aiming at the search for greater legitimacy, transparency and accountability of public entities. To achieve these goals, many countries have implemented changes in models of public sector management, based on the strategic and intensive use of new information and communication technologies (ICT). Through the advances in information systems, we are seeing an explosion in the number of technology-enabled participatory platforms for civic engagement.

In this regard, policy informatics promotes interactive policy-making by involving the public as co-producers of solutions to public problems (Kim and Johnston 2008). It is an emergent area of study that explores how information and technology can support policy-making and governance (Johnston and Desouza 2015). Indeed, the field of policy informatics is emerging to cross-fertilize between computational sciences and public administration and policy, and to advance the framework of and infrastructural support for public policy-making (Zeng 2015). Policy informatics is already using technological means to enabling informed policy-making processes (Zeng 2015) and fostering transparent, collaborative, and participatory public institutions (Krishnamurthy et al. 2013).

One of the main technological means used by policy informatics is the social media technologies because they provide great potential for policy makers to gauge public opinion (Zeng 2015). This way, in the last years, governments have implemented social media technologies into the governmental workplace as effective tools to engage citizens in public sector management and public goals (Rowe and Frewer 2005). The advent of social media using Web 2.0 technologies[1] has introduced new ways for governments to reach and interact with citizens (Jiang and Xu 2009), allowing citizens greater involvement in public affairs and creating more affordable, participatory and transparent models of public sector management (McMillan et al. 2008).

These new developments put pressure on government organizations to improve democratization, citizens' trust in the government (Chun et al. 2010), governmental legitimacy, governmental efficiency (Peedu 2011), and governmental responsiveness (Yang and Holzer 2006). In fact, Web 2.0 technologies have facilitated closer contact between public managers and citizens, which may produce a higher level of responsiveness in public policy design and delivery (Sørensen 2004). Therefore, Web 2.0 technologies could help information to be efficiently and effectively

[1] In this paper, Web 2.0 should be viewed as a networked platform, spanning connected devices to encourage collaboration, in terms of the creation, organization, linking and sharing of content (O'Reilly 2007). Thus, it is related to the technical platform on which social media applications are built to create and exchange user-generated content.

mobilized to enable evidence-driven policy design, implementation, and analysis, which is a basic premise of policy informatics (Krishnamurthy et al. 2013).

In addition, while the potential impact of social media technologies on the functioning of government is expected to be "profound", it will come with "challenges in the areas of policy development, governance, process design, and conceptions of democratic engagement" (Bertot et al. 2010). Therefore, the analysis of the improvement of the relational links between governments and their constituencies with the use of social media tools could be of high interest into the field of policy informatics, because it could shed some light regarding the potential of these technological tools for citizen engagement in public policies.

Also, social media applications have a great potential analytical value (Dawes and Helbig 2015). In this regard, Social Network Analysis (SNA) puts emphasis on explaining social phenomena using the structural and relational features of the network of actors involved (Wonodi et al. 2012). It is based on the assumption that relationships among interacting units are important and provides an understanding of the goals, characteristics and relative influence of relevant groups and individual actors (Wasserman and Faust 1994). This way, it could be interesting to analyze the role of the governmental actors regarding public service delivery in social networks and the involvement of citizens in the co-production of public services. This knowledge could shed some light regarding the SNA in the co-production of public services.

Despite previous comments, research into the field of policy informatics and SNA remains at an incipient, dispersed stage, lacking a clear literature base or research approach. In addition, little research has been conducted in the field of local governments to analyze the use of Web 2.0 technologies to transform the relational links between citizens and local governments in public service delivery. Also, it is necessary to analyze the capacities that these technologies provide to governmental actors and stakeholders in some relevant tasks such as accountability, transparency or legitimacy.

Therefore, this chapter contributes to the current debate on Web 2.0 technologies and its implication for policy informatics and SNA, aiming at identifying the perceptions of policy makers responsible of e-government in local governments about the influence of Web 2.0 technologies on: (a) the relational benefits of social networks in the public service delivery; (b) the capacities of different actors for leading these networks and their influence on the design, construction, and management of public sector services; and (c) the capacities of social networks for improving government legitimacy and accountability.

This analysis is especially relevant in local governments due to the tradition of greater citizen participation at the local level (Berry et al. 1993) and the tradition of the use of mechanisms to permit direct citizen involvement, in part because they are more manageable at that scale (Peters 2001) and the wide variety of services they provide (Russell and Bobko 1992). Inside local governments, the perception of policy makers responsible of e-government is of great interest taking into account not only their significant role in the policy-making process within local government, but also their direct involvement in the possible implementation of Web 2.0 technologies in public sector delivery.

To achieve this aim, a questionnaire was designed and sent to all policy makers responsible of e-government in large Spanish municipalities (those with a population of over 50,000 inhabitants) in order to capture their perceptions to answer each one of the research questions previously posed in this chapter.

Relational Benefits and Stakeholders' Capacities with the Use of Web 2.0 Technologies for Public Services

Society is becoming increasingly difficult to govern due to the growing complexity and fragmentation of social, political, and economic processes (Edelenbos et al. 2011). Public policies are to be employed on a number of issues which are intertwined, asynchronous, and spatially superposed, which suggests that most objects of public policies can be viewed as complex systems (Furtado et al. 2015), and public managers and elected politicians have ambitions in terms of the quality of public governance and its ability to solve problems (Sørensen and Torfing 2011). Under these conditions of complexity, policy informatics is becoming relevant because more intensive and creative use of information and technology can improve policy-making processes and lead to better policy choices (Dawes and Helbig 2015).

The advent of social media tools using Web 2.0 technologies has allowed the two-way communication and rich data exchange among different actors for purposes of communication to the network, knowledge exchange, and problem solving (Welch 2012). Indeed, these technologies have opened up unprecedented possibilities for engaging the public in government work and has changed public expectations about how government work should be done (Chun et al. 2010). In this regard, Web 2.0 applications will require that governments rethink how they design their content and services (Chang and Kannan 2008).

Opportunities to receive feedback from residents, inform them of government-provided opportunities, and increase engagement with the governance process have all been proposed as ways that social media tools can play a role in Governance 2.0 (Lampe et al. 2011). With social media tools, governments have the potential to be simultaneously more reachable, available, and relevant to users, while offering users more opportunities to become actively engaged in government (Bertot et al. 2012). In fact, social media tools have the potential to increase inclusion in political discourse, accessibility of information, and the ability to form and maintain relationships to strengthen advocacy efforts (Edwards and Hoefer 2010).

This way, the relational links between governments and citizens are being transformed and social media tools are altering the nature of social relations (Li 2011) and changing the nature of political and public dialogue (Osimo 2008). Indeed, evolving Web 2.0 applications will demand a new environment of collaborative culture within government agencies and organizations (Chang and Kannan 2008). Such novel approaches of connecting with citizens through Web 2.0 technologies create conditions for improving transparency and fostering innovation (Meijer and Thaens 2010).

An experiment conducted in four focus groups in early January 2008 in U.S. citizens (groups 1 and 2 for younger people –from 18 to 25– and groups 3 and 4 for elder people –from 30 to 40-), concluded that the more frequent the interaction, the more the desire to participate and change the world, the more likely they were to participate in the social computing engagements (Chang and Kannan 2008). Also, many policy makers feel that capturing the voice of citizens in local matters is very important, and it becomes easier with convenient online access and virtual town hall meetings. Such uses could provide a structured discussion, clear reasoning and asynchronous mode of interaction, and could lead to better-informed decision making (Chang and Kannan 2008).

This means that institutions will have to engage citizens and customers at sites where they are (in social network sites and online communities) rather than create portals and all-purpose websites and expect citizens and customers to approach them (Chang and Kannan 2008). This way, local governments are increasingly embracing Web 2.0 technologies to encourage the use of means of bidirectional communication to change how they interact with stakeholders and to become more efficient in their response to stakeholders' demands, thus providing the greater accountability demanded (Leighninger 2011). In addition, a push towards government co-production of services with citizens has been very clear in behavioral public policy fields, the 'nudge' territory of changing life choices (Thaler and Sunstein 2009), where even more interventionist European governments acknowledge that government-only interventions are unlikely to be successful (Margetts and Dunleavy 2013).

Nonetheless, recent research indicates that the use of the Web 2.0 tools is still very recent, and the information disclosed on them has an essentially ornamental focus. In this regard, it is necessary to increase the content disclosed at the relational level, and especially at the information level (Gandía et al. 2016). Therefore, governments must take a reflection regarding the role that social media tools could play in their organizations. Indeed, social media tools may have the potential to serve as a 'weapon of the weak' and diminish the differences between minor and major players in the interest groups worldwide, in a similar manner as one view that argues that ICT could help cultivate democracy more generally (Larsson 2013). Thus, the following research question is derived:

RQ1. How can Web 2.0 technologies affect the relation between citizens and governments with the aim at improving citizen engagement in providing public services?

In addition, the relational impact of social media tools on the relationship between governments and citizens also make the need to analyze the different role they can play in the governance of the cities and in the co-production of public services. Indeed, the exchange of information in social networking sites modifies participants' attitudes (Soderlund and Rosegren 2007). Citizens feel that they would have "more say" in decision making and thus have more influence in their government's policies and actions (Chang and Kannan 2008). More importantly, Web 2.0 technologies offer a platform that allows citizens' input to be integrated into the decision-making process, and increases transparency by sharing information (Mergel 2013), which offers justifications for the process and consequently increases citizen satisfaction (Verdegem and Verleye 2009).

In a study conducted by van der Graaf et al. (2015), results detect a strong role for interest groups representing citizens and workers who use social media tools in order to be communicated with each other to address the political decision-makers. In this framework, government interactions with citizens will also become less formal in such settings, which may have a negative impact on the power of the government to wield authority (Chang and Kannan 2008). This way, some government agencies that appear to be socially active by using Web 2.0 technologies are often unwilling to fully interact, fearing that their control will be weakened (Brainard and Derrick-Mills 2011).

Therefore, in contrast to state-centric perspectives on policy-making, the network approach assumes that informal decision-making arrangements and the involvement of non-state actors are crucial aspects (Adam and Kriesi 2007). This relevance of actors has made authors to focus their studies on structural patterns between actors and the assumption that relationships among interacting units are important (Wasserman and Faust 1994). This approach, called SNA approach, combines a relational measure of the organizations' centrality with a measure based on collective perceptions of the organization's powers in terms of decision-making (Dörry and Decoville 2013). Thus, under the framework of our study, the following research questions are derived:

RQ2. Would it be expected the governmental actors of their respective local jurisdictions to be the most important decision-makers in the policy network regarding public sector services delivery?

RQ3. Do social networks improve the involvement of stakeholders in the co-production of public services?

Finally, regarding public services, in the Web 2.0 era, users have become important actors in almost all aspects of online services (Huijboom et al. 2009), because they are no longer conceived as 'end-users', as they have moved into the heart of the value chain (Tuomi 2002), and are expected to provide insight and intelligence that will improve public services. Thus, citizens need to play a distinct and more direct role in designing public services with the aim at obtaining more citizen-oriented services, and this has led to a blurred distinction between production and consumption (Valtysson 2010). This has involved a wider process of governing through constructing and reconstructing ideas of the public, community and individual citizen-consumers who take on a role in their own governance (Morison 2010).

Besides, the involvement of governments in the process of implementing social media tools in e-services can improve democratization, citizens' trust in the government (Chun et al. 2010), governmental legitimacy, governmental efficiency (IDA 2011), and governmental responsiveness (Yang and Holzer 2006). In fact, appropriately designed initiatives to engage citizens in their own settings will also enhance the trust that citizens have in their government and will help government build citizen loyalty (Chang and Kannan 2008).

Shkabatur (2011) argued that those who participate in citizen-sourcing may help government obtain legitimacy and political support to adopt new policies or test novel objectives. Even if the final outcomes that participation generates do not represent participants' individual preferences, studies have demonstrated that participants positively evaluate processes in which they are permitted to participate and in which decision-makers actually consider their views (Tyler 2006). This positive view of citizens could be due to the increase of information transparency involved in the use of Web 2.0 technologies. In fact, these technologies can perform a key role in the enhancement of the administrations' transparency, because of the potential synergies that can develop with Web 1.0, through a broader circulation of the contents disclosed in the websites, improving their visibility and fostering a more extensive interaction with citizens (Gandía et al. 2016).

Nowadays, government organizations are using social media tools as a public affairs communication medium to increase transparency by sharing content that citizens are otherwise not aware of or cannot easily access through a government organization's website or another offline media (Rodríguez Bolívar 2015a). This new form of representation can be seen as the lowest degree of online engagement and is oftentimes misinterpreted as true citizen participation (Mergel 2016). So, the final research question of this study is derived:

RQ4. How can Web 2.0 technologies affect government legitimacy and accountability?

Policy Makers Perceptions Regarding Relational Benefits and Stakeholders' Capacities with the Use of Web 2.0 Technologies for Public Services

Sample Selection

Local governments are considered the closest tier (level) of government to citizens (Cegarra Navarro et al. 2012), manage very large budgets and provide a wide variety of services (Saiz 2011), which leads local governments to be a central focus on public sector reforms (Smith 2004). In this context, local government is an important subject for the study of social media tools and interactivity due to the tradition of greater citizen participation at the local level (Berry et al. 1993; Oakerson 1999). Indeed, Peters (2001) argues that local governments tend to use more mechanisms that permit direct citizen involvement, in part because they are more manageable at that scale -see also Briggs (2008) and Sirianni (2009)-. Therefore, social networks are becoming increasingly relevant in the local government context (Gibson 2010) to encourage the use of means of bidirectional communication with the aim at changing how they interact with stakeholders and to become more efficient in their response to stakeholders' demands, thus providing the greater accountability demanded (Leighninger 2011). And the largest cities, which are examined in this paper, have generally been at the forefront in the adoption of e-government innovations (Moon 2002; Ho 2002; Scott 2006).

This paper focuses on Spanish local governments in particular. This approach could be of interest in view of the managerial devolution process implemented in Spain in the 1990s (Gallego and Barzelay 2010) and the rapid introduction of new technologies by these local governments, which has been fostered with the promulgation of legislation in Spain in the last years (for example, the Information Society Services and E-Commerce Act No. 34/2002, the Local Government Modernization Act No. 57/2003, or the Electronic Access to Public Services Act No. 11/2007). Therefore, the various levels of public administration have to develop a wide range of Web-delivered services. In addition, according to recent studies, the e-services provided by local administrations in Spain account for 66% of all public services (Orange Foundation 2014) and the 79% of Internet users in Spain use some type of social network (IAB 2014) mainly as a means to chat with friends or organizations as well as to generate content –this figure is over the mean of European Union (54%)- (Orange Foundation 2014).

In accordance with numerous prior empirical studies of the use of new technologies in local governments (Bonsón et al. 2012), we chose to examine exclusively municipalities with relatively large populations. These were selected because they are usually among the first to adopt new technologies (Bonsón et al. 2012) with the aim at providing efficient services to the public (Cegarra Navarro et al. 2012) and their delivery of services is more complex. In addition, the quantity and variety of services delivered by these administrations are very comparable.

Therefore, the present empirical study is based on a sample of large Spanish municipalities, defined as those with a population of over 50,000 inhabitants, together with those which, although smaller in terms of numbers are classified as "large population" under Article 121 of Local Government Regulatory Act 7/1985, amended by the Local Government Modernization Act 57/2003, i.e. municipalities that are provincial capitals, regional capitals or in which the headquarters of regional institutions are located. In total, 148 Spanish municipalities meet these conditions, and account for over 50% of the total population of Spain (SNSI 2014).

Data were obtained by sending a link to perform an e-survey, and this was sent to the policy makers of all the local authorities studied, via email. The contact details were obtained from the Spanish central government's website. Of the 148 municipalities that comprised the survey sample, seven stated that the municipality had not yet introduced communication channels such as social networks, and thus neither had experience of Web 2.0 nor dedicated human resources to this area. Therefore, the questionnaire was sent to 141 local governments and 46 complete replies were received from policy makers (thus there were 107 incomplete responses to the questionnaire). To date, therefore, the minimum response rate is 32.62%. In order to characterize the municipalities that answered all the questions in the e-survey, this paper begins its analysis with the number of Web 2.0 applications used in sample municipalities. All these official web pages were re-visited to obtain this information (see Table 1). Also a geographic distribution of the respondents can be seen in Fig. 1.

Table 1 Characterization of the sample by population and Web 2.0 applications embraced by the local governments that answered all the e-survey questions

	Number of sample municipalities by population						
Number of municipalities, by population categories	Total number of municipalities in Spain (Dec 2014)	Total number of inhabitants (Dec 2014)	Total sample municipalities	Completed responses	%	Incomplete responses	Accumulated %
Small local governments (fewer than 50,000 inhabitants)	7972	22,304,187	–	–	–	–	–
Medium-sized (50,000–249,000 inhabitants)	129	13,624,612	124	40	32.26%	8	38.71%
Large-sized (more than 249,000 inhabitants)	16	10,842,542	15	5	33.33%	1	40%

	Web 2.0 applications embraced by the local governments that answered all the e-survey questions					
	Mean	Median	Mode	Maximum	Minimum	Standard deviation
Number of inhabitants of sample municipalities	131,141	88,7	99,32	442,203	51,719	89,538.77
Number of web 2.0 applications used in sample municipalities, by population categories						
Medium-sized (50,000–249,000 inhabitants)	3	3	3	8	1	1.68
Large-sized (more than 249,000 inhabitants)	2	2	2	3	1	1.00
Number of web 2.0 applications used in sample municipalities (Total)	3	3	3	8	1	1.64

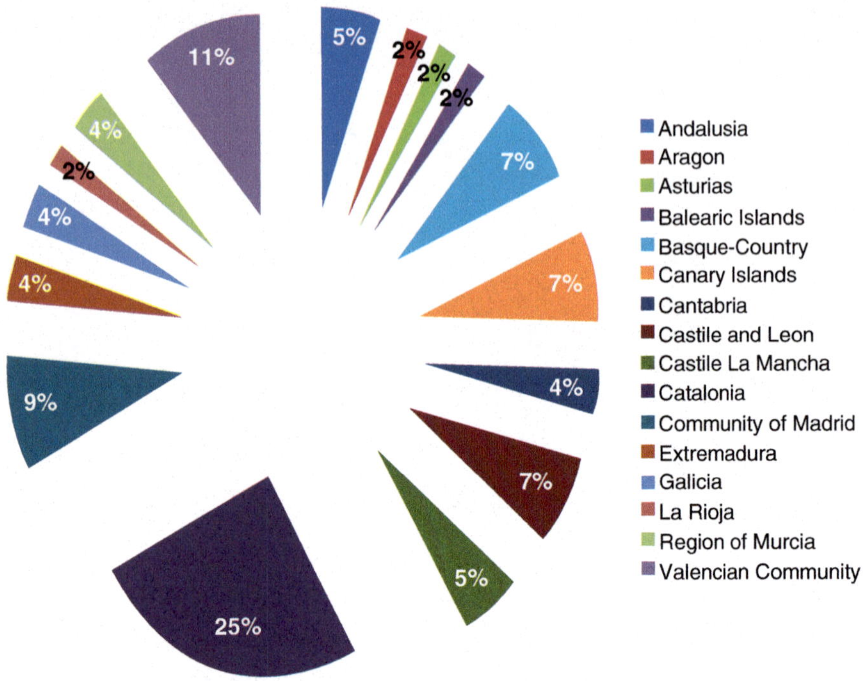

Fig. 1 Geographic distribution of the respondents

Nonetheless, some policy makers of local governments have responded some items without finishing the full e-survey. In consequence, for some questionnaire items, the response rate exceeded the above-mentioned minimum (see Table 2). This sample size is reasonable; according to Roscoe (1975), a sample size between 30 and 500 is considered satisfactory. Data were compiled over the research period utilizing an appropriate sampling technique.

Methodology of Research

To achieve the study aims and address the research questions established, a questionnaire was designed and sent to all policy makers in the municipalities in question in order to capture their perceptions on the issues that are analyzed in this paper. The questionnaire was made up of 19 questions covering the relational benefits (7 items), role of governmental actors (3 items), involvement of stakeholders in the co-production of public services (4 items) and the improvement of legitimacy and accountability with the implementation of Web 2.0 technologies (5 items) (see Table 2).

Policy makers were addressed in this survey taking into account not only their significant role in the policy-making process within local government, but also their direct involvement in the possible implementation of Web 2.0 technologies in public

Table 2 Results for items related to relational benefits and using Web 2.0 technologies in providing public services

Questionnaire	Frequency	Response rate	Median	Mean	Mode	Standard deviation	Max	Min
RQ1. Relational links between governmental and no-governmental actors under web 2.0 technologies in providing public services								
1.1. Web 2.0 technologies promote the creation of debates about public affairs in real time.	54	38.30%	4.00	3.72	5	1.12	5	1
1.2. Web 2.0 technologies enable the access to the information about public services.	54	38.30%	5.00	4.46	5	0.79	5	1
1.3. Web 2.0 technologies introduce tools for a better evaluation of the efficiency in the delivery of public services.	47	33.33%	4.00	3.74	4	0.99	5	1
1.4. Web 2.0 technologies allow the contribution of suggestions for improving the quality of public services.	53	37.59%	4.00	3.96	5	1.14	5	1
1.5. Web 2.0 technologies improve the capacity of the local government to be communicated with non-governmental actors.	55	39.01%	5.00	4.44	5	0.79	5	1
1.6. Web 2.0 technologies contribute to a better social cohesion and development of cities.	48	34.04%	4.00	3.79	4	0.90	5	2
1.7. Web 2.0 technologies foster the effective collaboration between citizenry and governments in the delivery of public services.	54	38.30%	4.00	3.67	4	0.95	5	1
RQ2. The role of governmental actors in the policy network regarding public services delivery								
2.1. Local governments must always play the lead role in the implementation of web 2.0 technologies regarding the delivery of public services.	46	32.62%	5.00	4.42	5	1.43	5	1
2.2. Local governments should designate a senior official responsible for the coordination of their web 2.0 activities.	48	34.04%	5.00	4.35	5	1.08	5	1
2.3. The implementation of web 2.0 in providing public services gives an essential role to users of these services.	54	38.30%	4.00	3.72	4	1.00	5	1
RQ3. The involvement of stakeholders in co-production of public services								
3.1. Social networks foster the co-production of public services.	53	37.59%	4.00	3.38	4	1.06	5	1
3.2. Citizens may participate in the generation of content and information.	53	37.59%	3.00	3.15	4	1.20	5	1

(continued)

Table 2 (continued)

Questionnaire	Frequency	Response rate	Median	Mean	Mode	Standard deviation	Max	Min
3.3. The local government opens up a problem or activity for resolution or co-execution by citizens in order to tap into the unique skills, talents, and knowledge of the population.	54	38.30%	3.00	3.13	4	1.20	5	1
3.4. Web 2.0 technologies gives capacity citizens to transform public services.	54	38.30%	4.00	3.98	4	0.92	5	2
RQ4. The improvement of legitimacy and accountability under web 2.0 technologies in public services delivery								
4.1. Policymakers need to use web 2.0 technologies for legitimacy purposes.	46	32.62%	4.00	3.83	4	1.08	5	1
4.2. Local governments use web 2.0 technologies with the aim at disclosing information for accountability purposes.	48	34.04%	4.00	3.92	4	0.85	5	2
4.3. The use of web 2.0 technologies allows the achievement of greater credibility for local governments.	48	34.04%	4.00	3.71	4	0.90	5	2
4.4. Enables a greater openness of information.	48	34.04%	4.50	4.31	5	0.88	5	1
4.5. Web 2.0 technologies contribute to changing the government's image, making it more friendly.	46	32.62%	4.00	3.98	4	0.95	5	1

sector delivery. Before the e-survey was sent out, every policy maker in the sample population was contacted and asked to participate in the study, after being informed of the study goals and of what was required by the questionnaire. They were also assured of its strictly scientific and confidential nature, and of the global, anonymous treatment of the data to be obtained.

Two draft versions of the survey were pre-tested on a selected group of stakeholders, to refine the design of the questionnaire items. First, the research team drafted a preliminary version based on the conclusions of previous work in the field of Web 2.0 technologies (Picazo-Vela et al. 2012; Gomes and Sousa 2012; Linders 2012). Based on this analysis, 19 items have been selected to analyze the issues involved in the objective of this chapter (see Table 2). These items were driven to answer the research questions previously posed (7 items for RQ1, 3 items for RQ2, 4 items for RQ3 and 5 items for RQ4). Second, the initial text was presented to two specialists on Web 2.0 technologies and to ten policy makers, to ascertain their opinions on: (a) the understandability of the questionnaire; (b) the clarity of the questions posed and possible ambiguities; (c) the possible inclusion of other questions relevant to the study aims. The comments and suggestions made were analyzed and, when considered appropriate, incorporated into the text of the questionnaire. The following changes were made in order to improve the understandability and clarity of the questionnaire: (a) lengthier textual description was given for some items; (b) some terms were replaced by others with a more specific meaning; (c) some alterations were made to the word order.

Then, the link to the second version of the questionnaire was provided to the policy makers of each local government in our sample. The introductory letter to the policy makers in each organization stressed the need for them, before answering the questions, to be quite certain of having understood the meaning of each questionnaire item and the ultimate goals of this study. Also, they were offered the possibility of clarifying any remaining doubts before completing the questionnaire. Thus, some e-mails were received concerning the exact meaning of some items; these questions were answered, and thus we may be reasonably sure that the questions measured the intended constructs.

Based on prior studies on attitude analysis (Collison et al. 2003; Emerson et al. 2007), the questionnaire was designed in which respondents were asked to describe their degree of agreement with each statement on a five-point Likert scale (ranging from strongly disagree, "1" to strongly agree, "5"). Although the Likert scale has some limitations, such as its inability to approximate intervals of ordinal data (Russell and Bobko 1992), and its closed response format (Hodge and Gillespie 2003), these limitations do not invalidate conclusions about the numbers (Norman 2010) and Likert scale was used in this research due to its suitability for attitude studies. In this regards, the measures are simple to administer, quantify and code (Spector 1992), and the results obtained have proven to be reliable and valid (Li 2013). In addition, Likert scale allows obtaining numerical measurement results directly used for statistical inference and has been proved to be "robust" when this scale is used for parametric statistics (Norman 2010). In brief, the Likert scale is one of the most popular rating scales for measuring attitudes (Dunn-Rankin et al. 2004), which is the key aspect addressed in the present study. Therefore, a five-point Likert scale has been used in our research to capture the attitudes of policy makers regarding the implementation of Web 2.0 in public services.

After the questionnaire was completed, each item was analyzed separately. Unlike in other methods, in Likert scaling each statement becomes a scale in itself and a person's reaction to each statement is given a score (Likert 1932). Nonetheless, it cannot be presumed that participants perceive the difference between adjacent levels to be equal (Bertram 2007) and, therefore, data obtained from such responses could not be analyzed using the mean to comparing results between questions due to scale problems. In this milieu, the analysis of the central tendency summarized by median and the mode of the responses has been proved to be useful in order to analyze data obtained using Likert scale (Bertram 2007).

Analysis of Results

RQ1: How can Web 2.0 technologies affect the relation between citizens and governments with the aim at improving citizen engagement in providing public services?

According to Table 2, policy makers indicate that the main relational benefits of Web 2.0 technologies are those related to the improvement of the access to the information and to the improvement of the communication with non-governmental actors (median scores 5.00 and media scores 4.46 and 4.44 respectively). In addition, respondents indicate that Web 2.0 technologies could help them to foster effective collaboration with citizens in the delivery of public services but this is the lowest score item in the survey (median score 4.00 and media score 3.67). Therefore, these results seem to confirm prior research that indicate that Web 2.0 technologies are seen as a new form of representation for governments (Rodríguez Bolívar 2015a; Mergel 2016).

In addition, policy makers indicate that Web 2.0 technologies could help them to obtain suggestions from the public regarding the delivery of public services (see median score 4.00 and media score 3.96) and they also think that social media tools could help them to achieve a better social cohesion and development of cities (median score 4.00 and media score 3.79). Therefore, respondents think that social media tools could have the capacity for a better communication with the citizenry in order to integrate them into the Local governments' decisions. Nonetheless, this involvement of citizenry does not seem to be effective. Only information collection from them seem to be valuable for policy makers.

Finally, the least scored items are those referred to the creation of debates about public affairs in real time (median score 4.00 and media score 3.72), the effective collaboration between citizenry and governments in the delivery of public services (median score 4.00 and media score 3.67), and the introduction of tools for a better evaluation of the efficiency in the delivery of public services (median score 4.00 and media score 3.74). These results confirm that policy makers are not willing to fully interact with the citizenry because they are not prone to introduce social media tools for improving e-democracy or accountability.

This way, respondents indicate that they are not sure of involving citizens into the public decision-making process regarding public affairs or regarding the design and execution of public services. They think that they are the leaders in the decisions in these areas. Also, they are afraid regarding the use of social media tools as a means to be accountable for the public. As Web 2.0 technologies could require government organizations to give up significant control over content and applications or over the way in which communications and relationships with stakeholders are handled (Graells-Costa 2011), policy makers could perceive the higher participation of citizens as the source of additional "noise".

RQ2. Would it be expected the governmental actors of their respective local jurisdictions to be the most important decision-makers in the policy network regarding public sector services delivery?

The results indicate that policy makers think that local government must play a leading role in the implementation and management of the Web 2.0 technologies in providing public services (see median and media scores for items 2.1 and 2.2). According to the results, local government should designate public officials for the management and coordination of the flow of communication with citizen through social media tools. This result is in the line of the results obtained previously for RQ1 of this study because it means the intention of policy makers to be the leaders of the network created for providing public services.

In addition, although relevant, policy makers think that the implementation of Web 2.0 in providing public services does not give an essential role to users of these services (see median and media scores for item 2.3). This result confirms that the policy makers' perception is about using the collective intelligence of citizenry to collect ideas regarding the design and production of public services but they are not prone to give up control regarding the management and monitoring of these services.

RQ3. Do social networks improve the involvement of stakeholders in the co-production of public services?

Results of this section of the study are very clear. According to the respondents, web 2.0 technologies have the capacity to make citizens principal actors for transforming public services (see median and media score of item 3.4). Nonetheless, they think that citizens must not participate actively in the generation, co-production or co-execution of public services (see median and media scores of item 3.1, 3.2 and 3.3).

This result is striking because web 2.0 technologies should be viewed as a networked platform, spanning connected devices to encourage collaboration, in terms of the creation, organization, linking and sharing of content (Chang and Kannan 2008). Therefore, sample policy makers are not prone to engage citizens in the management and co-production of services. Perhaps, they think that they cannot monitor or play a key role if this possibility is opened.

RQ4. How can Web 2.0 technologies affect government legitimacy and accountability?

Regarding the legitimacy and accountability issues derived from the use of Web 2.0 technologies, the most scored items are those referred to the legitimacy of the local government (see median and media scores of item 4.1, 4.4. and 4.5). The openness of information is the most scored item (median score 4.50 and media score 4.31) and it reflects the intention of policy makers to use Web 2.0 technologies as a unidirectional communication channel. It seems that Web 2.0 technologies are thought to be used as a unidirectional portal of information where citizens can access to have knowledge regarding public services but it is not addressed to foster citizen collaboration or engagement in public policies.

On the other hand, accountability purposes are also highly-scored (median score 4.00 and media score 3.92), but it seems that it is not a preference for policy makers due to the results obtained in item 1.3 of the survey.

Conclusion and Discussions

The popular maxim that "power comes with responsibility" would suggest new responsibilities for the newly empowered citizen (Linders 2011). But this reworking of the social contract is not without controversy. Policy informatics and SNA are arising as new theories that seek to explain why governments are embracing new technologies as well as the relational effects that the implementations of these technologies could have in the relation with citizens.

This chapter contributes to the current debate on Web 2.0 technologies and its implication for policy informatics and SNA with the aim at identifying the perceptions of policy makers responsible of e-government in local governments about the influence of Web 2.0 technologies on the relational benefits, on the capacities of actors for leading social networks and on the design, construction, and management of public sector services, and on the capacities of social networks for improving government legitimacy and accountability. A main limitation of our research is the limited number of responses (32.62% as the minimum response rate). Thus, our findings must be interpreted taking into account the level of responses obtained. In this regard, future research should increase the number of responses to the questionnaire to obtain richer data with the aim at undertaking wider analysis of the issue analyzed. Nonetheless, the methodology of research used in this paper provides valid and robust results.

Findings indicate that policy makers do not take advantage of the capabilities that have Web 2.0 technologies to enhance the participation of citizens in public sector management. They only seem to advocate the introduction of social media tools as new channels for unidirectional communication and representation of the local government, which limits the capacity of these new technologies for citizen engagement in public policies and, by this way, in policy informatics. They seem to be unwilling to fully interact with citizens, fearing that their control will be weakened (Brainard and Derrick-Mills 2011). Perhaps this is the result of the current inexperience of local governments in Spain in managing social media tools, in providing public sector services with Web

2.0 technologies and in the way of interaction with individuals through these technologies (Meijer et al. 2012). Indeed, experience has been shown to be a highly significant factor for networking and network management (Edelenbos et al. 2011) and, in Spain, we are viewers of the early stage in the development and implementation of social media tools into governments.

Findings also indicate that policy makers think that Web 2.0 technologies could work as a crowd-brainstorming that could help governments to improve the quality of public services and to meet better the citizens' expectations regarding public services. Policy makers think that social media tools could be used for collecting ideas from the public that will improve public services. Also, with the use of Web 2.0 technologies, governments could know better the needs of citizens and focus on those operations that create public value for the citizenry, which could make governments more efficient. This could help local governments to achieve the goal of financial sustainability.

Nonetheless, local governments should also foster an improvement of their relational links with citizens allowing the creation of debates regarding public services and public policies. In brief, they should improve citizen engagement in public policies with the use of social media tools (policy informatics), but it does not seem to be a preference of the survey respondents. In addition, governments should implement social media tools to fully interact with citizens who should be actively involved in the process of design, co-production and execution of public services. Again, respondents are not prone to introduce these social media strategies into their local governments.

Perhaps, the existence of a clear regulatory framework for the activities related to social networks or the establishment of a process to combat unauthorized or fraudulent postings could mitigate the perceived risk of the inappropriate use of social networks by stakeholders (Rodríguez Bolívar 2015). So, it could make policy makers to be more prone to the effective involvement of citizens in the co-production of public services. Therefore, regulation about the use of these technologies could be a relevant aspect to solve these problems. This way, future research in policy informatics could analyze if policy makers in countries with legal environments regarding the use of social media tools in the interaction with governments are more prone to use these technologies to debate public policies with the citizenry.

Also, the training of employees regarding the use and monitor of Web 2.0 technologies could be another main aspect to improve citizen engagement. The training of public officials is especially relevant because findings indicate that, nowadays, policy makers perceive the need to keep the leading role in the implementation and management of Web 2.0 technologies in providing public services. Besides, the hiring of community managers that actively work with the citizenry in order to collect thinking and feelings of the citizenry regarding public services could help local government to foster effective collaboration of the citizenry in providing public services.

According to our findings, local governments want to play the role of commissioner (executor) rather than that of co-producer or facilitator. This result confirms prior research that indicates that there is currently little evidence to support claims that e-government 2.0 has radically changed government (Dixon 2010). Perhaps, the institutional change needs another time-scale different from that needed by technological changes. So, it is possible that this perception could be changed in the future when policy makers have experience in using Web 2.0 technologies. Therefore, future research could replicate our study in order to compare institutional changes provoked by the technological changes introduced by Web 2.0 technologies in local governments.

In addition, recent research has indicated that social media tools have their own logic, but it is only manifest when it encounters fertile ground within a government bureaucracy (Meijer and Thaens 2013). In fact, ICT has been sometimes used to support and enable bureaucratic practices in favor of government reforms and service delivery improvements (Cordella and Tempini 2015). In this milieu, governments that strictly manage public services under the Web 2.0 era only seems to use these new technologies with the aim at improving transparency in public services but not at obtaining citizen participation in the design, solution of problems or for monitoring purposes in public services (Rodríguez Bolívar 2015).

Another main finding of our study indicates that policy makers are not prone to engage citizens in the generation of content and information, co-production or co-execution of public services. In brief, the technology has not yet changed the interactions considerably. Citizens that are involved in the co-production of public services will probably make greater demands for some form of reward for their involvement, making it necessary to ensure the perception of user-added value and greater control by citizens of the e-services process, as has been achieved in the private sector (Wirtz et al. 2013). Therefore, this finding of our study probably shows the fear of policy makers to give up control of the public services if citizens get a key role in the decision-making process of public services, perhaps due to the lack of training of public officials to monitor this process or because policy makers have not yet figured out how to best use the technology to engage citizens in more effective ways.

Finally, prior research has demonstrated that Web 2.0 technologies hold an enormous potential to enhance the effectiveness and legitimacy of government and, therefore, Government 2.0 is presented as the appropriate reaction to changes in society (Meijer et al. 2012). Indeed, a municipal presence on social networks may convey the message that government is more responsive, open, and democratic, by allowing citizens to express their views via this channel (Hibbing and Theiss-Morse 2002). Our findings confirm these assertions and clearly indicate that policy makers perceive Web 2.0 technologies as a means to enhance the government's image and to acquire greater legitimacy for its actions.

In conclusion, local governments must make greater efforts to improve their relational strategies regarding the use of Web 2.0 technologies in providing public services. Else, these technologies will only play the role of another unidirectional communication channel without capacities for improving e-democracy, public accountability or public services. Are the findings of this paper the results of the early stage of the implementation of Web 2.0 technologies?; Are these findings the result of the no capacity of governments to monitor and control the activity of social media tools?. These questions remain without answer and could be the aim of future research.

Acknowledgments This research was carried out with financial support from the Regional Government of Andalusia (Spain), Department of Innovation, Science and Enterprise (Research project number P11-SEJ-7700).

References

Adam S, Kriesi H (2007) The network approach. Theories Policy Process 2:189–220
Berry J, Portney K, Thomson K (1993) The rebirth of urban democracy. Brookings Institution Press, Washington
Bertot JC, Jaeger PT, Hansen D (2012) The impact of polices on government social media usage: issues, challenges, and recommendations. Gov Inf Q 29(1):30–40
Bertot JC, Jaeger PT, Munson S, Glaisyer T (2010) Social media technology and government transparency. Computer 11:53–59
Bertram D (2007) Likert scales. Retrieved June 28, 2014, from the University of Calgary, Department of Computer Science Web Site: http://poincare.matf.bg.ac.rs/~kristina/topic-dane-likert.pdf
Bonsón E, Torres L, Royo S, Flores F (2012) Local e-government 2.0: social media and corporate transparency in municipalities. Gov Inf Q 29(2):123–132
Brainard LA, Derrick-Mills T (2011) Electronic commons, community policing, and communication. Adm Theory Prax 33(3):383–410
Briggs XS (2008) Democracy as problem solving: civic capacity in communities across the globe. MIT Press, Cambridge
Cegarra Navarro JG, Córdoba Pachón JR, Moreno Cegarra JL (2012) E-government and citizen's engagement with local affairs through e-websites: the case of Spanish municipalities. Int J Inf Manag 32(5):469–478
Chang AM, Kannan PK (2008) Leveraging web 2.0 in government. IBM Center for the Business of Government, Washington
Chun S, Shulman S, Sandoval-Almazan R, Hovy E (2010) Government 2.0: marking connections between citizens, data and government. Inf Polity 15(1):1–9
Collison D, Lorraine N, Power D (2003) An exploration of corporate attitudes to the significance of environmental information for stakeholders. Corp Soc Responsib Environ Manag 19(4):199–211
Cordella A, Tempini N (2015) E-government and organizational change: reappraising the role of ICT and bureaucracy in public service delivery. Gov Inf Q 32(3):279–286
Dawes SS, Helbig N (2015) The value and limits of government information resources for policy informatics. Governance in the Information Era: Theory and Practice of Policy Informatics, 25
Dixon BE (2010) Towards e-government 2.0: an assessment of where e-government 2.0 is and where it is headed. Public Adm Manag 15(2):418–454
Dörry S, Decoville A (2013) Governance and transportation policy networks in the cross-border metropolitan region of Luxembourg: a social network analysis. European Urban and Regional Studies, 0969776413490528
Dunn-Rankin P, Knezek GA, Wallace SR, Zhang S (2004) Scaling methods. Lawrence Erlbaum Associates, Mahwah
Edelenbos J, Klijn EH, Steijn B (2011) Managers in governance networks: how to reach good outcomes? Int Public Manag J 14(4):420–444
Edwards HR, Hoefer R (2010) Are social work advocacy groups using web 2.0 effectively? J Policy Pract 9(3–4):220–239
Emerson TLN, Conroy SJ, Stanley W (2007) Ethical attitudes of accountants: recent evidence from a practitioners' survey. J Bus Ethics 71(1):73–87
Furtado BA, Sakowski PAM, Tóvolli MH (2015). A complexity approach for public policies. Discussion Paper, Institute for Applied Economic Research, Brasilia

Gallego R, Barzelay M (2010) Public management policymaking in Spain: the politics of legislative reform of administrative structure, 1991–1997. Governance 23(2):277–296
Gandía JL, Marrahí L, Huguet D (2016) Digital transparency and web 2.0 in Spanish city councils. Gov Inf Q 33(1):28–39
Gibson A (2010) Local by social. How local authorities can use social media to achieve more for less. NESTA, London
Gomes R, Sousa L (2012) Contributions to the development of local e-government 2.0. Future Internet 4(4):882–899
Graells-Costa J (2011) Administración colaborativa y en red. El profesional de la información 20(3):345–347
Hibbing JR, Theiss-Morse E (2002) Stealth democracy: Americans' beliefs about how government should work. Cambridge University Press, Cambridge
Ho ATK (2002) Reinventing local governments and the e-government initiative. Public Adm Rev 62(4):434–444
Hodge DR, Gillespie D (2003) Phrase completions: an alternative to Likert scales. Soc Work Res 27(1):45–55
Huijboom N, Van den Broek T, Frissen V, Kool L, Kotterink B, Nielsen M, Millard J (2009) Public Services 2.0: The Impact of Social Computing on Public Services. Institute for Prospective Technological Studies, Joint Research Centre, European Commission, Office for Official Publications of the European Communities, Luxembourg
IAB Spain Research (IAB) (2014) V Estudio Anual de Redes Sociales. IAB, Madrid
International Development Association (IDA) (2011) Information and communication technology agency of Sri Lanka (ICTA), vol 3. Colombo-05, Sri Lanka
Jiang M, Xu H (2009) Exploring online structures on Chinese government portals: citizen political participation and government legitimation. Soc Sci Comput Rev 27(2):174–195
Johnston EW, Desouza KC (eds) (2015) Governance in the information era: theory and practice of policy informatics. Taylor & Francis, Routledge, New York
KimY, Johnston E (2008) Policy informatics v1. 0. In: Proceedings of the Minnowbrook 3 Conference
Krishnamurthy R, Bhagwatwar A, Johnston EW, Desouza KC (2013) A glimpse into policy informatics: the case of participatory platforms that generate synthetic empathy. Commun Assoc Inf Syst 33(21):365
Lampe C, LaRose R, Steinfield C, DeMaagd K (2011) Inherent barriers to the use of social media for public policy informatics. Innov J Public Sect Innov J 16(1):1–17
Larsson AO (2013) Bringing it all back home? Social media practices by Swedish municipalities. Eur J Commun 28(6):681–695
Leighninger M (2011) Using online tools to engage – and be engaged by – the public. IBM Center for The Business of Government, IBM Center for The Business of Government, Washington
Li D (2011) Online social network acceptance: a social perspective. Internet Res 21(5):562–580
Li Q (2013) A novel Likert scale based on fuzzy sets theory. Expert Syst Appl 40(5):1609–1618
Likert R (1932) A technique for the measurement of attitudes. Arch Psychol 140:1–55
Linders D (2011) We-government: an anatomy of citizen co-production in the information age. In: Proceedings of the 12th annual international conference on digital government research, DG.O 2011, College Park, MD, USA
Linders D (2012) From e-government to we-government: defining a typology for citizen co-production in the age of social media. Gov Inf Q 29(4):446–454
Margetts H, Dunleavy P (2013) The second wave of digital-era governance: a quasi-paradigm for government on the web. Philos Trans R Soc A 371:20120382
McMillan P, Medd A, Hughes P (2008) Change the world or the world will change you: the future of collaborative government and web 2.0, Deloitte Touche Tohmatsu www.deloitte.com
Meijer A, Thaens M (2010) Alignment 2.0: strategic use of new internet technologies in government. Gov Inf Q 27(2):113–121
Meijer A, Thaens M (2013) Social media strategies: understanding the differences between north American police departments. Gov Inf Q 30(4):343–350

Meijer AJ, Koops BJ, Pieterson W, Overman S, Tije S (2012) Government 2.0: key challenges to its realization. Electron J e-Gov 10(1):59–69
Mergel I (2013) Social media adoption and resulting tactics in the US federal government. Gov Inf Q 30(2):123–130
Mergel I (2016) Social media institutionalization in the US federal government. Gov Inf Q 33(1):142–148
Moon MJ (2002) The evolution of e-government among municipalities: rhetoric or reality? Public Adm Rev 62(4):424–433
Morison J (2010) Gov 2.0: towards a user generated state? Mod Law Rev 73(4):551–577
Munro HA, Roberts M, Skelcher C (2008) Partnership governance and democratic effectiveness community leaders and public managers as dual intermediaries. Public Policy Adm 23(1):61–79
Norman G (2010) Likert scales, levels of measurement and the "laws" of statistics. Adv Health Sci Educ 15(5):625–632
O'Reilly T (2007) What is web 2.0: design patterns and business models for the next generation of software. Commun Strateg 65:18–37
Oakerson RJ (1999) Governing local public economies: creating the civic metropolis. ICS Press, Richmond
Orange Foundation (2014) eEspaña. Informe anual 2014 sobre el desarrollo de la sociedad de la información en España. Fundación Orange, Madrid
Osimo D (2008) Web 2.0 in Government: Why? and How? Institute for Prospective Technological Studies, Joint Research Centre, European Commission. Luxembourg: Office for Official Publications of the European Communities
Peedu G (2011) Enhancing public service user experience in information society. Master Thesis. Tallinn University, Estonia
Peters BG (2001) The future of governing. University Press of Kansas, Lawrence
Picazo-Vela S, Gutiérrez-Martínez I, Luna-Reyes LF (2012) Understanding risks, benefits, and strategic alternatives of social media applications in the public sector. Gov Inf Q 29(4):504–511
Rodríguez Bolívar MP (2015) Policy makers' perceptions on the transformational effect of web 2.0 technologies on public services delivery. Electron Commerce Res, 1–28
Roscoe JT (1975) Fundamental research statistics for the behavioural sciences, 2nd edn. Holt, Rinehart and Winston, New York
Rowe G, Frewer L (2005) A typology of public engagement mechanisms. Sci Technol Hum Values 30(2):251–290
Russell CJ, Bobko P (1992) Moderated regression analysis and Likert scales: too coarse for comfort. J Appl Psychol 77(3):336–342
Saiz MP (2011) La Ley de Economía Sostenible: La sostenibilidad financiera del Sector Público. Revista de Contabilidad y dirección 13:21–42
Scott JK (2006) "E" the people: do U.S. municipal government web sites support public involvement? Public Adm Rev 66(3):341–353
Shkabatur J (2011) Cities @ crossroads: digital technology and local democracy in America. Brooklyn Law Rev 76(4):1413
Sirianni C (2009) Investing in democracy: engaging citizens in collaborative governance. Brookings Institution Press, Washington
Smith KA (2004) Voluntary reporting performance measures to the public: a test of accounting reports from U.S. cities. Int Public Manag J 7(1):19–48
Soderlund M, Rosegren S (2007) Receiving word-of-mouth from the service customer: an emotion-based effectiveness assessment. J Retail Consum Serv 14(2):123–136
Sørensen E (2004) Democratic governance and the role of public administrators. In: Bogason P, Kensen S, Miller HT (eds) Tampering with tradition: the unrealized authority of democratic agency, pp 107–130
Sørensen E, Torfing J (2011) Enhancing collaborative innovation in the public sector. Adm Soc 43(8):842–868
Spanish National Statistics Institute (SNSI) (2014) Internet document available at: http://www.ine.es/inebmenu/mnu_padron.htm. Accessed 1 June 2014

Spector PE (1992) Summated rating scale construction: an introduction. Sage, Newbury Park
Thaler M, Sunstein C (2009) Nudge: improving decisions about health, wealth and happiness. Penguin, London
Tuomi I (2002) Theory of innovation: change and meaning in the age of internet. Oxford University Press, Oxford
Tyler TR (2006) Why people obey the law. Princeton University Press, Princeton
Valtysson B (2010) Access culture: web 2.0 and cultural participation. Int J Cult Policy 16(2):200–214
van der Graaf A, Otjes S, Rasmussen A (2015) Weapon of the weak? The social media landscape of interest groups Eur J Commun, 0267323115612210
Verdegem P, Verleye G (2009) User-centered E-government in practice: a comprehensive model for measuring user satisfaction. Gov Inf Q 26(3):487–497
Wasserman S, Faust K (1994) Social network analysis: methods and applications, vol 8. Cambridge University Press, Cambridge
Welch EW (2012) The rise of participative technologies in government. In: Transformational government through eGov practice: Socioeconomic, cultural, and technological issues, pp 347–367
Wirtz BW, Piehler R, Ullrich S (2013) Determinants of social media website attractiveness. J Electron Commerce Res 14(1):11–33
Wonodi CB, Privor-Dumm L, Aina M, Pate AM, Reis R, Gadhoke P, Levine OS (2012) Using social network analysis to examine the decision-making process on new vaccine introduction in Nigeria. Health Policy Plan 27(suppl 2):ii27–ii38
Yang KF, Holzer M (2006) The performance-trust link: implications for performance measurement. Public Adm Rev 66(1):114–126
Zeng D (2015) Policy informatics for smart policy-making. IEEE Intell Syst 30(6):2–3

Manuel Pedro Rodríguez Bolívar is Professor at the University of Granada. He has authored numerous articles in international journals, among them we can highlight *Public Money & Management, Government Information Quarterly, Public Administration and Development, Online Information Review, International Review of Administrative Sciences, American Review of Public Administration, ABACUS, International Public Management Journal, Environmental Education Research, Transylvanian Review of Administrative Sciences, Journal of Environmental Policy & Planning, Lex Localis, Journal of Environmental Planning and Management, Information Technology for Development, Electronic Commerce Research, Internet Research* or *Administration & Society*. He has been also the author of several book chapters published in Routledge, Kluwer Academic Publishers, Springer, Nova Publishers and IGI Global, and is author of full-length books published in Spain. He is also editor of books in Springer, IGI Global and Palgrave, member of the Editorial Board of Government Information Quarterly and Associate Editor in other leading international journals.

Policy Analytics Tool to Identify Gaps in Environmental Governance

Julia A. Ekstrom, Gloria T. Lau, and Kincho H. Law

Abstract Persistent and emerging social and environmental issues require new approaches and tools to help develop policies that address the complexities inherent in these problems. Here a policy informatics tool, developed iteratively with the feedback of ocean and coastal domain experts, is presented. The tool relies on two user inputs: a conceptually modelled ecosystem and a compilation of policy documents that covers a given jurisdiction of interest. With these inputs the user can explore what ecosystem components and linkages are potentially acknowledged in policy and which are not. The linkage acknowledgement is based on a co-occurrence of key terms. When viewing highly complex ecosystems (or other systems), the tool offers an efficient way to systematically identify potential gaps in policy and where relevant policies do exist that could be leveraged to cover emerging environmental issues. While the tool is simple in its design, the development with the potential user-community (policy makers, government staff, and ecosystem scientists) demonstrates the usefulness even in such a modest form and the desire for creative, but transparent and accessible, policy analytic tools.

Abbreviations

MINOE	Management Identification for the Needs of Ocean Ecosystems
EBM	Ecosystem-based management
DFO	Department of Fisheries and Oceans
DPSIR	Driver-Pressure-State-Impact-Response
ICES	International Council for Exploration of the Sea

J.A. Ekstrom (✉)
Policy Institute for Energy, Environment and the Economy, University of California, Davis, One Shields Avenue, Davis, CA 95616, USA
e-mail: jekstrom@ucdavis.edu

G.T. Lau • K.H. Law
Department of Civil and Environmental Engineering, Engineering Informatics Group, Stanford University, 473 Via Ortega, Stanford, CA 94305-4020, USA

J.R. Gil-Garcia et al. (eds.), *Policy Analytics, Modelling, and Informatics*, Public Administration and Information Technology 24,
DOI 10.1007/978-3-319-61762-6_13

Introduction

As early as the mid-1800s, scientists began exploring the idea of viewing the earth from above. This "remote sensing," termed in 1950s to describe the science and practice of identifying, observing, measuring an object without coming into direct contact with it, became instrumental in international relations and international environmental policy (Campbell and Wynne 2011). Photographs, initially generated from rockets and space crafts for military purposes and, more recently, for understanding land use and global and regional environmental change have begun to provide dramatically improved understanding of environmental change. Since its inception remote sensing's technological and analytical advancements have led to its powerful use by scientists to inform policies, from measuring snowpack that guides flood and water supply management (Edberg et al. 2016) to documenting deforestation that guides international treaties (de Sherbinin et al. 2002), identifying and tracking the ozone hole (Farman et al. 1985) and monitoring ecological impacts of global climate change (Kerr and Ostrovsky 2003).

In stark contrast to the rapid application of remote sensing to understand the biophysical systems, no equivalent of an objective "remote" analysis for *policies* exists. Policies, including those formed by and composed of laws, regulations, court cases, management documents, and others, largely govern our use, abuse, and interaction with each other and the biophysical world. In an information age where policies are often digitally accessible, the equivalent of remote sensing for policy is now possible. Here we address *policy informatics* in regards to the application of information retrieval on policy documents to create a remote broad view of the patterns of policy across management of different resources (air, water, biodiversity, fishing, transportation, etc.). In this chapter we present a tool developed to demonstrate and test the power and acceptability of analyzing policy from a "remote" view for ocean resource management.

Policies, including laws, regulations, and other types of management frameworks and plans, are fundamental to guiding human use and impacts on the environment. Government policies of all jurisdictional levels from international to municipal have traditionally been constructed by experts of a certain field—transportation, water, public health, land use planning, and civil engineering, among others. As population has grown and resources have degraded worldwide, the threat of this fragmented approach to governance has become apparent through reduced productivity and overlap of policies that may interfere with one another (Piñon Carlarne 2008; Biermann et al. 2009; Doremus 2009; Leal-Arcas and Filis 2013; Galster et al. 2001). The drivers, impacts, and solutions to these challenges inherently cut across traditional management sectors. A broad view of the full scope of existing policies can serve to inform where gaps in environmental policy exist, track effectiveness over time and examine potential interference of conflicting policy goals. In this chapter we present pilot applications of a methodology that can provide an analogous view of policies—spanning multiple jurisdictions simultaneously and observing how policies change over time. This provides a way to navigate the complex governance systems for a specific issue of interest.

There is a fundamental need to gain a better understanding of governance from a system perspective (Kim 2013; Fischer et al. 2015). A *system perspective* is viewing a problem as part of its connected drivers and implications, whereas most resource policy traditionally has focused more narrowly on the problem (or symptom) itself. Historically, fisheries management, for example, has been focused on single species. Modelling guiding catch limits account for how the environment (sea temperature, storms, etc.) affect that single species, but disregards the impact of removing that species will have on the environment, including other species (Fogarty and Rose 2014). An analysis tool that can help reveal where there are potential gaps in policy, overlapping jurisdictions, and opportunities for coordination, could be valuable not only for policy making purposes and for improving the management and protection of ecosystem services (Dawes and Janssen 2013), but also for addressing many other major complex problems that need a system approach (Janssen and Wimmer 2015).

This chapter is intended to review and to show cases that illustrate the utility of one policy analytics tool, MINOE, with insights into future needs and applications. The software tool, MINOE (Management Identification for the Needs of Ocean Ecosystems), was designed to pilot and demonstrate how to examine and identify disconnects between science and policy. A user inputs a scientifically understood ecological system, and then outputs those areas where policy potentially exists and where it is absent. At the same time through this analysis, the tool identifies overlapping decision space among governing agencies.

The chapter discusses three case studies that have used the MINOE tool to support investigations of governance and policies from ecological system perspective. The three case studies include: (1) an analysis of policies for California Aquatic Invasive Species; (2) an analysis of the science-policy disconnect in international policy of the marine systems (Treml et al. 2015); and (3) evaluation of federal Canadian coastal environmental laws and management plans. For each of these cases, MINOE is employed to help researchers to gain a better understanding of existing policy from a system perspective, and to evaluate how topics were linked within extant policies and where gaps potentially persisted. The purpose of the chapter is to illustrate how this application of policy informatics can inform policy-making. Last but not least, the chapter concludes with a discussion on the benefits of this approach to support regulative and legislative analysis, the limitations of the tool, and the needs for further research and development.

System Analysis of Governance

Fragmented Governance, Jurisdictional Overlap, and Fit

This section provides the background describing the problem of fragmented governance, and the concepts developed to better understand, analyze and overcome gaps and misfits of policies, providing context for the need to develop and apply policy informatics tools. MINOE was developed to help understand and inform how to provide a first-order analysis with a focus on ocean policy challenges. To set the context for the ocean-related case applications, this section describes the challenges of fragmented governance within the context of ocean policy.

Coastal and ocean health is rapidly deteriorating from overuse, land development, pollution and other stressors (Halpern et al. 2015). No area of the oceans worldwide has been untouched by human influence (Halpern et al. 2012). Fish populations that support many coastal community economies globally, nearshore ecosystems of coral reefs, mangroves and shellfish beds that provide flood protection (and other important ecosystem services) continue to be threatened. Policies governing ocean uses (and abuses) were traditionally made in piecemeal without oversight of the potential unintended consequences that could affect other users of the ocean (Crowder et al. 2006; Fogarty and Rose 2014). Globally, management of the marine environment traditionally has been divided into individual sectors, such as transportation, mining, and fishing (USCOP 2004; Ban et al. 2014; Cormier et al. 2013). Laws were created to address symptoms of problems on a case by case basis, which has resulted in a piecemeal approach (Visbeck et al. 2014). Due to increased coastal populations and improved technology in recent years, uses affecting the ocean have dramatically increased as well, resulting in overlapping laws that regulate different activities. Inconsistent management and lack of coordination across political jurisdictions and between sectors have resulted in a collection of fragmented, sector-based management systems (Cook 2014). When one sector makes a policy decision, without due diligence and insights into other existing policies, it can result in unintended negative consequences for other sectors. For instance, a shipping lane through the Santa Barbara Channel off Southern California geographically overlapped with a major blue whale feeding grounds. As cargo traffic increased, ship strikes became frequent, killing the whales protected under international and national marine mammal policies (Redfern et al. 2013; Monnahan et al. 2015).

To address the complexity of environmental challenges, over a decade ago the United States and Pew ocean commissions (Pew 2003, USCOP 2004) and marine scientists (McLeod et al. 2005) called for the implementation of marine and coastal ecosystem-based management (EBM). This approach has been adopted widely by European nations for land-based resource management (Council of Europe 2000) and more recently as part of the international agreement, Convention on Biological Diversity (CBD 2014). It requires the integration of laws and regulations so that ocean-related activities can be managed under an overarching holistic goal to consider the entire ecosystem rather than the individual sector (guiding principles defined in McLeod et al. 2005). Transitioning to the EBM approach requires an improved understanding of existing ecological environments and policy systems (Borgström et al. 2015; Rosenberg and McLeod 2005; Juda 2003). Over a decade later, this transition is still in process. Essentially, the transition requires a paradigm shift in management so that both policy makers and operational staff embrace the EBM principles (Borgström et al. 2015).

The transition to a new management approach across the multiple management sectors cannot occur without greatly improved knowledge of the extant institutions involved in governance (Imperial and Hennessey 1996; Cortner et al. 1998; Sutinen et al. 2000). Knowledge of the linkages among laws and other management policies is as fundamental to EBM as understanding the relationships among ecological components, such as those among species, their habitats, and stressors (Olsen et al. 2006). A better understanding of the type and degree of interactions among management systems can help set priorities for EBM programs in filling gaps and preventing unhelpful overlaps in coordination. Marine sectors off the coasts of North America include, but are not limited to, shipping, tourism, oil mining, aquaculture and fishing. Implementing EBM requires greater coordination and consistency within and between these sectors, as well as within and between political jurisdictions.

As part of adopting an EBM approach, it is important to understand the landscape of existing problematic fragmented governance by answering the following questions:

- Where are the gaps in ocean management, such as relationships between issues that should be acknowledged and addressed?
- Do policies overlap unintentionally that could or do interfere with one another?
- Where can coordination be improved between policies and government agencies?

Jurisdictional Overlap

To assess, mitigate and prevent environmental impacts of increasing severity and scale requires coordinated and improved governance at all levels of society. Laws and policies dealing with the same issues and connected topics can overlap and interact with one another. These interactions can have a complementary or conflicting effect on each institution's performance (Young 2002; Raakjaer et al. 2014). Once we better understand the interactions, this knowledge can be applied towards addressing the gaps and conflicts, or where appropriate, coordinating and integrating policies. Creating conflict resolution procedures, where possible and necessary, may also improve effectiveness of marine management (Young 1996). Whereas policy analysis typically cover a dozen or fewer policies, the text analysis methods of laws, policy informatics, make it possible to analyze *all* of the policies and their formal written interactions (Cookey et al. 2017). By identifying potential priority areas (of concern), policy analytics can inform the direction of further empirical and legal analysis.

The Ecological-Institutional Problem of Fit

Research on sustainability is increasingly focused on an integrative systems perspective that acknowledges complex social-ecological interdependencies (McGinnis and Ostrom 2014). One of the core problems that can impede effective governance in social-ecological systems is referred to as the "problem of fit". This problem centers on the idea that to be effective, management systems need to reflect and be aligned with the structure, properties, and processes of the ecosystem within their scope (Crowder et al. 2006; Ekstrom and Young 2009; Folke et al. 2007; Young 2002). Poor alignment can lead to ineffective governance and thus reducing the ability for meeting long-term ecological and social benefit goals. Few studies have quantitatively evaluated the problem of fit, as most have been descriptive case studies, though creative techniques are advancing (Bergsten et al. 2014; Vatn and Vedeld 2012; Bodin and Tengö 2012; Cookey et al. 2017). MINOE represents an attempt designed to allow users to evaluate fit quantitatively through the lens of their own ecosystem of interest, and also to identify where the gaps are in the policy system that are causing any poor alignment (or poor fit).

MINOE: A Tool for Analyzing Management Across Sectors

MINOE is a desktop application designed as a pilot to flexibly generate policy information about an ecosystem of interest. The analysis is based on two main pieces of information: (1) a model of a system of interest or concern, e.g., ecosystem, socio-ecological model, or any other system of elements linked in some way to one another; and (2) a collection of text documents representing policies. While sophisticated technological tools for document search and text analysis exist, MINOE is designed to specifically deal with ecosystem models and to address the issues from fragmented governance and jurisdiction. The tool uses the conceptually modeled ecosystem as defined by the user, and performs text analysis of a set of policy documents to determine which ecosystem linkages are potentially acknowledged in the policies. In essence, this method can highlight where links and components in the ecosystem model are discussed in the given set of policies. The tool analyzes the potential policy gaps but also provides some information about overlapping or redundant policies. In addition, MINOE includes initial transparent metrics and visualization tools to synthesize information derived from the documents.

The basic steps for running an analysis in MINOE are as follows:

1. User constructs an ecosystem model by defining the elements and their relationships, and enters the model in the form of a Boolean matrix.
2. The user defines the search criteria for analysis (such as year of documents), jurisdictions (such as location and scale), document type (regulations, laws and statutes), or specific inclusion and exclusion of policy documents for analysis.
3. MINOE performs text analysis based on the ecosystem model and outputs an interactive matrix displaying the analysis results.
4. The user may view list and text of the documents containing the ecosystem elements, a bar graph of the agencies responsible for those documents, as well as graphic visualization of the output results. Two modules are implemented to facilitate the viewing of the output results; namely, the matrix module, which

shows the results in the same form as the user-defined ecosystem model; and the visualization module, which creates an interactive network diagram of the results.

The following describes the system as the user sets up an ecosystem model and ends in the policy visualization module.

Ecosystem Model

Matrices are common formats for developing and quantitatively portraying system relationships. This approach is especially common as characterization of ecosystem models and as a way to organize the complexity of stresses and activities and their associated linkages and relationships (e.g. see Boehlert et al. 2015; Newton et al. 2000; Jacobs et al. 2015). MINOE uses a matrix format as the basic input structure for the program's analysis. First, the user enlists the elements that make up the ecosystem of interest. Each element is defined by a single term or phrase or a set of terms and phrases. The ecosystem elements are the key concepts that users define to represent the ecosystem model that they are interested in. However, a single term or phrase alone is generally too limited in defining an ecosystem element. Often domain experts can define a list of related terms to a concept. For example, when *climate change* is one of the ecosystem elements, the user may also define a related concept as *global warming*. As another example if *coho salmon* is one of the ecosystem elements, the user may also allow to input related terms such as *coho* and *Onchorhynchus kisutch*, *O. kisutch*, or *silver salmon*. If the user is interested in more laws and regulations dealing more generally with *fish*, he or she might choose to include *salmon* and *salmonid*. To facilitate user's input, MINOE provides a simple wizard as shown in Fig. 1 to define terms as well as a list of synonymic terms and concepts.

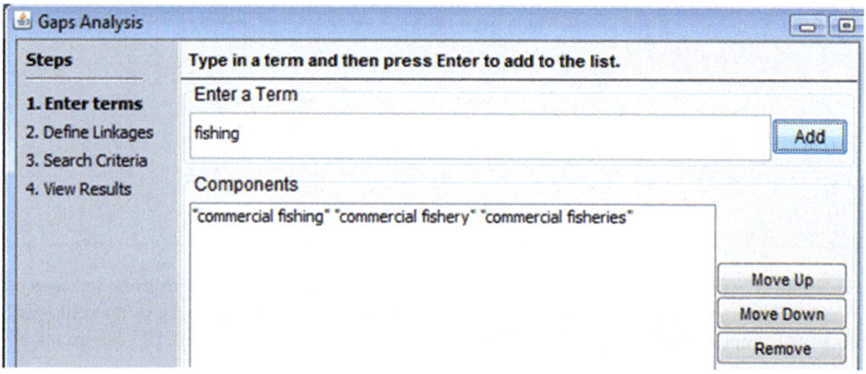

Fig. 1 GUI to define terms and synonymic terms and concepts

Once all the elements are entered, MINOE uses them to generate a symmetric matrix on a spreadsheet. The user then inserts a one (or zero) representing the existence (or non-existence) of a relationship between the corresponding elements. For elements unrelated in the ecosystem model, the user inserts a zero in the corresponding cell. Figure 2 shows an example of an ecosystem model of *Estuary* represented graphically and in a matrix format. Figure 2a shows the ecosystem network model where the nodes and edges depict, respectively, the ecosystem elements (or concepts) and the linkages between the elements. A Boolean matrix of ones and zeroes can be constructed to represent the presence and absence of linkages between the ecosystem elements in the network. As shown in Fig. 2b, a "1" in the corresponding cell of "salmon" and "ocean" rep-

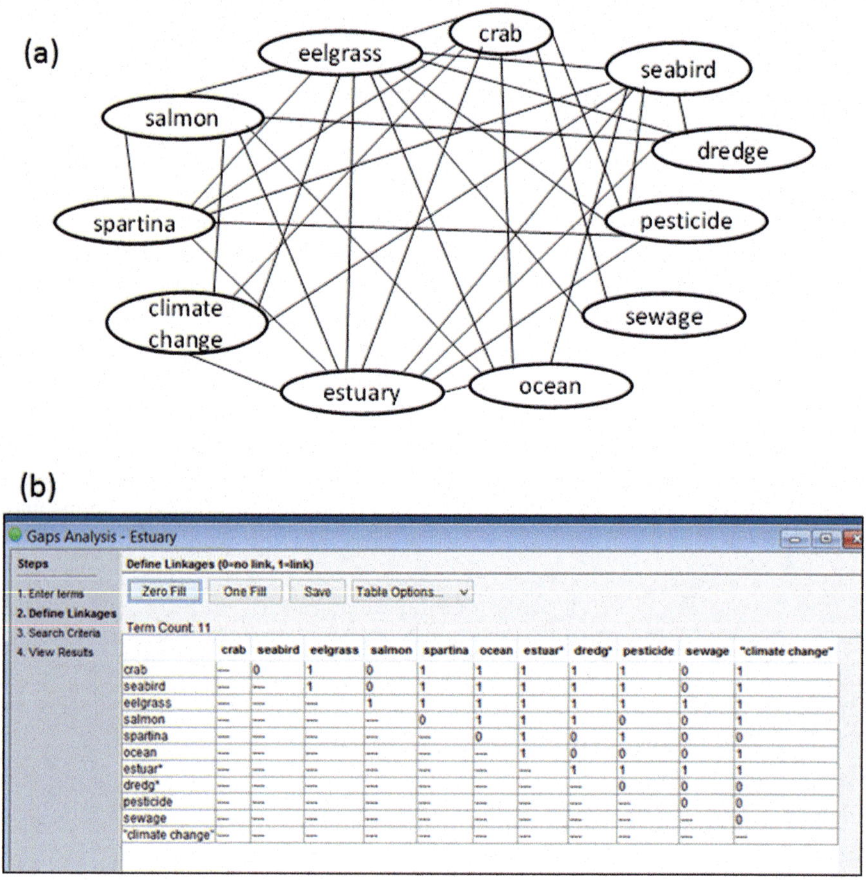

	crab	seabird	eelgrass	salmon	spartina	ocean	estuar*	dredg*	pesticide	sewage	"climate change"
crab	—	0	1	0	1	1	1	1	1	0	1
seabird	—	—	1	0	1	1	1	1	1	0	1
eelgrass	—	—	—	1	1	1	1	1	1	1	1
salmon	—	—	—	—	0	1	1	1	0	0	1
spartina	—	—	—	—	—	0	1	0	1	0	0
ocean	—	—	—	—	—	—	1	0	0	0	1
estuar*	—	—	—	—	—	—	—	1	1	1	1
dredg*	—	—	—	—	—	—	—	—	0	0	0
pesticide	—	—	—	—	—	—	—	—	—	0	0
sewage	—	—	—	—	—	—	—	—	—	—	0
"climate change"	—	—	—	—	—	—	—	—	—	—	—

Fig. 2 An ecosystem model represented in a matrix format. (**a**) Example of terms and linkages of an ecological system; (**b**) Matrix representation of linkage between elements in an ecological model ("1" represents presence of a linkage between two elements, whereas "0" represents no linkage between elements)

resents an ecological linkage. On the other hand, a "0" in the cell of "seabird" and "salmon" represents the lack of such a relationship between the two elements. Modeled links noted in the cells can help organize what scientific relationships exist so that management personnel can account for such system relationships. These matrices are used in MINOE to evaluate the laws, regulations, other policy documents, and government agencies that deal with the issues defined in the ecosystem models.

The matrix representation is a flexible structure for defining and organizing concepts and linkages in a system. A user may input ecosystem models that incorporate species relationships in the form of food web ecosystem models. Incorporation of human dimensions is also a critical part of ecosystem-based management; therefore, users may choose to input ecosystem models with elements and relationships representing not only ecological linkages, but also human activities, societal values, and economic industries. Models including human elements likely will be generated qualitatively based on economic activities, human impacts, and ecological elements and the relationships among one another (dependencies, positive and negative impacts). In the current implementation, neither the strength of linkages between concepts (for example, corresponding to the degree of ecological relationship) nor the indications of positive or negative dependencies or impacts are captured in MINOE. Such relationships would require a non-Boolean matrix representation.

Document Repository and Scope for Analysis

MINOE uses an ecosystem model to analyze a collection of policy documents to determine which ecosystem elements and linkages are potentially acknowledged in the documents. For instance, MINOE contains in its repository a variety of sample documents that include statutes and regulations from four geopolitical jurisdictions (States of Washington, Oregon, and California, and Federal United States). To function properly with MINOE's features, a metadata file documenting the title or description of each document and responsible agency or agencies is included. Other metadata such as document type (statute, regulation, management plan, etc.), year, and geographic location are incorporated in the indexing (for search filter and results display purposes) through this metadata file. Users can also import their own policy document compilations.

Governance varies widely in scope and scale based on the context and nature of the question. The variance in scope and scale also plays a role when analyzing governance in the context of an ecosystem. One user may be more interested in federal U.S. statutes, while another user may need to investigate the regulations and authoring agencies for a specific state. Another user may want to access regulations from multiple states simultaneously. Therefore, MINOE allows users the flexibility of setting the scope and scale of analysis by specify-

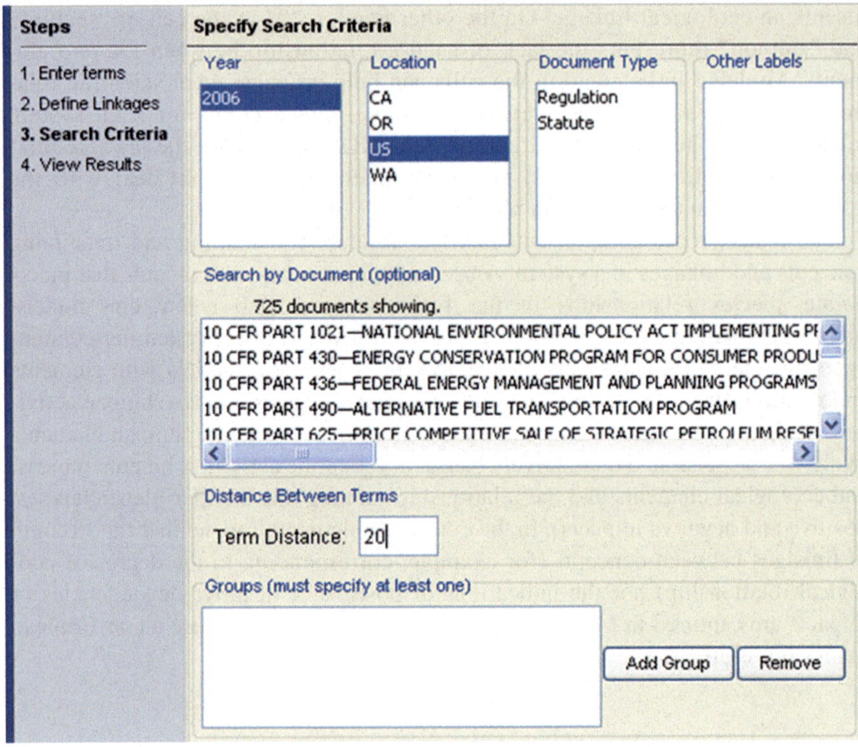

Fig. 3 Scope definition by specifying search criteria for document inclusion

ing a set of search criteria as shown in Fig. 3. For instance, using this interface, a user may want to analyze and group both Federal law and California State law simultaneously for a single ecosystem. In this case, the user would select the year of interest and then the jurisdiction of interest, and add these documents as a single group for analysis. Alternatively, as developed in Ekstrom and Crona (2017), a user may want to look at a single jurisdiction over multiple years by importing documents into MINOE representing the different years of interests. Such multi-year analyses generate a time series data set to evaluate how policy has changed over time to address complex issues. Analysis of policy documents from single or multiple jurisdictions could be informative in tracking ecosystem stressors such as climate change. Lastly, a user may select the specific documents to be included for analysis. MINOE indexes all the policy documents in the repository and any other documents imported by the user. The most basic functionality of MINOE is to capture the occurrence of the ecosystem elements in the various sections of the law and its metadata. For instance, using a 2006 collection of policy documents from different federal agencies, Fig. 4 shows the number of sections that contain the ecosystem element *salmon* (defined by the term and the list of synonymic terms) in their documents.

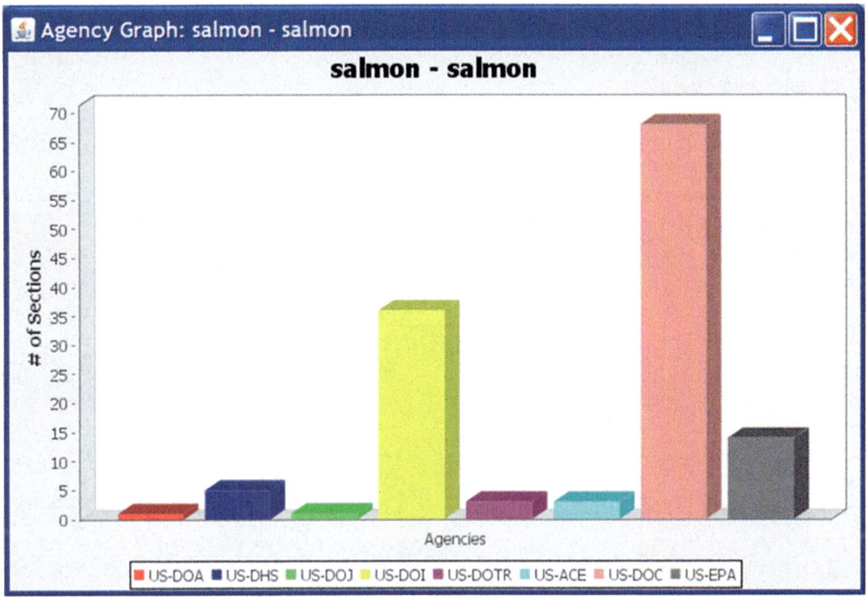

Fig. 4 Federal documents, by responsible government agency, containing ecosystem element "salmon"

Analysis Results

Once the policy documents of interest are selected (into relevant groupings), the system performs the analysis by comparing the text of the documents with the ecosystem model. As discussed, a concept or an ecosystem element is defined by term or phrase, related terms and synonyms. MINOE, through text analysis, parses the documents of interest and keeps track of the frequency of all terms and computes the concept frequency by summing the occurrences of all the terms that define the concept. Results are presented in a matrix format that matches the user's ecosystem model elements. Users then can access the information displayed in cells, which corresponds to the documents (such as laws and regulations) and the issuing agencies.

As an example, Fig. 5 shows the results for comparing the ecosystem model as shown in Fig. 2 using the 2006 collection of laws and regulations in the State of Washington. The results are displayed in a matrix synonymous to the ecosystem model. The diagonal cells of the matrix contain the frequency for the single corresponding term in the collection. For instance, the terms *eelgrass* and *dredge* occur 14 and 76 times, respectively. Each non-diagonal cell contains the co-occurrence frequency of each dyad of terms. For example, the cell *eelgrass* and *dredge* shows seven co-occurrences of the two terms within a given "distance" (such as separating sentences, sections, paragraphs or chapter within a document) in the collection.

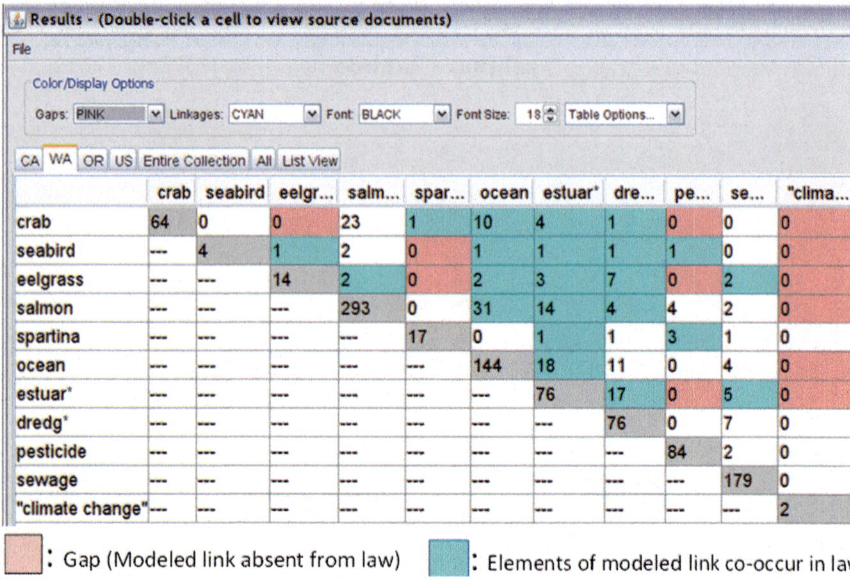

	crab	seabird	eelgr...	salm...	spar...	ocean	estuar*	dre...	pe...	se...	"clima...
crab	64	0	0	23	1	10	4	1	0	0	0
seabird	---	4	1	2	0	1	1	1	1	0	0
eelgrass	---	---	14	2	0	2	3	7	0	2	0
salmon	---	---	---	293	0	31	14	4	4	2	0
spartina	---	---	---	---	17	0	1	1	3	1	0
ocean	---	---	---	---	---	144	18	11	0	4	0
estuar*	---	---	---	---	---	---	76	17	0	5	0
dredg*	---	---	---	---	---	---	---	76	0	7	0
pesticide	---	---	---	---	---	---	---	---	84	2	0
sewage	---	---	---	---	---	---	---	---	---	179	0
"climate change"	---	---	---	---	---	---	---	---	---	---	2

Fig. 5 Analysis results of an ecosystem model

As shown in Fig. 5, the cells are highlighted with colors to indicate existing linkages in the ecosystem model while those cells that were zero in the ecosystem model remain white. Each cell contains the number of times the corresponding terms (representing ecosystem elements) occurs (within a given distance of one another) in the selected collection of laws and regulations. The matrix representation, although simple, can be very informative to review gaps and linkages in the laws and regulations with respect to the ecosystem model.

1. If a cell shows a linkage (i.e. with an entry >0) in the ecosystem model but contains zero frequency, it is marked as a "gap." Gaps are highlighted and colored pink (see Fig. 5). For example, a linkage exists in the ecosystem model between *salmon* and *climate change* (see Fig. 2); this same cell in the matrix results for the collected 2006 regulations in the State of Washington contains zero (Fig. 5), indicating that there are no regulations in this collection of documents that reference the terms *salmon* and *climate change* together. Therefore, despite the understanding of the large threats to salmon from climate change, no regulation or law explicitly acknowledges this relationship, according to this 2006 collection of laws in the State of Washington.
2. If a cell represents a linkage (i.e. with an entry >0) in the ecosystem model and the corresponding elements co-occur in one or more laws, it is marked as a potential linkage in law. These linkages are highlighted with cyan color. An example of a modeled linkage acknowledged in law is the relationship between *eelgrass* and *dredge*. This same cell in the matrix results indicates that there are seven incidences in which the terms *dredge* and *eelgrass* co-occur together. That is, the link between *eelgrass* and *dredge* modeled in the ecosystem is potentially acknowledged in the collection of laws and regulations being examined (to what extent, however,

requires further investigation). The user then may opt to view what laws and regulations contain the co-occurrence of these terms and view the text of the documents.

Using the metadata defined, the user may retrieve more specific information for each cell. Right clicking on the cell of interest, the user can opt to view (1) what laws and regulations contain the ecosystem elements and (2) what agencies are responsible for the laws and regulations that contain the elements and relationships of interest. The first option brings up directly a list of laws and regulations from all the policy documents included in the analysis, each of which the user can open and view the text of individual documents for the highlighted terms that represent the relevant ecosystem elements. The second option displays the responsible agencies in a bar graph as shown in Fig. 6a. In this figure, the number of sections with co-occurrence

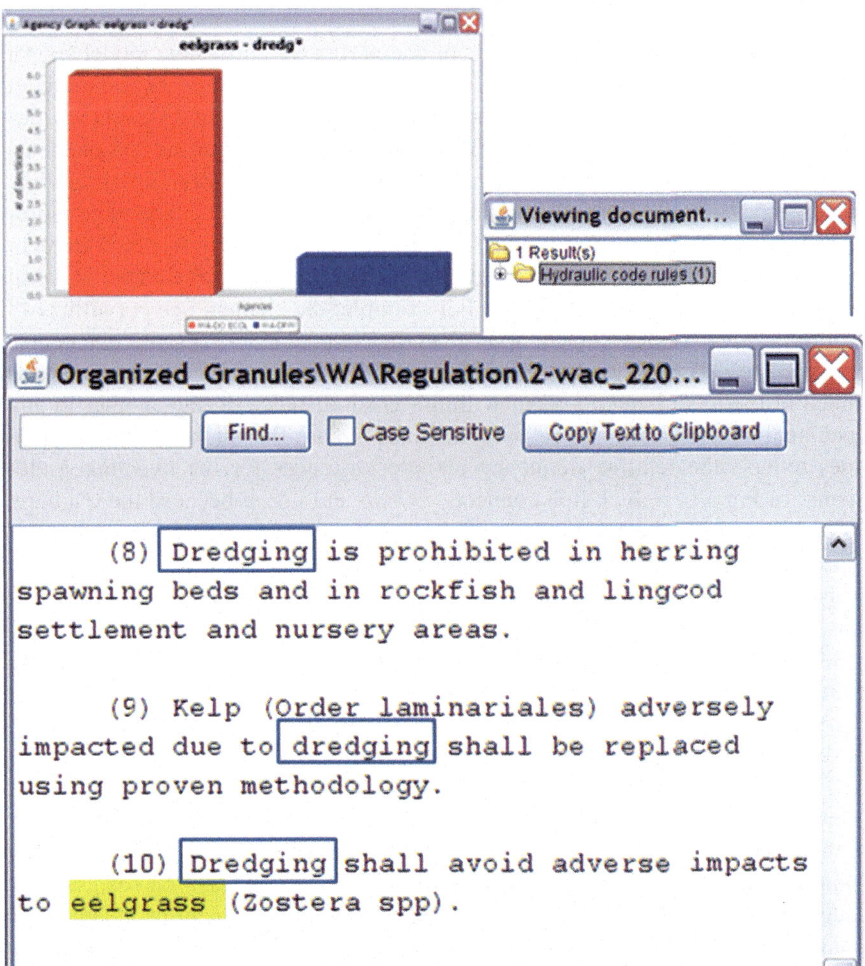

Fig. 6 Retrieval of documents with co-occurrence of ecosystem elements. (**a**) Agency involvement (with documents showing term co-occurrence); (**b**) Document showing co-occurrence of terms

of the terms in the documents of each individual agency is also shown. The user can then click on the bar for the agency to open and view the text of the document that contains co-occurrence of the ecosystem elements as shown in Fig. 6b.

Visualization of Output Results

As the number of elements in the ecosystem model increases, the complexity of the model increases exponentially, making interpretation of results more challenging. The increased complexity makes the matrix view of the results difficult. To assist the user to visualize gaps and linkages between the ecosystem model and the institutional documents of interest, MINOE provides a network diagram to display the data graphically so that the user can further synthesize and explore the analysis results.

Figure 7 shows the analysis results of comparing the ecosystem model shown in Fig. 2 with a 2013 collection of US federal policies. For an ecosystem model with only 11 elements, the matrix representation as depicted in Fig. 7a provides a concise presentation of the results. Obviously, as the number of elements increases, displaying the results in matrix form can be difficult to interpret visually. Figure 7b shows the network diagram which uses the same information presented in Fig. 7a. By activating this visualization module, each modeled element is displayed in the visualization window where the size of each element is based on the frequency in which it occurs. For example, the element *ocean* (with 1148 occurrences) will be displayed larger than the element *eelgrass* (with one occurrence) since *ocean* occurs more frequently than *eelgrass*. The elements that are linked in the model and co-occur within a given distance of one another in any document analyzed are shown by solid lines, where the thickness of the solid lines reflects the relative frequency of co-occurrences for the two linked elements. In Fig. 7b, a thick line connects *salmon* and *ocean* because there are six co-occurrences of these two elements in the 2013 collection of federal regulations, while a thin line connects *dredg** and *salmon* because there is only one co-occurrences of the two ecosystem elements. MINOE draws dotted lines between the elements that are linked in the ecosystem model but do not co-occur in any law, such as between *eelgrass* and *salmon*. This represents a potential policy gap (and continued ecological threat), an instance of disconnect between two species that scientists have found to be ecologically interconnected (Levin et al. 2015), but is not reflected in policy.

The user can access the corresponding list and text of laws and the chart of agency involvement directly through the visualization module. As with the results in the matrix module, the user can right click on any ecosystem element to view additional information. This includes a list of relevant laws and the bar graph of gauging involvement of agencies to managing a specific element or linkage. For instance, Fig. 8 shows the number of occurrences of the term *climate change* in the 2013 collection of the US regulations issued by the different agencies.

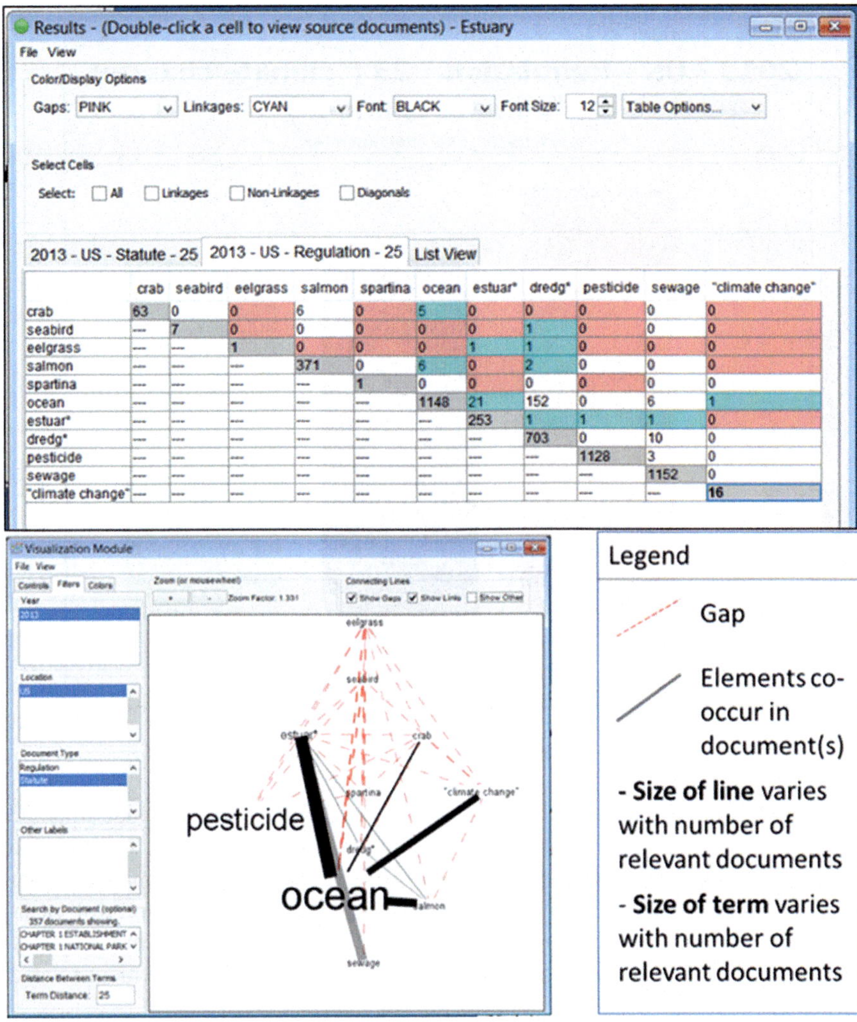

Fig. 7 Results modules showing the frequency of ecosystem linkages found in document collections (*top*) and the interactive graphic of the linkages (*lower*)

Case Studies

MINOE can support analysis of management documents flexible to the needs of the user. It allows users to input their ecosystem model of choice and conduct the analysis on any scope and scale of documents. Lastly, MINOE provides features to help users analyze and synthesize the data generated. Users who may find potential applications for MINOE in governance and policy analysis include governmental and nongovernmental organization personnel, policy advocates, resource users, concerned citizens, and policy course instructors and students. As a tool for

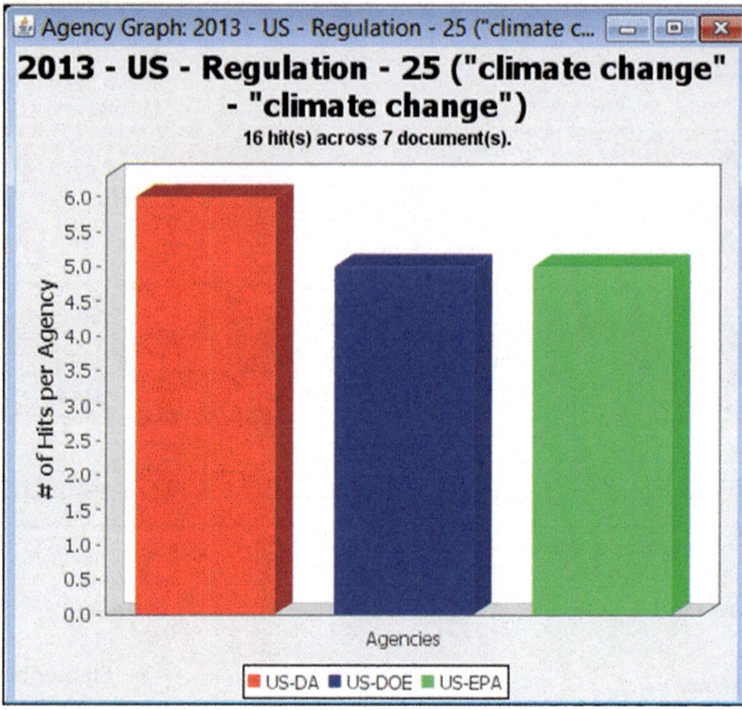

Fig. 8 Number of occurrences on the term *climate change* in regulations

agencies, it could be used to assist on improving collaboration, to enhance strategic resource sharing, and to increase strategic policymaking. The technique could also be useful for individual agencies in writing new regulations to determine whether there are resource-sharing opportunities with other agencies in order to fulfill mandates. In the following, we illustrate ways MINOE has been used by resource agency staff and by interdisciplinary scientists for addressing research management-related questions.

Invasive Species California

To demonstrate how MINOE can be used to assist regulatory and legislative analysis, our first case study example involves a regulatory analysis conducted for an ocean policy agency in California on aquatic invasive species. In 2008, a primary goal set by the California Aquatic Invasive Species Management Plan was to conduct an analysis of existing management, identifying what laws and regulations that the State already had were relevant to each specific pathway and invasive species. Additionally, one of the plan's primary tasks was to identify which agencies were and should be involved in management of invasive species. Given the complexity and long list of pathways through which non-native species were introduced into the state waters, the task could be a time consuming project. Tasked with documenting all California State legislation relevant to aquatic invasive species, a California State Research Fellow used MINOE, in combination with more traditional law databases, to carry out this project.

The user found the unique results output format useful in that it provided the same matrix structure so she could investigate and compare the usage of the different invasive species and vector pathways simultaneously, but still independently of one another. Producing this, with the graphics module (as illustrated in Fig. 7), and the clickable agency involvement bar graphs (as illustrated in Fig. 8), provide a clear benefit over the traditional legal database systems she queried, which at the time provided output in the form of simple lists. The other benefit was that the user could view multiple jurisdictions through a single query (three states' and federal law), whereas otherwise she had to go to several different websites to investigate her target jurisdictions. The primary reported drawback of MINOE relative to the more widely used legal databases was that the compilation of laws and regulations provided through MINOE was not current. This case study has clearly identified there is a need to develop a practical strategy to facilitate compilation of policy documents and a way to alleviate propriety legal database restrictions where they exist.

Coral Triangle and International Governance

This second case example of a project using MINOE is set in the Coral Triangle (Treml et al. 2015), a marine area in the Western Pacific Ocean that includes Indonesia, Australia, Papua New Guinea, Philippines, East Timor, and Malaysia. The Coral Triangle is widely regarded as the home to the most biodiverse coral reef ecosystems, which support (and can be threatened by) large industry and small local level fisheries and tourism. The region was chosen by several conservation organizations as a target to help protect the region's threatened biodiversity. In 2013 one of these organizations commissioned a study to evaluate the existing governance, whether it adequately covered the ecological systems, and where potential gaps could be in transboundary conservation efforts. This project used a database of 200 documents (e.g., treaties, conventions, policy declarations, action plans, and memoranda of understanding) representing international environmental agreements involving two or more of the 13 countries of interest was developed. The documents were analyzed using MINOE, updated from the international policy analysis conducted in Fidelman and Ekstrom (2012). The analysis included counting the number of times key terms appeared in each of the documents within the database. Term frequency was used as a proxy indicator of the extent to which the documents related to one or more ecological topics of interest (defined through the key terms).

The linkage data for the region were represented as institutional networks where the nodes consisted of individual countries and the links among them showed the strength of the connection (i.e., the number of linkages in common between a pair of countries) for each concept. Although 14 concepts in total were analyzed, results focused on the four that are most central to marine connectivity and conservation issues: Coral Reefs, Marine Protected Areas (MPAs), Fisheries, and a composite (MPAs + Fisheries) network.

In this case study (published in Treml et al. 2015), the nodes are countries and the relationships between countries represent co-signatories of international agreements or ecological pathways, the former of which was done by Fidelman and Ekstrom (2012) and the latter done by Treml and Halpin (2012). Similar to this study, Bergsten et al. (2014) have compared networks of the same structure (though at the county/sub-national level), but only for a single type of topic (i.e., wetlands). As a further development in analyzing fit, Treml et al. (2015) have evaluated multiple topics of concern as a way to incorporate the notion of functional fit. Specifically, the study sought to reveal cases in which two countries are linked ecologically and whether these same countries were also connected through shared participation in international agreements. From this case study project, Treml et al. (2015) have also developed a typology of combined policy and ecological linkages based on the union of the two networks.

In integrating large-scale ecological-policy networks, we have identified the important geopolitical fit and misfit between the region's ecology and its governance. This analysis found that overall, the region appears to be largely well-aligned in terms of shared signatories of international agreements regarding coral reefs, but less so for those arrangements related to fisheries and marine protected areas. The study also found a potential gap between how East Timor and Australia are connected institutionally, which may interfere with the strong ecological ties the countries' coral reefs have with one another. In short, with MINOE as an enabling tool, Treml et al. (2015) have developed an innovative method to evaluate a complex set of international treaties against an equally complex tropical marine ecological system.

Southern Gulf of St. Lawrence, Canada

In this third case example, we describe here the use of MINOE by the federal government of Canada for its planning for ecosystem-related policies. The Federal government of Canada is mandated to transition towards an ecosystem-based management approach for the coastal and marine regions through the Oceans Act in 1996. Efforts to plan for and carry out activities relating to integrated ecosystem-related goals became a priority with the Canada Oceans Action Plan in 2005. A pilot project was initiated to assess the use of MINOE in identifying potential gaps in Canada's context of coastal and oceans in relation to adverse environmental effects (Ekstrom et al. 2011).

The MINOE software was used by the federal government (in consultation with the authors here) to conduct an analysis to identify the potential gaps in a set of Provincial and Federal laws, regulations and best management practices for the jurisdictions adjacent to the southern Gulf of St. Lawrence. Their document collection consisted of laws, regulations, and best management practices to represent management from Federal level and three provinces (Nova Scotia, New Brunswick, and Prince Edward Island). The Department of Fisheries and Oceans (DFO) has also developed a conceptual model on the basis of the DPSIR (Driver-Pressure-State-Impact-Response) structure to link activities to cumulative and adverse environmental effects for coastal and marine waters and the associated watersheds. This project served as a pilot to determine the utility of extending MINOE's application in the context of a collaborative process for identifying areas where management measures may not be present, enforced or sufficiently effective in mitigating cumulative effects to significant fish and fish habitat.

Following the evaluation of MINOE's application, in 2013 the International Council for Exploration of the Sea (ICES) Working Group on Marine Planning and Coastal Zone Management recommended use of MINOE to identify and locate management gaps as part of the Risk Analysis for implementing marine spatial planning (Cormier et al. 2013). The document provides governments and other organizations basic project planning guidance for any ecosystem-based management project. ICES is the oldest international intergovernmental organization concerned with fisheries and other marine resources.

Discussion

Benefits

As illustrated in the case studies, there are ample opportunities for decision support tool such as MINOE in policy analytics. Most directly, MINOE can be used to assist regulatory and legislative analysis, which is a common step for the development of new policy and amending existing policies for emerging and existing threats to ocean ecosystems. When compared to on-line law databases, which are not specifically designed for system-based analysis, unique advantages of using MINOE include:

- Access to an approximation of agency responsibility (through bar graphs)
- Matrix-based search
- Facilitating investigation of documents addressing linkages between elements
- Providing system-perspective of regulations across state and federal and across multiple states for comparison
- Identification of regulations containing overlapping concepts
- Identification of potential gaps in policy
- Ability to access multiple jurisdictions through a single application

As illustrated in the case studies with MINOE, quantitative and systematic techniques of text analysis can be applied to a broad set of management scenarios involving jurisdictional and functional overlaps. They can help navigate governance and help identify overlaps in different domains as seen already in application development for analysis and comparison of laws and regulations (Cookey et al. 2017; Lau et al. 2005, 2006; Law 2006, 2014; Law et al. 2015; Fidelman and Ekstrom 2012).

Limitation of Tool

As with any decision support tool, MINOE comes with limitations and opportunities for extensions, as described below.

User Knowledge

A frequently asked question by MINOE users in early stages of an analysis is for advice on how to define the boundaries and concepts included in the ecosystem—or otherwise defined—conceptual model. Although we send advice and suggestions based on our experience, it is important to recognize that the user needs to have some prior knowledge of the system of interest in order to conduct a useful analysis. This is typical of any decision-support tool (and in fact, any tool) that the tool relies heavily on the user's input and likely would benefit if it is used in conjunction with multiple support tools and some initial knowledge to help guide the study.

Compiling Policy Text

The most substantial limitation of MINOE thus far is the absence of having a clear and easy way to create management document compilations for any given jurisdiction. Particularly in the United States, online "live" databases of codified statutes and regulations have grown tremendously over the last decade. However, some barriers to developing an API to pull from these sites persist. Most are not aligned with one another across jurisdictions. For example, the federal United States makes its laws for public access (including easy download) in a different format from the State of Washington laws and other states. For some jurisdictions, the digital versions of the laws are considered proprietary because they are organized and maintained by private companies contracted by the state. This is the case in California, a state that is known for its complicated and massive system of law. Furthermore, no central public repository exists to coordinate local government policies (municipal or other local level written policy). Local level may benefit immensely from the policy analytic tools given they can have high turnover (lowering institutional knowledge) and often are economically constrained.

Search Algorithm

The method thus far has used a basic keyword matching algorithm. In presenting early results using more advanced text mining methods (for example, using singular value decomposition for clustering concepts and finding potential overlapping policies), experts in the ocean and coastal ecosystem management domain strongly preferred the basic keyword matching because of its simplicity and clarity. Resource managers and other policy makers preferred the transparent search algorithm for precise interpretability (Ekstrom et al. 2010). Given the growth in information science, social acceptability, and interest in "big data analytics" in recent years, advanced text analysis techniques, particularly those with well-designed interface to communicate the relevant results, can have values to ecosystem analyses. Advancements in information science could be harnessed to develop analytics that accurately represent how concepts are acknowledged in policies and where gaps may exist. For instance, text mining software, such as QDAMiner (developed by Provalis Research, see https://provalisresearch.com/), could be adapted to explore the utility of advanced text analytics and to investigate how effective and acceptable such tools are for policy makers' uses. Last but not least, continuing advancements in natural language processing, multi-dimensional modeling, and machine learning could potentially lead to better technique and algorithm for both the policy analytics in relation to the ecosystem models, as well as for the agency jurisdiction exploration.

Conclusion

Taking a broad system view of policy can have major policy implications in the long term for environmental governance and ecosystem health. Current policy analysis is typically conducted either on an ad-hoc or qualitative basis, which we argue could benefit from an earlier evaluation of the larger policy landscape to develop environmental solutions. As a policy maker, government agency staff or scholar, searching for concepts in regulations, statutes and management documents is an arduous task, especially given that one may have to run a number of related queries in several databases. A broad scale, semi-quantitative view, if made transparent, accessible, and easily interpretable can be useful for initial policy-related research by government agencies, policy scholars, and even resource users wanting to participate in decision-making but that do not have a full grasp on the policy environment surrounding their resource or issue. For example, government agency staff developing new policy, for example, need to be familiar with existing laws and regulations that relate to the new proposed law in order to avoid unintentional conflicting rules. Staff also can use a semi-quantitative policy analysis to discover aspects of environmental issues that are not covered in law or policy so far. This is especially important to promote policy making across sector-focused agencies, meaning as an example that transportation planning or management of oil mining be done with consideration of drinking water permitting. As another example, many environmental issues cross jurisdiction boundaries triggering governments to forge collaborative policies across states and nations. Developing region-wide policies that are consistent with existing national policy is important to avoid obstructing implementation. Though simple in its current state, the methodology and the tool presented here could save a tremendous amount of time to target policy priorities for further examination once the collection procedure of policy documents is either automated or otherwise made more efficient. Offering a unique landscape view of policy, it is analogous to the remote sensing widely used for understanding landscapes and biophysical systems. Landscape views of policy using informatics tools will become increasingly important as the morass of policies continues to grow.

The methodology illustrated through MINOE in this chapter is simple and is designed specifically relevant to analyzing the intersection of governance and ecological system management. The tool is created with feedback from policy makers that have not previously been exposed to the landscape view of policy. Future research includes the development and testing of more rigorous quantitative text mining and analysis methods and then the pursuit of incorporating these methods either into MINOE or to partner with existing legal database organizations to adopt the matrix-view and gaps analysis techniques into part of publicly accessible (and free) legal databases. This would offer the advancements in the tool beyond academia to international and domestic government agency staff and policy-makers.

Acknowledgments Our thanks to the David and Lucile Packard Foundation (Ecosystem-Based Management Tools Initiative Fund), California Sea Grant and the California Ocean Protection Council for supporting this research. The authors would also like to acknowledge partial support by the National Science Foundation grant IIS-0811460 and the Canada Department of Fisheries and Oceans. Special thanks to Hannah Torres of University of South Florida for assistance in preparing the 2013 federal policy dataset. Any opinion expressed in this paper are those of the authors and do not necessarily reflect the opinions of the David and Lucile Packard Foundation, the National Science Foundation or their collaborators.

References

Ban NC, Maxwell SM, Dunn DC et al (2014) Better integration of sectoral planning and management approaches for the interlinked ecology of the open oceans. Mar Policy 49:127–136. doi:10.1016/j.marpol.2013.11.024

Bergsten A, Galafassi D, Bodin O (2014) The problem of spatial fit in social-ecological systems: detecting mismatches between ecological connectivity and land management in an urban region. Ecol Soc 19(4):6

Biermann F et al (2009) The fragmentation of global governance architectures: a framework for analysis. Glob Environ Polit 9(4):14–40. Available via http://www.mitpressjournals.org/doi/abs/10.1162/glep.2009.9.4.14#.VuckA_krKUk. Accessed 5 Nov 2016

Bodin Ö, Tengö M (2012) Disentangling intangible social–ecological systems. Glob Environ Chang 22:430–439. doi:10.1016/j.gloenvcha.2012.01.005

Boehlert B et al (2015) Climate change impacts and greenhouse gas mitigation effects on U.S. water quality. J Adv Model Earth Syst 7(3):1326–1338. Available via http://doi.wiley.com/10.1002/2014MS000400

Borgström S, Bodin Ö, Sandström A, Crona B (2015) Developing an analytical framework for assessing progress toward eco-system-based management. Ambio 44:357–369. doi:10.1007/s13280-015-0655-7
Campbell JB, Wynne RH (2011) Introduction to remote sensing, 5th edn. Guilford Publications, Inc., New York
Convention on Biological Diversity (CBD) (2014) Ecosystem approach sourcebook. Available via https://www.cbd.int/ecosystem/sourcebook/default.shtml. Accessed 5 Nov 2016
Cook C (2014) Governing jurisdictional fragmentation: tracing patterns of water governance in Ontario, Canada. Geoforum 56:192–200
Cookey PE, Rotchanatch D, Ratanachai C (2017) Text mining analysis of institutional fit of Lake Basin water governance. Ecol Indic 72:640–658
Cormier R et al (2013) Marine and coastal ecosystem-based risk management handbook. ICES Cooperative Research Report No. 317, Copenhagen, Denmark. Available via http://www.ices.dk/sites/pub/Publication Reports/Cooperative Research Report (CRR)/crr317/CRR317 Marine and coastal ecosystem based risk management handbook.pdf. Accessed 5 Nov 2016
Cortner HJ et al (1998) Institutions matter: the need to address the institutional challenges of ecosystem management. Landscape and Urban Planning, pp 159–166
Council of Europe (2000) European Landscape Convention, ETS No. 176
Crowder L et al (2006) Resolving mismatches in U.S. ocean governance. Science 313(5787):617–618. Available via http://www.sciencemag.org/cgi/content/full/sci;313/5787/617. Accessed 5 Nov 2016
Dawes SS, Janssen M (2013) Policy informatics: addressing complex problems with rich data, computational tools, and stakeholder engagement. In: Proceedings of the 14th annual international conference on digital government research, pp 251–253. doi:10.1145/2479724.2479759
Doremus H (2009) CALFED and the quest for optimal institutional fragmentation. Environ Sci Pol 12(6):729–732
Edberg SJ, Evans DL, Graf JE et al (2016) Studying earth in the new millennium: NASA Jet Propulsion Laboratory's contributions to earth science and applications space agencies. IEEE Geosci Remote Sens Mag 4:26–39. doi:10.1109/MGRS.2016.2525678
Ekstrom JA, Crona BI (2017) Institutional misfit and environmental change: a systems approach to address ocean acidification. Sci Total Environ 576:599–608. Available via http://dx.doi.org/10.1016/j.scitotenv.2016.10.114. Accessed 5 Nov 2016
Ekstrom JA, Young OR (2009) Evaluating functional fit between a set of institutions and an ecosystem. Ecol Soc 14(2):16. Available via http://www.ecologyandsociety.org/vol14/iss2/art16/. Accessed 5 Nov 2016
Ekstrom JA, Lau GT, Cheng JCP, Spiteri DJ, Law KH (2010) Gauging agency involvement in environmental management using text analysis of laws and regulations. I/S J Law Policy Inf Soc 6:189–210
Ekstrom JA et al (2011) Application of the MINOE regulatory analysis framework: case studies. In: Proceedings of the 12th annual international conference on digital government research. Universitiy of Maryland, College Park, Maryland
Farman JC, Gardiner BG, Shanklin JD (1985) Large losses of total ozone in Antarctica reveal seasonal ClOx/NOx interaction. Nature 315(6016):207–210
Fidelman P, Ekstrom JA (2012) Mapping seascapes of international environmental arrangements in the Coral Triangle. Mar Policy 36(5):993–1004. Available via http://www.sciencedirect.com/science/article/pii/S0308597X12000139. Accessed 5 Nov 2016
Fischer J, Gardner T, Bennett E, Balvanera P, Biggs R et al (2015) Advancing sustainability through mainstreaming a social-ecological systems perspective. Curr Opin Envir Sust 14:144–149. doi:10.1016/j.cosust.2015.06.002
Fogarty MJ, Rose K (2014) The art of ecosystem-based fishery management. Can J Fish Aquat Sci 71:479–490. doi:10.1139/cjfas-2013-0203
Folke C, Pritchard L, Berkes F, Colding J, Svedin U (2007) The problem of fit between ecosystems and institutions: ten years later. Ecol Soc 12:30

Galster G et al (2001) Wrestling sprawl to the ground: defining and measuring an elusive concept. Hous Policy Debate 12(4):681–717. Available via http://www.tandfonline.com/doi/abs/10.1080/10511482.2001.9521426. Accessed 5 Nov 2016

Halpern BS et al (2012) An index to assess the health and benefits of the global ocean. Nature 488(7413):615–620. Available via http://www.ncbi.nlm.nih.gov/pubmed/22895186. Accessed 5 Nov 2016

Halpern BS et al (2015) Spatial and temporal changes in cumulative human impacts on the world's ocean. Nat Commun 6:7615

Imperial MT, Hennessey TM (1996) An ecosystem-based approach to managing estuaries: an assessment of the national estuary program. Coast Manag 24:115–139

Jacobs S, Van Daele T, Staes J, Schneiders A (2015) "The Matrix Reloaded": a re-view of expert knowledge use for mapping ecosystem services. Ecol Model 295:21–30. doi:10.1016/j.ecolmodel.2014.08.024

Janssen M, Wimmer MA (2015) Introduction to policy-making in the digital age. In: Policy practice and digital science. Springer, Cham, pp 1–14

Juda L (2003) Obstacles to ecosystem-based management. In: UNESCO, Paris, pp 67–71

Kerr JT, Ostrovsky M (2003) From space to species: ecological applications for remote sensing. Trends Ecol Evol 18:299–305. doi:10.1016/S0169-5347(03)00071-5

Kim RE (2013) The emergent network structure of the multilateral environmental agreement system. Glob Environ Chang 23:980–991. doi:10.1016/j.gloenvcha.2013.07.006

Lau GT, Law KH, Wiederhold G (2005) Analyzing government regulations using structural and domain information. Computer 38(12):70–76

Lau GT, Law KH, Wiederhold G (2006) A relatedness analysis of government regulations using domain knowledge and structural organization. Inf Retr 9(6):657–680

Law KH (2006) A distributed information management framework (REGNET) for environmental laws and regulations. In: Proceedings of the 2006 national conference on digital government research - dg.o '06. ACM Press, New York, p 423. Available via http://dl.acm.org/citation.cfm?id=1146598.1146730. Accessed 5 Nov 2016

Law KH (2014) REGNET: regulatory information management, compliance and analysis. Govern Inf Q 31:S37–S48. Available via http://www.sciencedirect.com/science/article/pii/S0740624X14000549. Accessed 5 Nov 2016

Law KH, Taduri S, Lau GT, Kesan JP (2015) An ontology-based approach for retrieving information from disparate sectors in government: the patent system as an exemplar. In: 48th annual Hawaii international conference on system sciences, Kauai, HI, pp 2096–2105. doi:10.1109/HICSS.2015.252

Leal-Arcas R, Filis A (2013) The fragmented governance of the global energy economy: a legal-institutional analysis. J World Energy Law Bus 6(4):348–405

Levin PS, Williams GD, Rehr A, Norman KC, Harvey CJ (2015) Developing conservation targets in social-ecological systems. Ecol Soc 20(4):6. doi:10.5751/ES-07866-200406

McGinnis MD, Ostrom E (2014) Social-ecological system framework: initial changes and continuing challenges. Ecol Soc 19(2):30. doi:10.5751/ES-06387-190230

McLeod KL et al (2005) Scientific consensus statement on marine ecosystem-based management. Communication Partnership for Science and the Sea. Available via http://compassonline.org/?q=EBM. Accessed 5 Nov 2016

Monnahan CC, Branch TA, Punt AE (2015) Do ship strikes threaten the recovery of endangered eastern North Pacific blue whales? Mar Mamm Sci 31:279–297. doi:10.1111/mms.12157

Newton J et al (2000) A conceptual model for environmental monitoring of a marine system developed for the Puget Sound ambient monitoring program (PSAMP). Available via http://www.monitor2manage.com.au/userdata/downloads/p_/Puget%20Sound%20Environment%20monitoring%20model.pdf. Accessed 5 Nov 2016

Olsen SB et al (2006) A handbook on governance and socioeconomics of large marine ecosystems. Coastal Resources Center: University of Rhode Island. Available via http://www.iwlearn.net/abt_iwlearn/pns/learning/lme-gov-handbook.pdf. Accessed 5 Nov 2016

Pew Oceans Commission (Pew) (2003) America's living oceans: charting a course for sea change. Available via http://www.pewtrusts.org/en/research-and-analysis/reports/2003/06/02/americas-living-oceans-charting-a-course-for-sea-change. Accessed 5 Nov 2016
Piñon Carlarne C (2008) Good climate governance: only a fragmented system of international law away? Law Policy 30(4):450–480. doi10.1111/j.1467-9930.2008.00289.x
Raakjaer J, van Leeuwen J, van Tatenhove J, Hadjimichael M (2014) Ecosystem-based marine management in European regional seas calls for nested governance structures and coordination—a policy brief. Mar Policy 50:373–381. doi:10.1016/j.marpol.2014.03.007
Redfern J, McKenna M, Moore T et al (2013) Assessing the risk of ships striking large whales in marine spatial planning. Conserv Biol 27:292–302
Rosenberg AA, McLeod K (2005) Implementing ecosystem-based approaches for its management. Ocean Coast Manage 44:729–756
de Sherbinin A, Kline K, Raustiala K (2002) Remote sensing data: valuable support for environmental treaties. Environ Sci Policy Sustain Dev 44:20–31. doi:10.1080/00139150209605589
Sutinen JG et al (2000) A framework for monitoring and assessing socioeconomics and governance of large marine ecosystems. NOAA Technical Memorandum NMFS-NE-158. Northeast Fisheries Science Center
Treml EA, Halpin PN (2012) Marine population connectivity identifies ecological neighbors for conservation planning in the Coral Triangle. Conserv Lett 5(6):441–449. doi10.1111/j.1755-263X.2012.00260.x
Treml EA, Fidelman, PJ, Kininmonth S, Ekstrom, JA, Bodin O (2015) Analyzing the (mis)fit between the institutional and ecological networks of the Indo-West Pacific. Glob Environ Change 31:263–271. Available via http://www.sciencedirect.com/science/article/pii/S095937801500014X
US Commission on Ocean Policy (USCOP) (2004) An ocean blueprint for the 21st century final report of the U.S. Commission on ocean policy. Available via http://govinfo.library.unt.edu/oceancommission/documents/full_color_rpt/welcome.html. Accessed 5 Nov 2016
Vatn A, Vedeld P (2012) Fit, interplay, and scale: a diagnosis. Ecol Soc 17:12
Visbeck M, Kronfeld-Goharani U, Neumann B et al (2014) Securing blue wealth: the need for a special sustainable development goal for the ocean and coasts. Mar Policy 48:184–191. doi:10.1016/j.marpol.2014.03.005
Young OR (1996) Institutional linkages in international society: polar perspectives. Glob Gov 2:1–24
Young OR (2002) The institutional dimensions of environmental change: fit, interplay, and scale. MIT Press, Cambridge

Julia A. Ekstrom is the Climate Adaptation Initiative Director at the UC Davis Policy Institute. She is a social scientist with expertise in climate change adaptation and vulnerability, coastal and marine management, and institutional and governance analysis. Before joining UC Davis, she worked at Natural Resources Defense Council and UC Berkeley examining vulnerabilities and adaptation to climate change and ocean acidification. She was a postdoctoral scholar at Stanford University developing software for analyzing environmental governance. Her current research focuses on the perspectives and needs of practitioners for using climate science to plan for extreme weather events. She has a PhD in Marine Science from UC Santa Barbara and a BS in Conservation and Resource Studies from UC Berkeley.

Gloria T. Lau is an Adjunct Professor at Stanford University. She has over a decade of experience in data science and informatics research across multiple industries, from the legal sector to social network, and to the consumer internet space. She is an equity partner at Data Collective venture capital fund, and previously worked at Google, LinkedIn and Thomson Reuters. Her current work is primarily focused on artificial intelligence and machine learning applications to assist the professional workforce. She has an MS and PhD in Engineering and a PhD minor in Computer Science from Stanford, and a BS in Engineering from UCLA.

Kincho H. Law is a Professor of Civil and Environmental Engineering at Stanford University. His professional and research interests focus on the application of computational and information science in engineering. His work has dealt with various aspects of computational science and engineering, legal and engineering informatics, engineering enterprise integration, web services and Internet computing. His current research interests also include smart manufacturing, smart city, monitoring and control systems and large scale simulations on high performance computers. He obtained his MS and PhD in Civil Engineering from Carnegie Mellon University, and a BS in Civil Engineering and a BA in Mathematics from the University of Hawaii.

On the Spot and Map: Interactive Model-Based Policy Support Under Deep Uncertainty

Erik Pruyt, Tushith Islam, and Thomas Arzt

Abstract In this chapter, we discuss and demonstrate the use of 'on the spot' and 'on the map' scenario exploration and policy-support in workshop settings. First we justify the need for exploratory model-based policy workshops. Then we present some methods and techniques needed for these workshops. Special attention is paid to new techniques we believe are crucially needed for this kind of interactive workshop if time is of the essence, namely (1) techniques to quickly generate small but diverse ensembles of alternative scenarios, and (2) techniques to visualize whole-system dynamics on maps by means of geospatial animations. We subsequently describe a workshop related to the 2015–2016 European refugee crisis for which this approach and these techniques were developed and used. Finally, we discuss shortcomings and improvements to deal with these shortcomings and conclude.

Introduction

Although policy modelling and simulation is used to inform policy-makers before they enter the policy arena and engage in political negotiations, it is not commonly used interactively in the policy arena for scenario exploration and joint model-based policy-making. High-level policy negotiations often consist of verbal exchanges of

---

E. Pruyt (✉)
Faculty of Technology, Policy and Management, Delft University of Technology,
Jaffalaan 5, 2628BX Delft, The Netherlands
e-mail: e.pruyt@tudelft.nl

T. Islam
Greenwood Strategic Advisors, Zugerstrasse 40, CH-6314 Unteraegeri, Switzerland

Faculty of Technology, Policy and Management, Delft University of Technology,
Jaffalaan 5, 2628BX Delft, The Netherlands
e-mail: tushith.islam@greenwood-ag.com

T. Arzt
SAT Strategic Advisors for Transformation GmbH,
Badenweilerstr. 4, 79115 Freiburg, Germany
e-mail: arzt@sat-ag.com

J.R. Gil-Garcia et al. (eds.), *Policy Analytics, Modelling, and Informatics*, Public Administration and Information Technology 24,
DOI 10.1007/978-3-319-61762-6_14

ideology-inspired narratives—some(times) based on model-based background studies. Policy modelling and simulation is used even less often in crisis situations when time pressure is really high.

For modelling and simulation to have a bigger impact on the outcomes of high-level policy-making processes related to complex uncertain issues, we believe modelling and simulation should play a more central role in these processes. To bring policy modelling and simulation to the forefront of the policy-making arena, new approaches and techniques are needed that allow for more interactivity and instant systemic understanding.

In this chapter we discuss two recent methodological contributions that could help in bringing policy modelling and simulation to the forefront of policy-making processes, namely (1) an output-oriented sampling method to quickly generate small sets of strongly divergent scenarios, and (2) on the spot and on the map animation to visualize the dynamics of model-generated scenarios. These techniques were developed more specifically to support a high-level model-based policy-exploration workshop during the 2015–2016 European migration crisis. In this chapter, we will use this workshop and some of our work related to the 2015–2016 European refugee crisis as an illustration.

First, we discuss the problem (i.e., the reasons for developing these techniques). Then, we zoom in on the methodology, methods, and techniques used in the work described in this chapter: We introduce the overall exploratory modelling approach, the modelling method used, and the two recent contributions we believe are crucial for the successful use of modelling to support policy-making workshops, namely an approach for quickly generating small subsets of scenarios that are representative for many similar plausible scenarios, and geospatial visualizations and animations. We subsequently discuss the 2015–2016 European refugee workshop and how different techniques were used before, during and after the workshop. In the next section, we discuss problems encountered during the workshop as well as future developments inspired by it. Finally, we provide some conclusions.

Context and Problem

The job of policy-makers and politicians consists of—or should to a large extent consist of—identifying potential problems and making policies to address them. Many problems and grand challenges slowly evolve over a long period of time, which often means that there is much time for investigation, policy-making processes, and gradual implementation of agreed upon policies and measures. In many such cases, there is enough time for policy modelling to provide insights that can be used for policy-making. Energy policies to mitigate climate change and pension age policies to deal with double societal ageing (Auping et al. 2015) are examples of policies and issues of the slow-moving type for which much time was available to perform extensive model-based research. This is not always the case though: some long-term issues require rapid decision-making and implementation. Whether urgent or not, long-term dynamics are likely to be uncertain due to a lack of knowledge about the future state of the world, which is problematic from an 'evidence-based' policy-making point of view.

Other social problems and grand challenges suddenly emerge and require on the spot foresight, urgent policy-making, and immediate implementation before accurate information and data could be gathered. Recent outbreaks of infectious diseases like the 2014–2015 Ebola outbreak in West Africa (Pruyt et al. 2015) and the 2015–2016 refugee crisis in Europe are examples of crisis-like issues in which policies and decisions were urgently needed and time to gather data and information, develop accurate models, and perform extensive model-based research was very limited. Such crises are often characterized by a lack of data, information, thorough understanding about what is happening, and time to gather better data and information which is once more problematic from an 'evidence-based' policy-making point of view.

Many potential societal issues and grand challenges (both transition-like and crisis-like) are so complex, dynamic, and uncertain, that good policy-support or decision-making requires more than just information, data, and opinions. Since the late 1950s, modelling and simulation approaches have been used to support policy-makers and decision-makers facing complexity and dynamics, especially for long-term issues. Depending on the characteristics of an issue, the approach used, the level of detail, the desired degree of accuracy, and the level of expertise of the modellers involved, modelling and simulation of complex social problems and grand challenges can take hours, weeks, months, or years. The time needed to make and use models is in fact partly the result of methodological and modelling choices.

Traditionally, uncertainty has also been dealt with in different ways. Traditional approaches for handling uncertainty in modelling and simulation could be summarized as (1) reducing the uncertainty through more research (i.e., increased accuracy through more evidence), (2) marginally addressing uncertainty through a small set of alternative assumptions, or (3) most commonly by simply overlooking or ignoring uncertainty. In many pressing complex social issues and grand challenges, however, decisions have to be made presently and waiting for (more) evidence, only marginally addressing uncertainty, let alone overlooking it, could be expensive—even dangerous. In other words, evidence-based approaches may be insufficient or too late. In such circumstances, the challenge is to find ways to let planning and decision-making proceed despite the presence of complexity and uncertainty.

Since the early 1990s, exploratory methodologies and approaches have been developed for dealing with 'deep' uncertainty (Bankes 1993). These methods embrace uncertainty which could be seen as a fourth fundamental way of dealing with uncertainty. In essence, these exploratory approaches make use of computational models to generate a wide variety of what-if scenarios and test policies across these scenarios. These ensembles of scenarios are generated preferably such that they cover the uncertainty space spanned by all remaining uncertainties. In contrast to traditional ways of handling uncertainty in modelling and simulation, which almost exclusively focus on 'known knowns' through shared assumptions and best estimates, exploratory approaches require easily manageable models that enable one to generate a maximally diverse set of possible ways in which the various irreducible uncertain factors might play out. Smart computer algorithms as well as human intelligence are then used to identify which differences really make a difference, and what policies are effective and efficient across these differences.

To provide policy-makers with exploratory computer assisted reasoning requires smart and fast methods and techniques, among else from data science, policy analytics, and informatics. Data science techniques could be used to inform model building; Informatics could be used to merge data and (multiple) models; Data science, machine learning, and policy analytics applied to the model-generated ensembles could be used to generate insights (e.g., to identify high leverage policies) (Pruyt 2015, 2016).

However, just performing integrated analyses under deep uncertainty, including (near) real-time data, and producing policy insights and advice, is also insufficient. More real interaction is needed too. That is, exploratory computer assisted reasoning necessitates close interaction between analysists and policy-makers. Policy analysts need to be able, on the spot, to generate ensembles of plausible scenarios and test policies across these ensembles. Given the nature of interactive processes, speed is of the essence. Speed is even more important for issues that require urgent decisions to be made. Until recently, speed was a problem for exploring vast uncertainty spaces, for generating and dealing with large ensembles of simulation runs is computationally intensive and computing power was limited. Today, increased computational power and some recently developed techniques allow for real-time interactivity and visualization of the outcomes of in-depth analyses in visually attractive, comprehensive, and comprehensible ways.

Although interactive workshops are sometimes used to design robust policies for important long-term problems, they are not (yet) commonly used during high-level policy-negotiation processes to interactively explore the effects of possible scenarios and jointly design policies for dealing with crises. So, while on the spot model-based policy support may technically be possible today, in high-level crisis-related policy processes, it is still in its infancy. For it to break through and really impact high-level policy-making, new techniques are needed, including new sampling and simulation techniques as well as techniques for on the spot visualization and animation of (ensembles of) scenarios. Such techniques should enable one to simulate policy-makers' ideas and suggestions on the spot without reducing the remaining uncertainty (e.g., by quickly generating a small subset of exemplar scenarios that, together, are representative for the output space), and to visualize the effects in a way that enables one to get at grasps with the overall system dynamics over time and space as well as the consequences of uncertainty.

In the next section, we will discuss a new output-oriented sampling technique to generate such small ensembles of divergent scenarios as well as techniques for on the spot visualization and animation of scenarios in the light of the 2015–2016 European refugee crisis. Before introducing these techniques, we will first discuss the overall methodological framework within which this work is situated as well as the modelling approach used to model the 2015–2016 European refugee crisis.

Methodology and Methods Used

Deep Uncertainty, Exploratory Modelling and Analysis, Robust Decision-Making, and Scenario Discovery

Most socio-technical issues and grand challenges are characterized by many so-called "deep uncertainties" due to conflicting information, incommensurate perspectives and worldviews, conflicting value systems, contested knowledge, as well as variability, immeasurability, and indeterminacy. Deep uncertainty has been defined as situations in which the parties to a decision do not know or agree on the model(s) that describe the system, the probability distributions and other model inputs to be used, and the evaluation of the outcomes of the model(s) (Lempert et al. 2003). Deep uncertainty thus refers to, and encompasses, uncertainties due to conflicting information, irreconcilable perspectives, conflicting value systems, and contested knowledge. The presence of deep uncertainty does not mean that a model-based approach cannot be used. It means that, in such situations, using a single model with a single set of inputs and assumptions, and a single evaluative perspective is insufficient.

Model-based methodologies such as Exploratory Modelling and Analysis (Bankes 1993; Agusdinata 2008; Lempert et al. 2003), Robust Decision-Making (Lempert and Collins 2007), and Scenario-Discovery (Groves and Lempert 2007; Bryant and Lempert 2010) have been—and are being—developed for dealing with situations characterized by deep uncertainty.

Exploratory Modelling and Analysis (EMA) is a model-based scenario methodology which consists of the following overall steps: developing sets of plausible exploratory computational models, simulating these models across large uncertainty spaces, using smart algorithms to explore and analyse the resulting ensembles of scenarios in search for un/desirable types of scenarios and policy leverage, and designing robust adaptive policies using advanced policy analytical techniques. A more detailed version of the EMA process is displayed in Fig. 1a. Although these steps correspond to the steps in a traditional modelling and simulation process, they differ in terms of the techniques used and the way in which results are interpreted.

EMA is useful for in-depth exploration, which necessarily takes time (at least a couple of hours). It is therefore, in its current form, not suited for interactive on the spot policy-support. It is nevertheless useful in preparation for an interactive policy-support workshop and for ex-post in-depth analysis and for testing the outcomes of interactive policy-support workshops.

Robust Decision-Making (RDM) is an iterative decision analytic framework that helps to identify potentially robust strategies, find potential vulnerabilities of such strategies, and evaluate the trade-offs among them under deep uncertainty. RDM makes intensive use of exploratory modelling and scenario discovery techniques. Scenario Discovery mostly refers, within the field of decision-making under uncertainty, to the use of specific techniques/algorithms (i.e., PRIM (Friedman and Fisher 1999; Bryant and Lempert 2010; Kwakkel et al. 2014)) within an interactive RDM

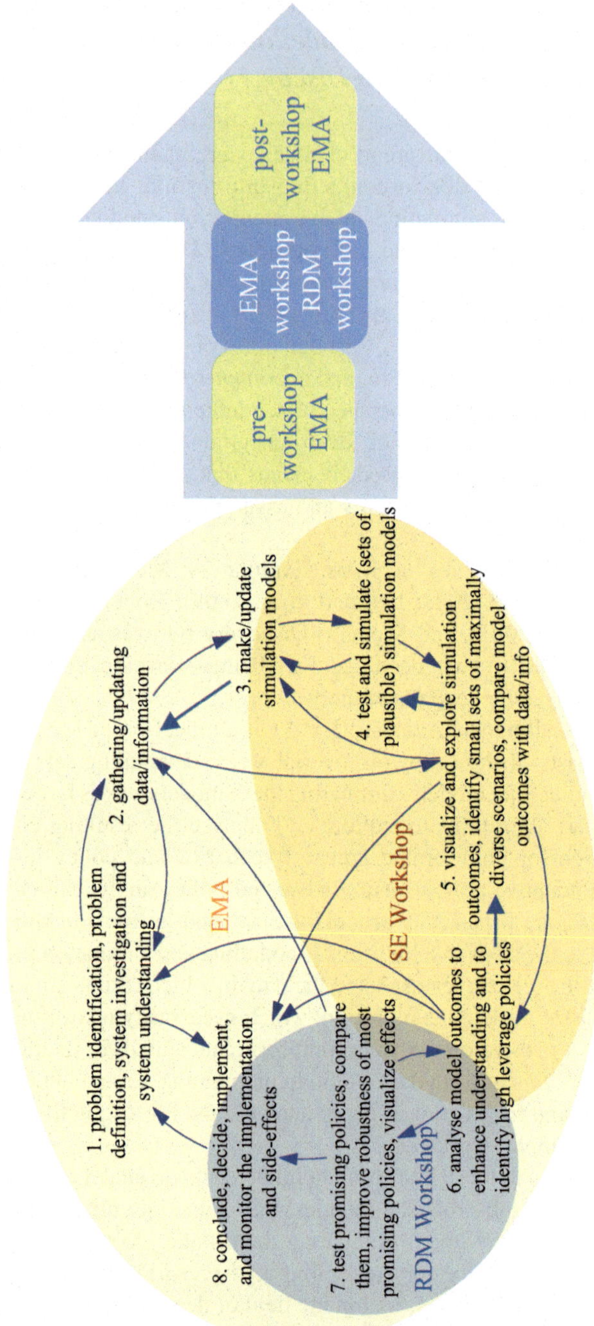

Fig. 1 Detailed EMA steps (*left*); locus of EMA and RDM workshops in the overall EMA process (*right*)

process to find subsets of model-based scenarios characterized by particular un/desirable properties and their joint causes in order to design overall robust policies that address these causes. Interactive RDM workshops thus focus on steps 6, 7, and 8 of the EMA cycle. Workshop settings are often used for steps 1 and 2 in Fig. 1a too (Vennix 1996). Workshop settings could also be used for the core "exploratory modelling and analysis" steps—namely steps 4, 5 and 6—if models can be adapted on the spot, ensembles can be generated rapidly, and results can be visualized and/or analysed on the spot. This is what we focus on in this chapter. We refer to this kind of workshop as "simulation and exploration" (SE) workshop, since these steps are all about figuring out what experiments to do, to simulate them, and explore the simulated consequences of these experiments. Figure 1b shows where RDM and SE workshops fit in the bigger EMA cycle: a SE workshop would typically be organized before or in junction with an RDM workshop.

In this chapter, we introduce and demonstrate both the SE workshop and the in-depth ex-post use of EMA. We show (1) how the combination of output-oriented sampling, scenario discovery, and dynamic visualization of simulation runs can be used for "on the spot and on the map" policy-support, and (2) how EMA can be used to test the outcomes of an interactive workshop across large uncertainty spaces.

The workshop approach presented here consists of the following steps: (1) simulation of scenarios and policies such that small but very diverse ensembles of simulation runs are generated, (2) clustering of the ensembles into subsets of similar runs and identification of exemplar runs for each of the subsets, (3) selection of exemplar simulation runs for each of these clusters/classes and visualizing the dynamics of the exemplars onto maps and other dynamic visualizations.

The full EMA approach used in conjunction to the workshop approach is rather—although not entirely—similar: Instead of using fast output-oriented sampling we use traditional input-oriented sampling, and instead of creating an animated visualization of the dynamics, we use graphs to visualize the effects of policies across large ensembles of scenarios over time.

During interactive processes, 'quick and dirty' approaches are most useful. After the workshop, crucial steps that require systematic research, like steps 3, 4, 5, 6, 7 and 8 can then be revisited, using traditional sampling approaches and simulation of (large) ensembles followed by in-depth analysis and policy robustness testing using methods and techniques that require more time than is available during workshops. These two approaches are thus complementary.

The Modelling Method: System Dynamics Modelling and Simulation

Exploratory modelling and model-based SE workshops require one or more manageable models. The modelling and simulation method used here is the System Dynamics modelling and simulation method (Forrester 1961; Sterman 2000; Bossel

2007; Ford 2009; Pruyt 2013; Rahmandad et al. 2015). System Dynamics (SD) is a modelling and simulation approach for dealing with dynamically complex issues that are characterized by feedback effects and accumulation effects. SD has been developed since the late 1950s (Forrester 1958, 1961, 1968, 1969, 1971). For several decades, the SD field has been a mature field characterized by a clear focus and stable state-of-the-art, as well as by strict diagramming conventions (e.g., boxed "stock variables" are used for underlying integral equations).

While leaving the core SD approach intact, recent innovations—including the use of SD models as scenario generators for exploratory modelling studies (Lempert et al. 2003; Kwakkel and Pruyt 2013, 2015)—significantly transform the state-of-the-art of SD modelling and simulation, and also the ways in which it could be applied (e.g., in workshop settings) (Pruyt 2015). Some of these innovations are combined in this chapter. That is, the work presented in this chapter combines a SD approach for dealing with issues characterized by deep uncertainty, innovations in sampling to generate a small set of very diverse exemplar scenarios, innovations in approaches to cluster and analyse the outcomes of SD models, and new ways to visualize outcomes of simulation models.

Although alternative modelling and simulation approaches could have been used here, SD was chosen because the European refugee issue is characterized by feedback effects, accumulation effects, and delays—which are the key components of SD models. Moreover, SD can be applied quickly for urgent issues when speed is of the essence—which was the case in the 2015–2016 European refugee crisis. Finally, it enables one to provide a big picture overview of grand challenges and complex issues. This is why SD has been used a lot to study grand challenges and complex issues, and test policies to address them. Examples of grand challenges addressed with SD include climate change mitigation (Fiddaman 1997) and health related issues (Thompson and Tebbens 2007).

Traditional Sampling and Exemplar Selection Versus Output-Oriented Sampling and Exemplar Selection

Two different sampling and exemplar selection approaches are used here—a 'quick-and-dirty' approach for workshop settings and a traditional 'slow-but-systemic' approach for in-depth back-room analyses. Quick-and-dirty output-oriented sampling and exemplar selection approaches like the ones in Pruyt and Islam (2015) and Islam and Pruyt (2016) are particularly useful for SE workshop settings. The multi-stage adaptive sampling approaches used there—referred to as Behavior Space (BS) sampling—are inspired by previous work on adaptive sampling including Bucher (1988), Bergot (2001) and Bishop et al. (2001). Here, we use the approach presented in Pruyt and Islam (2015) as depicted in Fig. 2. This particular adaptive sampler enables one to generate a wide variety of behaviors with sufficient density to identify the corresponding input spaces and select representative exemplars.

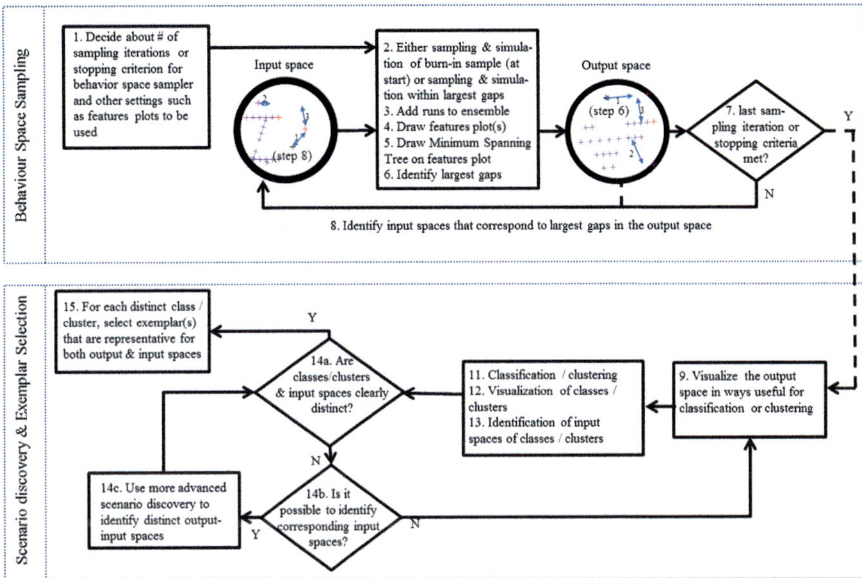

Fig. 2 Flow chart of Output-Oriented Sampling and Exemplar Selection process. Adapted from: Pruyt and Islam (2015)

Starting from a burn-in sample generated with a Latin Hypercube sampler, the BS sampler draws in each of p resampling iterations a minimum spanning tree (i.e., a graph that connects all vertices at minimum length) on a two-dimensional representation of the output space to determine the n largest gaps to be filled, each with m Latin-Hypercube samples (see Fig. 2). For a small ensemble of simulation runs, the resulting ensemble is extremely diverse, yet sampled densely enough for exemplars to be identified by means of behavior-based classification such that the resulting exemplars, together, are representative for the behavioral spectrum the model could generate. Here we use a simple classification approach to select exemplars. Alternatively, one could use time-series clustering based on dynamic pattern features based on Yücel (2012) and Yücel and Barlas (2011) or other clustering approaches like Dynamic Time Warping (Petitjean et al. 2011). Finally, simple visualization techniques are used to identify the corresponding areas in the input space that predominantly generate these behaviors.

If time and process permit to perform ex-post analyses, then it makes sense to also perform traditional input-oriented sampling (e.g., Latin Hypercube sampling (McKay et al. 1979)) to generate (tens of) thousands of scenarios, perform in-depth analyses, design adaptive policies and apply multi-objective robust optimization (Hamarat et al. 2013, 2014), *et cetera*. Although traditional sampling and analysis approaches may be too slow to quickly generate representative ensembles, they are more suitable for in-depth back-room analyses than BS sampling. Here, we use traditional Latin Hypercube sampling for post-workshop analysis under deep uncertainty.

Maps, Visualizations, and Animations

Modellers and model-based analysts often generate behaviour-over-time graphs like the one displayed in Fig. 3. These graphs are argued, at least by SD modellers and analysts, to be useful for studying the dynamics of complex models/systems over time. For those who are trained in reading such graphs, that may be true, for relatively simple/small cases. However, behaviour over time graphs are hard to read if more than a dozen dissimilar simulation runs are displayed or if more than a few dozen similar simulation runs are displayed. Moreover, in spite of the fact that behaviour-over-time graphs enable one to visualize the dynamics over time of one key performance indicator for a few subgroups (e.g., the dynamics of refugee accumulation in a few EU countries) or of a few key performance indicators for one subgroup (e.g., several key performance indicators for one EU member state), they do not enable one to grasp or communicate the big picture dynamics across many key performance indicators for many subgroups (e.g., the dynamics of refugee accumulation as well as the dynamics of political tensions across all EU countries).

Even more problematic, from a policy-support point of view, is that many policy-makers seem to be unable to properly read and interpret behaviour-over-time graphs. From experience, we know that many people even have a hard time understanding simple graphs that summarize accumulation over a period of time. In an ever more visually-oriented digital world, it is the task of modellers and model-based analysts to provide better visualizations and animations—visualizations and animations that are comprehensible to those whose decision-making is supposed to be supported (i.e., policy-makers and decision-makers, and in many public policy cases, the public at large). This requires an additional effort from the side of modellers and policy analysts who are used to generate and interpret graphs that are well suited for them but not for policy-makers or the public at large. The investment required to develop

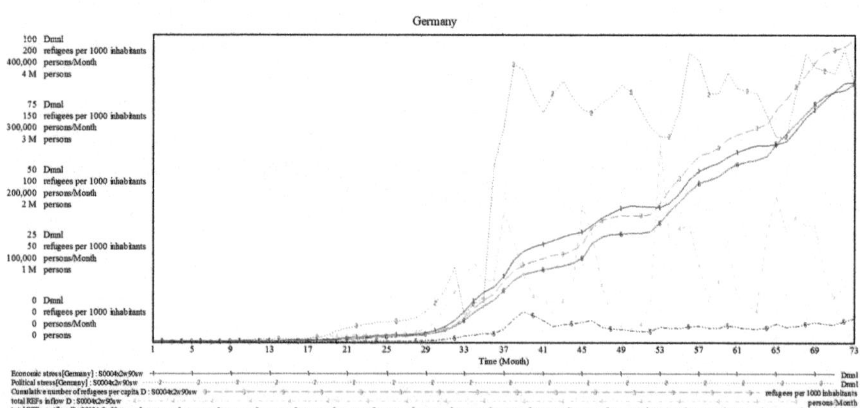

Fig. 3 Behaviour over time graph of a randomly selected simulation run (given particular values for uncertainties and decisions) generated with the refugee model

policy-maker-friendly graphs is relatively modest, since a large variety of visualizations and animations exist (see for example the matplotlib.org gallery which can be linked directly to (the outputs of) simulation models).

Two particular areas that deserve more attention are geo-spatial visualizations and geo-spatial animations. Many current policy problems have an important geo-spatial component to it (e.g., refugee crises, outbreak and spreading of diseases, adverse impacts from climate change across countries), which can be mapped on global, regional or local maps, or visualized with dynamic map-based animations.

For direct policy-support to policy-makers, visualizations and animations need to be visually attractive. To interact directly with high-level policy-makers in the case of the European refugee crisis, a visually attractive interface for showing animations of simulation runs (one at a time) on multiple key performance indicators was developed using serious gaming software. Figure 5 shows an animation related to the European problem of the 2015 refugee crisis developed for the high-level workshop. Development of fancy interfaces is time consuming though.

Most analysts do not necessarily want visually attractive graphs and animations. They want visualizations and animations that are useful for their analytical purpose and can be made or changed on the spot. More advanced visualizations and animations may nevertheless help to grasp the bigger picture, also for researchers and analysts. We believe it is possible to bridge both worlds. To that purpose, we have developed visualizations that are useful for analysts and decision-makers alike (see the Discussion). These figures have been generated in jupyter notebooks which means they can be adapted and connected to any model, and they can be rendered on the spot. They allow one to visualize data and simulation runs in any way desirable. Although less fancy and beautiful than the interface built for the high-level workshop reported on below, they are more versatile.

High-Level Policy-Exploration Workshop: The European Refugee Crisis

The 2015–2016 European Refugee Crisis

Between July 2015 and May 2016, several EU member states—including Austria, Germany, Sweden, Belgium and the Netherlands—have been dealing with large cumulative inflows of refugees from Syria, Afghanistan, Iraq, Eritrea and several other countries. Although these cumulative inflows were large, they were not unprecedented: large migration flows have occurred every decade or every few decades.

Countries on the main routes connecting Turkey to these Northern European countries were confronted to large—albeit mostly temporary—flows of refugees passing through their territories. Images of the refugees making their way across Europe and stories about their journeys were all over the news, invoking highly emotional responses—positive as well as negative. Political tensions in many European countries were rising dramatically, not only in transit and final reception countries, but also within and between other EU member states. Many blamed Merkel for her "*Wir schaffen das*" statement. While some politicians first claimed we should be able to manage, many claimed the flows and accumulations of refugees and migrants were unmanageable or would soon become unmanageable. In political circles, many claimed that the future of the European Union depended upon its capacity to deal with this "*European*" refugee crises. While successive EU policies did not seem to have the desired effects. Europe was not only trying to deal with a crisis—because of the refugee crisis, Europe was in crisis.

The model-based workshop and in-depth analysis reported on in this chapter were performed in November 2015 while tensions in Europe were still on the rise and it was uncertain what would happen in 2016. At the time of writing this chapter, end of April 2016, the deal between the EU and Turkey[1] seemed to work for Europe—but not necessarily for displaced Syrians.[2] That is, the Balkan route was closed and mass migration from Turkey to Greece had ended. With the EU-Turkey deal, hot spots on the Greek Islands and the Balkan borders closed, it is practically impossible for refugees to make their way from Turkey via Greece to Northern Europe. However, if the situation deteriorates, then refugees are likely to take alternative, more risky and expensive, routes like the middle-Mediterranean route.

The Simulation Model

In November 2015, we developed a SD simulation model regarding the European refugee crisis for a high-level workshop. The simulation model connects Turkey—via Greece, the so-called Balkan route, Hungary, and Austria—to reception countries in North-Western Europe (Germany, Sweden, the Netherlands). Figure 4a displays the connections between countries of transit and destination (the ones modelled are displayed in bold).

The simulation model consists of an interlinked set of country sub-models with modules to simulate refugee movements and decisions, reception capacity and asylum seeker dynamics, and potential economic and political consequences of refugee transit and reception. Figure 4b shows a simplified stock-flow diagram of a country sub-model. In these sub-models, refugees entering a country either pass through, ask for asylum, or remain in the country illegally. Part of those whose cases are rejected also join the illegal ones. They may also join those who seek fortune in another country. The more refugees that are staying in a country, the more housing is needed. This may cause financial and political stress. Large streams of refugees passing may cause political stress too. Political stress and frustration of basic needs from the part of refugees makes a country less attractive to migrate to, which is—in the smartphones era—fed back to those *en route* or about to migrate.

[1] See http://europa.eu/rapid/press-release_MEMO-16-963_en.htm.

[2] See for example http://www.spiegel.de/international/europe/the-refugee-deal-between-the-eu-and-turkey-is-failing-a-1094339.html.

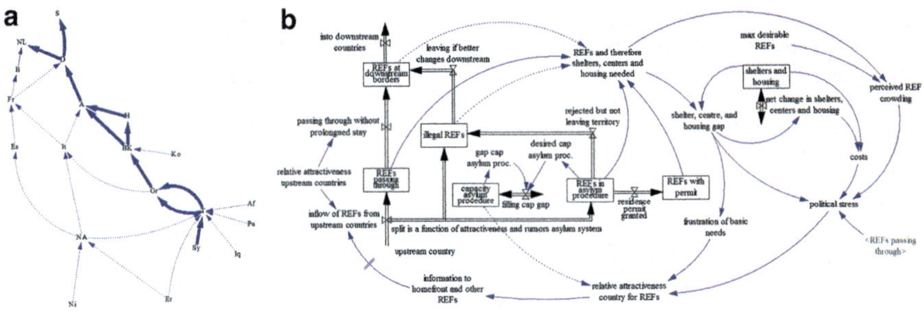

Fig. 4 (**a**) Flows between countries (*left*); (**b**) extremely simplified SFD of the country sub-models

Table 1 Four scenarios simulated before and at the start of the workshop together with high-level decision-makers

	S1: War in Syria ends soon	S2: War in Syria will go on
Policy 1: Angela Merkel's "Wir Schaffen das"	Ensemble simulation 1a	Ensemble simulation 1b
Policy 1: Perfect implementation of EU-17-point-plan	Ensemble simulation 2a	Ensemble simulation 2b

Simulations, Analyses, and Visualizations

Setting the Stage: Pre-Workshop Simulations

Before the workshop, the simulation model was used to generate base ensembles, as well as basic policies (what could possibly be changed by EU policy-makers) and scenarios (what cannot not be changed by EU policy-makers). These combinations are displayed in Table 1. The first ensemble of simulations, simulation 1a, combines the simulated effects of Merkel's "*Wir schaffen das*" policy with an abrupt (unrealistic) ending of the armed conflict in Syria. The second ensemble of simulations, simulation 1b, combines the effects of Merkel's "*Wir schaffen das*" policy with a "normal" continuation of the armed conflict in Syria. The third simulation/ensemble, simulation 2a, combines the effects of a perfect implementation of the EU17-point plan of action on top of Merkel's "*Wir schaffen das*" policy in combination with an abrupt (unrealistic) ending of the war armed conflict in Syria by 2016. The fourth simulation/ensemble, simulation 2b, combines the effects of a perfect implementation of the EU17-point plan of action on top of Merkel's "*Wir schaffen das*" policy in combination with a "normal" continuation of the armed conflict in Syria.

Fig. 5 Animation based on outputs of a simulation model regarding the European impact of the refugee crisis

The Workshop

During the workshop, we showed the effects of these scenarios and policies using a fancy gaming interface (displayed in Fig. 5) to visualize the simulated dynamics over time for multiple countries and on multiple key performance indicators. This enables one to assess the effects of scenarios and policies on the spot and on the map, although only for 1 exemplar scenario at a time.

Using the example of the *what-if* simulation/ensemble combining the (unrealistic) perfect implementation of the EU17-point plan of action with the (unrealistic) scenario in which the armed conflict would be over by the end of 2015, we showed that—even in case of a short-term resolution of these conflicts and a perfect implementation of the EU17-point plan—the European refugee crisis would keep affecting reception countries due to accumulation effects.

During the workshop, participants proposed several what-if scenarios and policies to be simulated, such as (1) a policy to provide housing, education (and work) in the region (e.g., in Turkey, Jordan and Lebanon), (2) the closure of internal EU borders with the external EU borders remaining open, (3) the continuation and slight intensification of current armed conflicts, (4) the outbreak of more armed conflicts and related refugee flows to Europe, (5) the establishment of safe zones, (6) the sudden closure of German borders, (7) the announced closure of German borders with an anticipatory wave of refugees, and (8) the establishment of safe zones in the region and a policy of sending back all refugees who are not already officially in the actual asylum process to such safe zones.

For each of these suggested scenarios/policies, we used adaptive sampling (Pruyt and Islam 2015) to generate ensembles of 100 plausible simulations runs, and subsequently reduced these ensembles to small sets of representative scenarios. Starting from an initial Latin-Hypercube (LH) burn-in sample of 20 runs, we resampled twice, each time sampling five LH samples within the six largest distances of a minimum spanning tree drawn on a features plot of roughness (i.e., the stretched length) of a simulation run and the maximum value for the 'Cumulative number of refugees in D'. This resulted for example in 100 simulation runs displayed in Fig. 6a for the "*Wir schaffen das*" policy. These runs were subsequently plotted on a features plot with the value of the square root of the maximum value divided by the mean value on the x-axis and the roughness on the y-axis as displayed in Fig. 6b. Breaking this features plot into different classes as in Fig. 6b and selecting the run with the median terminal value in each of these classes as an exemplar, results in the five exemplar runs in Fig. 6c. The corresponding outcomes on another key performance indicator are displayed in Fig. 6d.

Looking at the exemplars in Fig. 6c, four exemplar scenarios represent cases with further increasing cumulative numbers of refugees in Germany. One of the exemplar scenarios (the bottom-most run, in blue) represents the situation in which the cumulative number of refugees would remain more or less constant after 2015. The red and purple run represent situations in which the rise in cumulative refugees can be dealt with. That is, the stress level is not at its maximum. Note that all scenarios show a temporary drop in political stress. This is part of the way in which the "*Wir schaffen das*" policy was implemented. A higher stress level results in two of the cases (blue and green, above the previous bottom-most run) with fewer refugees in Germany. In these cases, feedback from refugees in Germany to others who might consider migrating to Germany makes Germany less attractive, even without additional policies (i.e., policies are not activated in the "*Wir schaffen das*" ensemble). Note, however, that this is not always the case: the yellow-green scenario (with the top-most terminal value) shows at least one situation in which the stress level oscillates while the cumulative number of refugees substantially increases.

Post-Workshop: In-Depth Analysis

Right after the workshop, these scenarios and policies were simulated thousands of times, using LH sampling, for each of the policies. These resulting ensembles are displayed in Fig. 7. It displays the envelopes-over-time graphs until month 50 for each of the policies (on the left) and kernel density estimates of the terminal values after 72 months (on the right), split out for the three main exogenous scenarios (S1: the war in Syria ends; S2: the war in Syria continues; S3: the war intensifies and spreads to other regions). Reference policies P1 ("*Wir schaffen das*") and P2 ("*Wir schaffen das* + perfect implementation of the EU 17 point plan") are displayed in each of the graphs.

Figure 6 shows that the "*Wir schaffen das*" and "EU17 point plan" mean that Germany would have to be prepared to accommodate any inflow of refugees that would make it to Europe and would prefer Germany above other reception countries. If wars in the Middle East and Africa would intensify and extend, then refugee inflows could dwarf the 2015 inflow of refugees. Border closures, especially unanticipated and early border closures, would strongly reduce the refugee inflow into Germany (Fig. 7a, c, e), but at the detriment of upstream countries and refugees alike. Providing safe zones with housing, schools and jobs in the region outperforms anticipated border closures, at least if minimizing the number of refugees in

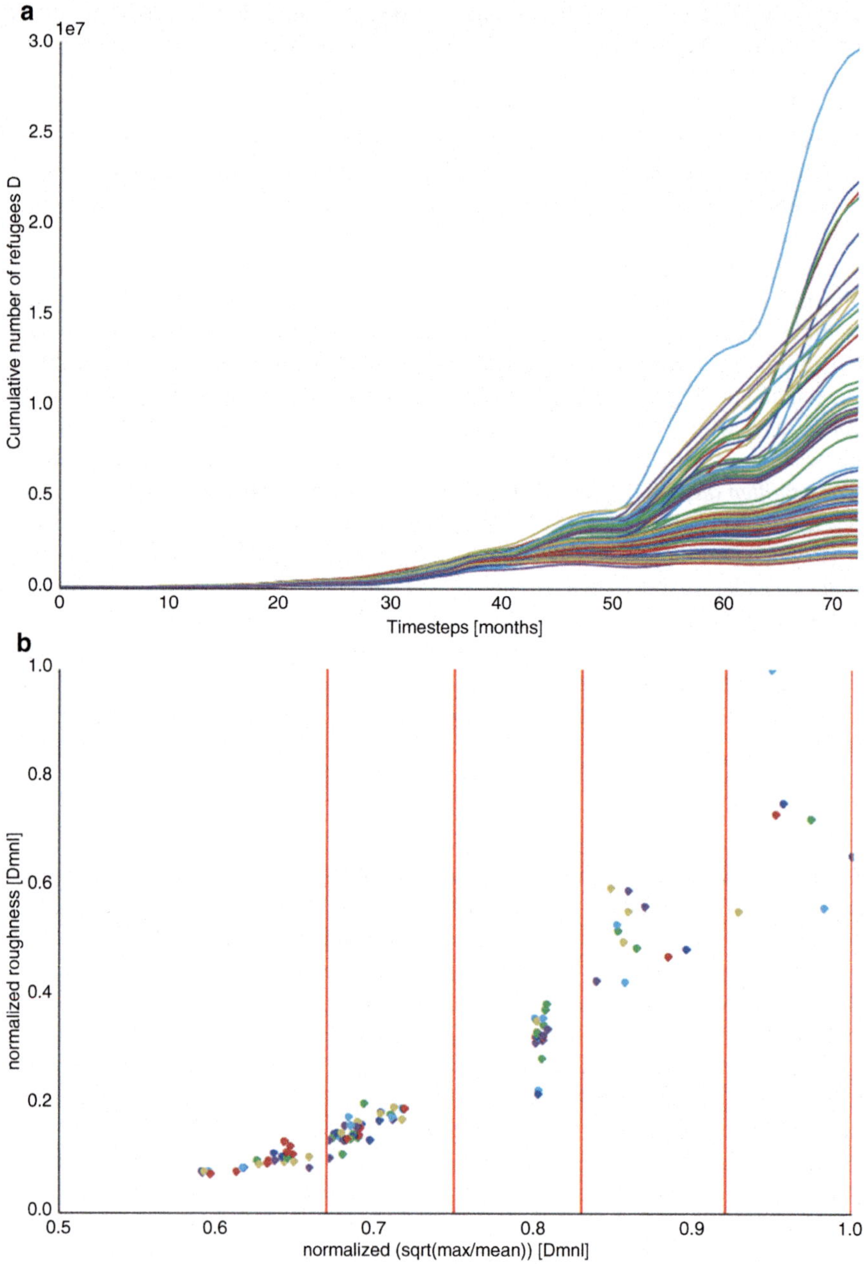

Fig. 6 Turning an ensemble of simulation runs generated with the adaptive sampler into a small set of exemplars. (**a**) 20 + 2 × 8 × 5 adaptive sampler runs. (**b**)Features plot roughness × max. breaks. (**c**) Exemplar per class for the cumulative refugees in D (median terminal value). (**d**) Corresponding exemplar behaviours for refugee-related political stress in D

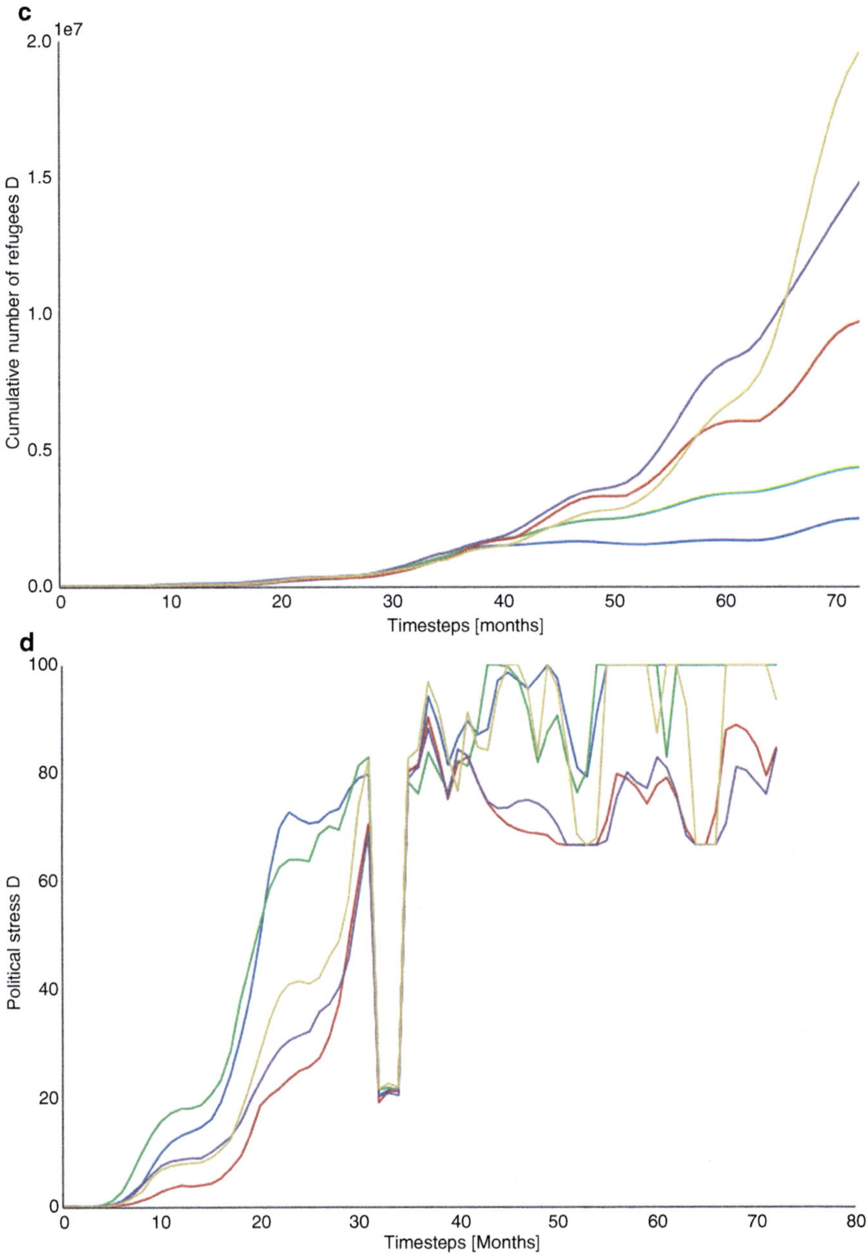

Fig. 6 (continued)

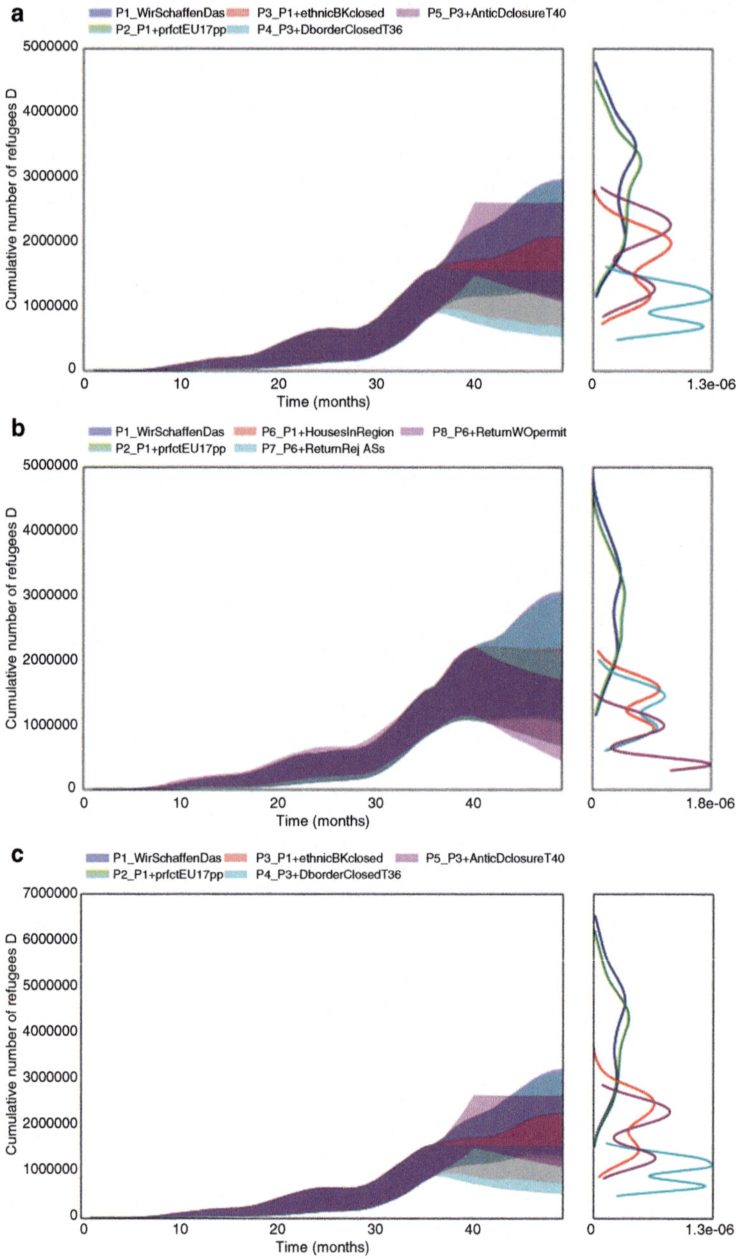

Fig. 7 Envelopes-over-time graphs for each the policies (*on the left*) and Kernel density estimates of the terminal values (*on the right*) for the three main exogenous scenarios (the war in Syria ends, the war in Syria continues and the war intensifies and spreads to other regions). (**a**) The war ends soon. (**b**) The war ends soon. (**c**) The war goes on. (**d**) The war goes on. (**e**) The war intensifies and extends. ( ) The war intensifies and extends

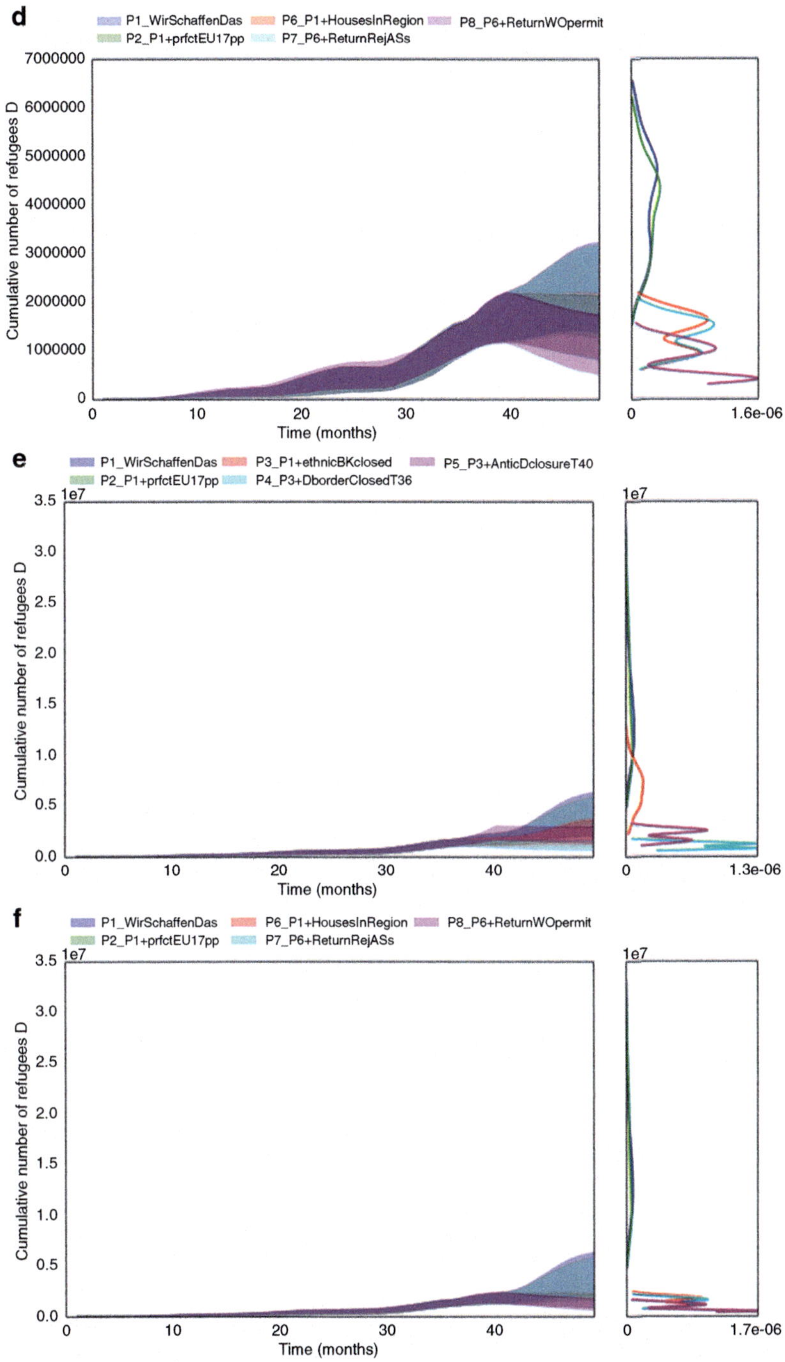

Fig. 7 (continued)

Germany would be the goal (see Fig. 7b, d, f). Returning rejected asylum seekers in addition to providing housing, schools and jobs in the region would further reduce the cumulative number of refugees. The biggest reduction would be achieved if all those who do not have a residence permit would actually be returned to these safe zones in the region too (see Fig. 7b, d, f).

The current EU-Turkey action plan corresponds more or less to border closures between Turkey and Greece (and along the Balkan route) in combination with the funding of safe zones with housing, schools and jobs in the region. Significant investments are needed though to provide housing, schools and jobs in the region, especially if armed conflicts in the Middle East and Africa intensify and extend. At the time of writing, the EU was also pursuing talks with nations in the Middle East and Africa to take back rejected asylum seekers. That would mean a shift from policy P6 to policy P7. Finally, several European countries were stalling the asylum and integration process to such an extent that it may take so much time for refugees to get a permanent residence permit and start the integration process that the conflict may be solved by then, which would correspond to P8, albeit delayed. The best situation would of course be an end to the armed conflict.

Discussion

A problem encountered in this and other studies is that behaviour-over time graphs and envelopes graphs are hard for people to read and grasp the big picture dynamics across the entire system. Decision-makers often prefer simple results, for example numbers that summarize the whole dynamics, like average values or cumulative values across a time span. At the workshop we were asked to provide a table that summarized the effects of these ensembles of simulation runs, like the Table for Germany in Fig. 8. The top-most table shows the minimum value, maximum value, and median value of the cumulative number of refugees received by Germany (in millions) between 01/01/2013 and 31/12/2016, predominantly via the Balkan route, for different sets of scenarios and policies, for a particular version of the model. The bottom table shows the resulting internal political tensions as included in the model, again only for Germany. With just a few external scenarios and policies, this already results in an overwhelming amount of information. Such tables are also problematic due to the fact that they lead to undesirable focus on crisp numbers, especially on the highest numbers, without fundamental knowledge of what is behind these numbers. Moreover, these numbers relate to modelled simulations only. Finally, such tables do not enable to visualize fine-grained dynamics over time, which in this case—and in many other cases—matters. It is not about terminal values, it is as much about how these numbers build up over time and the stress they cause.

Moreover, even without focussing on crisp numbers—just by comparing the ensembles of scenarios—it is possible to draw conclusions from Figs. 7 and 8. First, the "*Wir schaffen das*" policy/situation alone was not sustainable. Second, much more than the EU-17 point plan alone was needed, especially in case of intensifica-

Impact policies & scenarios on Germany 2013-2017 in terms of refugees in Germany (in millions - from/via Balkan Route)	Scenarios wrt further development of the war in Syria (and beyond): S1: War in S ends soon			S2: War in S will go on			S3: More wars than S&I			Outcomes across uncertain war scenarios (S1-S2-S3)		
Policies	Minimum	Median	Maximum	Minimum	Median	Maximum	Minimum	Median	Maximum	Minimum	Median	Maximum
P1: "Wir schaffen das" without further policies	1.02	1.93	3.07	1.28	2.22	3.22	1.87	3.21	6.40	1.02	2.45	6.40
P2: P1 + perfect implementation EU-17 point plan	0.97	1.88	3.01	1.17	2.14	3.12	1.48	2.88	5.85	0.97	2.30	5.85
P3: P1 + partial (ethnic) closure of Balkan borders	1.02	1.81	2.60	0.69	1.44	2.23	0.95	1.79	3.67	0.69	1.68	3.67
P4: P3 + sudden internal border closures	0.51	1.00	1.54	0.51	1.00	1.54	0.50	0.95	1.58	0.50	0.98	1.58
P5: P3 + anticipated internal border closures	0.64	1.36	2.06	1.05	1.84	2.63	1.10	1.92	2.87	0.64	1.71	2.87
P6: P1 + more/better housing|edu|work in region	0.67	1.38	2.16	0.78	1.44	2.15	0.71	1.47	2.24	0.67	1.43	2.24
P7: P6 + sending back (part of) failed asylum seekers	0.65	1.35	2.11	0.76	1.41	2.11	0.69	1.43	2.19	0.65	1.40	2.19
P8: P6 + sending back all refugees without temp permit	0.40	0.94	1.66	0.43	1.01	1.69	0.43	0.97	1.70	0.40	0.97	1.70

Impact policies & scenarios on Germany 2013-2017 in terms of political stress (0-100) due to refugees from Balkan Route)	Scenarios wrt further development of the war in Syria (and beyond): S1: War in S ends soon			S2: War in S will go on			S3: More wars than S&I			Outcomes across uncertain war scenarios (S1-S2-S3)		
Policies	Minimum	Median	Maximum	Minimum	Median	Maximum	Minimum	Median	Maximum	Minimum	Median	Maximum
P1: "Wir schaffen das" without further policies	[illegible]	[illegible]	[illegible]	[illegible]	[illegible]	[illegible]	[illegible]	[illegible]	[illegible]	[illegible]	[illegible]	[illegible]
P2: P1 + perfect implementation EU-17 point plan	[illegible]	[illegible]	[illegible]	[illegible]	[illegible]	[illegible]	[illegible]	[illegible]	[illegible]	[illegible]	[illegible]	[illegible]
P3: P1 + partial (ethnic) closure of Balkan borders	[illegible]	[illegible]	[illegible]	[illegible]	[illegible]	[illegible]	[illegible]	[illegible]	[illegible]	[illegible]	[illegible]	[illegible]
P4: P3 + sudden internal border closures	47	[illegible]	[illegible]	47	[illegible]	[illegible]	43	[illegible]	[illegible]	43	[illegible]	[illegible]
P5: P3 + anticipated internal border closures	[illegible]	[illegible]	[illegible]	[illegible]	[illegible]	[illegible]	[illegible]	[illegible]	[illegible]	[illegible]	[illegible]	[illegible]
P6: P1 + more/better housing|edu|work in region	54	[illegible]	[illegible]	[illegible]	[illegible]	[illegible]	57	[illegible]	[illegible]	[illegible]	[illegible]	[illegible]
P7: P6 + sending back (part of) failed asylum seekers	53	[illegible]	[illegible]	[illegible]	[illegible]	[illegible]	[illegible]	[illegible]	[illegible]	53	[illegible]	[illegible]
P8: P6 + sending back all refugees without temp permit	35	[illegible]	[illegible]	39	[illegible]	[illegible]	37	[illegible]	[illegible]	35	[illegible]	[illegible]

Fig. 8 *Top*: Ranges of cumulative numbers of refugees in Germany (via the Balkan route). *Bottom*: Resulting internal political tension

tion or extension of armed conflict in the region. Third, although sudden (full or partial) border closures may look like a solution, they are really problematic from the point of view of upstream countries (not shown here) and if they would be closed in the wrong order—resulting in accumulations of refugees along the Balkan route. Limiting the analysis to these options and looking from a systems perspective for options to minimize the cumulative number of refugees in Germany (and Austria, Sweden, the Netherlands, …), the best policy from a Northern-European perspective seemed to be to provide good housing, education and work in the region and in safe zones and returning as many refugees without official residence permit as possible to the region. This would require very significant investments of the EU in the region though. Else, refugees could be expected to try to make their way to Europe, via more dangerous and more expensive routes, or would pay the price by being trapped in supposedly safe and human zones, that are not so safe and human.

During the workshop, we did not have the means nor the time to show the effects of a multiplicity of scenarios. Providing insights about potential system-wide dynamics under deep uncertainty, nevertheless requires ways to generate and communicate a multiplicity of futures. For generating a multiplicity of futures, one may either generate a large ensemble of scenarios that covers the uncertainty space and communicate these ensembles (or their bounds) or specific runs within the ensembles (e.g., the minimum, median and maximum run at some point in time), or one may generate small sets of maximally distinct simulation runs, identify exemplars, and communicate these representative runs. The latter approach requires ways to generate simulation runs and to identify exemplars that are representative of diverse types of behaviours and their origins in the input space. One such approach was used here. The large-ensemble approach and the small-set-of-exemplars approach are complementary and can be used for different tasks within model-based interventions. The latter approach is particularly interesting for use in workshop settings, the former for detailed analysis.

Communicating a multiplicity of potential system-wide dynamic consequences is a challenge that still needs to be tackled, especially when communicating results to high-level policy-makers. One way to do so may be to visualize and communicate the system-wide dynamics of key simulation runs, preferably several at the same time, using geo-spatial visualizations and animations. Compelling visualizations and animations may be crucial for conveying insights regarding the dynamic effects of complex uncertain issues and policies applied to them. In view of better communicating the outcomes of dynamic models to decision-makers and analysts alike, we developed scripts based on work by Ramiro Gomez[3] to plot the outcomes of pre-selected simulation runs directly onto choropleth maps and animations thereof. Choropleth maps are thematic maps to display data that varies across geographic regions. Figure 9a displays a choropleth map of the political stress from Turkey across the Balkan route to some reception countries at month 41 for one of many scenarios/simulation runs.

Dynamic choropleth maps can also be linked directly to simulation models (of any type and of any simulation model). Simple choropleth animations can display the big-picture dynamics of a single key performance indicator of one simulation run. This limitation is a good reason for identifying and selecting as few representative scenarios as possible. However, visualizing just one key performance indicator is almost always insufficient. Options for dealing with multiple key performance indicators or multiple runs are to (1) simultaneously display multiple choropleth animations (e.g., next to each other), or (2) use colour shading in a single map (e.g., for minimum values, median values, and maximum values). Alternatively, more advanced choropleth animations could be made, for example by combining a geo-spatial basemap[4] based on information related to one key performance indicator or run with bar graphs or icons related to other key performance indicators or runs. Stacking different basemaps on top of each other, it is possible to deal with multiple levels of geo-spatial aggregation and multiple key performance indicators. Stacking too much information in such maps may not improve their communicability though.

Different types of information may also have to be dealt with in different ways. Adding lines between different regions with line thicknesses relative to the flows between the countries (or provinces or towns) as in Fig. 9b allows for adding flow-related information, possibly on top of a basemap coloured relative to the stock value of each of the countries. Alternatively, one could display the flows between countries as in Fig. 9c, possibly with the size of the country nodes relative to their stock value, or with a Sankey plot.

The distinct behaviour of stocks and flows and the difficulty most people have in understanding their joint dynamics, may have to be addressed with different types of graphs. Figure 10 displays the dynamics of stocks of refugees in countries (top—on two different scales) and the main flows into and out of countries in relation to the countries of origin and destination (bottom). Alternatively, stacked graphs may provide information about the change in the stocks, and therefore implicitly about the flows.[5] In sum, outcomes of simulation models can be displayed in many differ-

---

[3] See: http://ramiro.org/notebook/basemap-choropleth/.

[4] See: http://matplotlib.org/basemap/index.html.

[5] The scripts can be obtained from the authors.

ent ways. They should be displayed in ways that are comprehensible and useful to the users of the information.

Conclusions

This chapter was all about bringing simulation under uncertainty in workshop modus to decision-makers, and related to that, about alternative sampling approaches and about creating attractive and comprehensible visualization and animation of simulation results. We combined two exploratory modelling and simulation

a

Fig. 9 (**a**) Choropleth map of Balkan route (here, political stress beginning 2016); (**b**) map of the Balkan route with line thickness for size of flows between countries; (**c**) line thickness to represent flows without underlying map

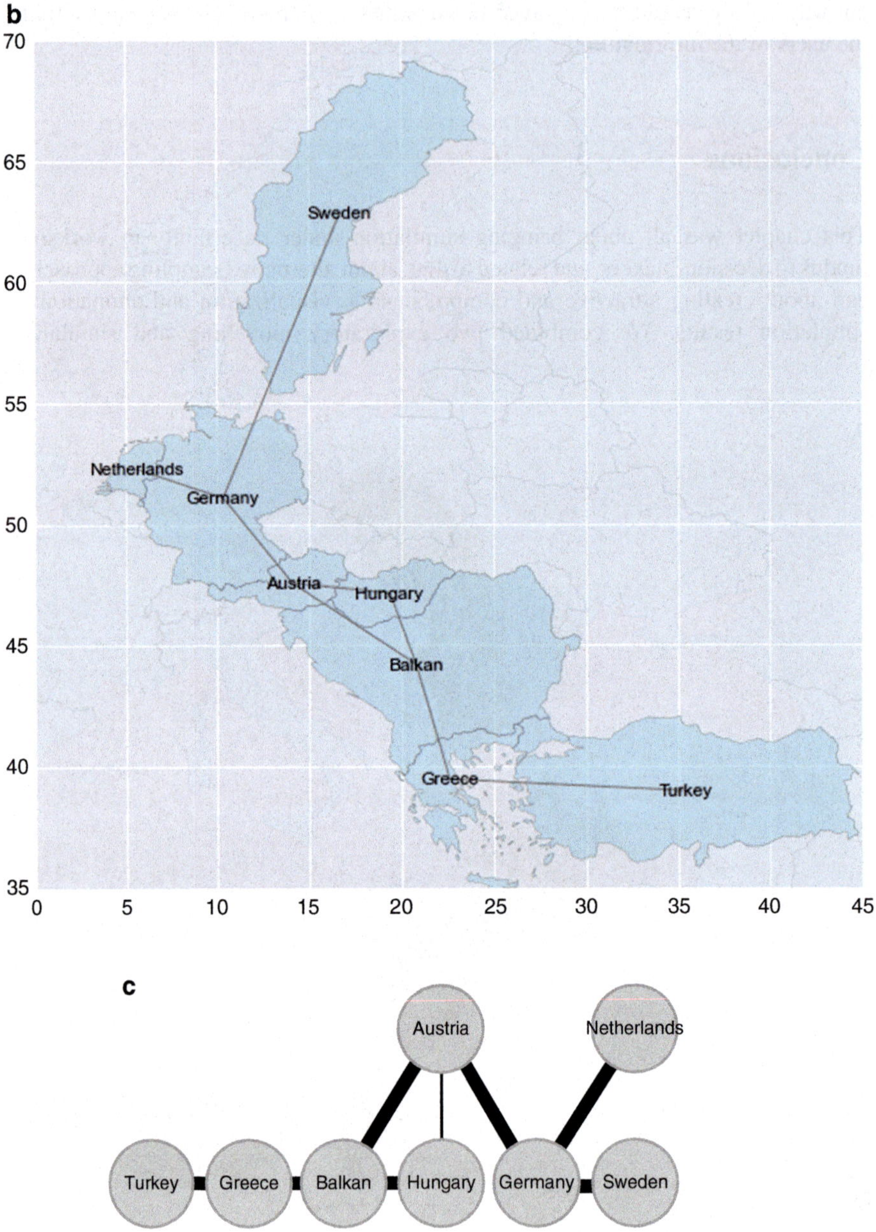

Fig. 9 (continued)

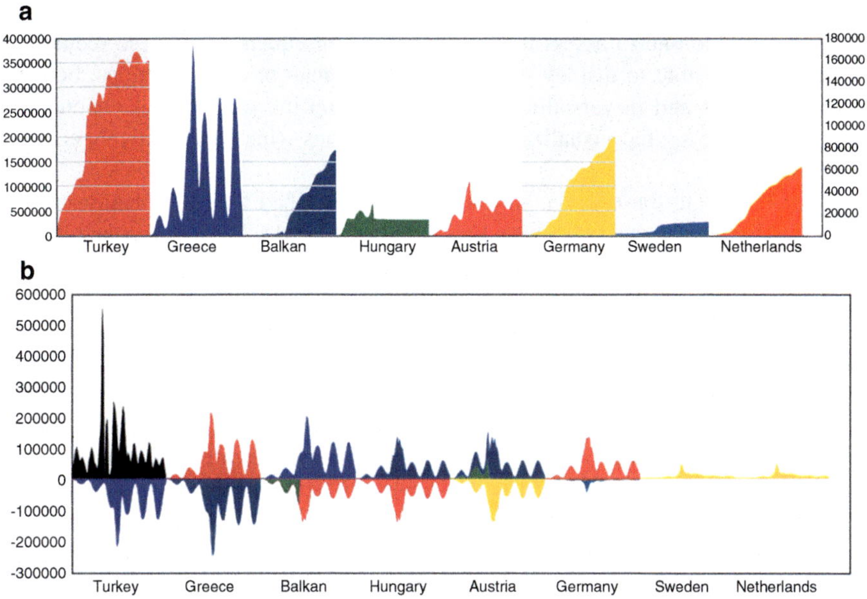

Fig. 10 Stock dynamics of the number of refugees in transit and destination countries on two scales (**a**) and inflows and outflows of countries by country colour code (**b**)

approaches: (1) to provide on the spot and on the map model-based policy-support in workshop settings, and (2) to provide in-depth model-based policy-support.

Although in-depth model-based policy-support under deep uncertainty is not new, it is not widely used yet. On the spot and on the map model-based scenario exploration and policy-making in workshop setting is new and requires new tooling. We focussed specifically on two recent contributions we believe are crucial for the successful use of modelling to support policy-making workshops on complex and uncertain issues, namely (1) approaches to quickly generate subsets of scenarios that are representative for large ensembles of scenarios, and (2) on the spot geospatial visualization and animation.

Quickly generating small subsets of simulated scenarios that are representative for large ensembles of simulated scenarios could either be done by using sufficient computing power or by using smart sampling and exemplar selection approaches. We used an innovative approach for sampling and exemplar selection which uses information from the output space to direct subsequent sampling and simulation. In terms of adaptive output-oriented sampling approaches, there is still room for further improvement. Using such approaches in workshop settings enables one to quickly generate a very small set of exemplars that are representative for the entire output space (and possibly also the input space). In other words, using such approaches, it is possible to—on the spot—generate and selected exemplar scenarios that are representative for large ensembles of scenarios.

Quick appraisal of these exemplar scenarios requires visualizations and animations that enable one to assess the system-wide consequences of these scenarios. Today, it is possible to display simulation runs as map-based animations, both in fancy interfaces and in versatile interfaces, again on the spot. In this chapter we focussed on geo-spatial visualizations and animations which are relatively easy to generate.

To demonstrate these two contributions, we described their application to the 2015–2016 refugee crisis in Europe. In the case of the 2015–2016 European refugee crisis, we used several approaches for generating sets of scenarios and developed interactive maps of Europe that show the dynamics of refugee flows, the political stress, and other key performance indicators over time. The animations that were developed for the workshop proved crucial for communicating the dynamics of refugee related indicators simultaneously across multiple countries. There is still much room for improving visualizations and animations of simulation results though. We believe good visualizations and animations are a precondition for modelling and simulation to become useful for on the spot and on the map scenario exploration and policy-support. There is still much work to do in that respect.

Integrative visualizations and animations may also help analysts to get at grasps with complex dynamics of disaggregated systems. Information-rich but easy-to-use visualization and animation techniques should be developed that can help both analysts and decision-makers. Such visualizations and animations may also help in bridging both worlds.

Acknowledgements This simulation project was conducted in partnership with the Munich Security Conference. The results were shown at the Munich Strategy Forum in November 2015 in Schloss Elmau (Germany). We greatly acknowledge Datenflug for the visualizations developed for the workshop. Finally, we want to thank Jan H. Kwakkel and Willem L. Auping for their contributions to TU Delft's EMA Workbench.

References

Agusdinata DB (2008) Exploratory modeling and analysis. A promising method to deal with deep uncertainty. PhD thesis, Next Generation Infrastructures Foundation, Delft

Auping WL, Pruyt E, Kwakkel JH (2015) Societal ageing in the Netherlands: a robust system dynamics approach. Syst Res Behav Sci 32(4):485–501. doi:10.1002/sres.2340

Bankes SC (1993) Exploratory modeling for policy analysis. Oper Res 41(3):435–449

Bergot T (2001) Influence of the assimilation scheme on the efficiency of adaptive observations. Q J R Meteorol Soc 127(572):635–660

Bishop CH, Etherton BJ, Majumdar SJ (2001) Behavior space sampling with the ensemble transform Kalman filter. Part I: theoretical aspects. Mon Weather Rev 129(3):420–436

Bossel H (2007) Systems and models: complexity, dynamics, evolution, sustainability. Books on Demand GmbH, Norderstedt

Bryant BP, Lempert RJ (2010) Thinking inside the box: a participatory, computer-assisted approach to scenario discovery. Technol Forecast Soc Chang 77(1):34–49

Bucher CG (1988) Behavior space sampling—an iterative fast Monte Carlo procedure. Struct Saf 5(2):119–126

Fiddaman T 1997 Feedback complexity in integrated climate-economy models. Ph.D. Thesis, MIT Sloan School of Management
Ford A (2009) Modeling the environment, 2nd edn. Island Press, Washington, D.C
Forrester JW (1958) Industrial dynamics. A major breakthrough for decision makers. Harv Bus Rev 36(4):37–66
Forrester JW (1961) Industrial dynamics. MIT Press, Cambridge
Forrester JW (1968) Principles of systems. Wright-Allen Press, Inc., Cambridge
Forrester JW (1969) Urban dynamics. Productivity Press, Cambridge
Forrester JW (1971) World dynamics. Wright-Allen Press, Inc., Cambridge
Friedman JH, Fisher NI (1999) Bump-hunting in high-dimensional data. Stat Comput 9:123–143
Groves D, Lempert RL (2007) A new analytic method for finding policy-relevant scenarios. Glob Environ Chang 17:73–85
Hamarat C, Kwakkel JH, Pruyt E (2013) Adaptive robust design under deep uncertainty. Technol Forecast Soc Chang 80(3):408–418
Hamarat C, Kwakkel JH, Pruyt E, Loonen ET (2014) An exploratory approach for adaptive policymaking by using multi-objective robust optimization. Simul Model Pract Theory 46:25–39
Islam T, Pruyt E (2016) Scenario generation using adaptive sampling: The case of resource scarcity. Environ Model Softw 79:285–299. doi:10.1016/j.envsoft.2015.09.014
Kwakkel JH, Pruyt E (2013) Exploratory modeling and analysis, an approach for model-based foresight under deep uncertainty. Technol Forecast Soc Chang 80(3):419–431. doi:10.1016/j.techfore.2012.10.005
Kwakkel JH, Pruyt E (2015) Using system dynamics for grand challenges: the ESDMA approach. Syst Res Behav Sci 32:358–375. doi:10.1002/sres.2225
Kwakkel JH, Cunningham SW, Pruyt E (2014) Improving scenario discovery by bagging random boxes. In Proceedings of the 2014 Portland international conference on management of engineering and technology conference
Lempert RJ, Collins MT (2007) Managing the risk of uncertain threshold response: comparison of robust, optimum, and precautionary approaches. Risk Anal 27(4):1009–1026
Lempert RJ, Popper SW, Bankes SC (2003) Shaping the next one hundred years: New methods for quantitative, long-term policy analysis. RAND report MR-1626, The RAND Pardee Center, Santa Monica, CA
McKay MD, Beckman RJ, Conover WJ (1979) Comparison of three methods for selecting values of input variables in the analysis of output from a computer code. Technometrics 21(2):239–245
Petitjean FO, Ketterlin A, Gancarski P (2011) A global averaging method for dynamic time warping, with applications to clustering. Pattern Recog 44(3):678–693
Pruyt E (2013) Small system dynamics models for big issues: triple jump towards real-world complexity. TU Delft Library, Delft. Available from http://simulation.tbm.tudelft.nl/smallSD-models/Intro.html
Pruyt E (2015) From building a model to adaptive robust decision making using systems modeling. In: Janssen M, Wimmer MA, Deljoo A (eds) Policy practice and digital science: integrating complex systems, social simulation and public administration in policy research. Volume 10. Series: Public administration and information technology. Springer International Publishing, Switzerland, pp 75–93
Pruyt E (2016) Integrating systems modelling and data science: the joint future of simulation and 'big data' science. Int J Syst Dyn Appl 5(1):1–16. doi:10.4018/IJSDA.201601010
Pruyt E, Islam T (2015) On generating and exploring the behavior space of complex models. Syst Dyn Rev 31(4):220–249. doi:10.1002/sdr.1544
Pruyt E, Auping WL, Kwakkel JH (2015) Ebola in West Africa: model-based exploration of social psychological effects and interventions. Syst Res Behav Sci 32:2–14. doi:10.1002/sres.2329
Rahmandad H, Oliva R, Osgood NA (eds) (2015) Analytical methods for dynamic modelers. MIT Press, Cambridge
Sterman JD (2000) Business dynamics: systems thinking and modeling for a complex world. Irwin/McGraw-Hill, Boston

Thompson K, Tebbens RJ (2007) Eradication versus control for poliomyelitis: an economic analysis. Lancet 369(9570):1363–1371

Vennix JAM (1996) Group model building: facilitating team learning using system dynamics. Wiley, Chichester

Yücel G (2012) A novel way to measure (dis)similarity between model behaviors based on dynamic pattern features. In: Proceedings of the 30th international conference of the system dynamics society, St.-Gallen, CH, 22–26 July 2012

Yücel G, Barlas Y (2011) Automated parameter specification in dynamic feedback models based on behavior pattern features. Syst Dyn Rev 27(2):195–215

Erik Pruyt is Associate Professor of Model-based Policy Analysis at Delft University of Technology. His methodological research focusses on developing methods and techniques for dealing with dynamically complex and deeply uncertain policy issues. His applied research interests include (but are not limited to) model-based policy support related to safety and security, health policy, humanitarian and development aid, energy and environmental policy, and public policy in general. He teaches modelling and simulation and advanced System Dynamics, and supervises many students in (exploratory) modeling, simulation and analysis. Some 150 of his teaching cases are available online through his free e-books (Small System Dynamics Models for Big Issues and the Systems Modelling and Simulation for Everyone series). Dr. Pruyt is a member of the Balaton Group and Vice President Meetings of the System Dynamics Society. Recently, Dr. Pruyt hosted two International System Dynamics Conferences and Summer Schools in Delft.

Tushith Islam currently works at Greenwood Strategic Advisors which focus on creating solutions that enhance management and organizational performance in complex and dynamic environments. At the time of writing, Dr. Islam was a Post-doctoral researcher at the Policy Analysis section of the Faculty of Technology, Policy and Management of Delft University of Technology (TU Delft) in the Netherlands. He worked on the application of robustness and adaptivity to multifunctional flood defences. Other research interests and contributions relate to adaptive sampling approaches, data visualization, geo-spatial simulation and analysis, and modelling socio-technical issues. His doctoral research explored the strategic planning of co-located infrastructure (road, water, sewers).

Thomas Arzt holds a Ph.D in physics and is the General Manager and President of SAT Strategic Advisors for Transformation GmbH in Freiburg, Germany. Dr. Arzt has built and led teams of simulation and optimization consultants over the past 20 years—nearly 150 consulting and advisory projects for clients across multiple industrial sectors. All Projects applied "Simulation-based Complexity Management", a methodology set up by SAT GmbH to allow for decision support and risk assessment in complex and volatile business environments. His recent projects apply "Stategic Foresight under Deep Uncertainty" to topics on the level of geostrategic and geopolitical importance.

An Event-Driven Lens for Bridging Formal Organizations and Informal Online Participation: How Policy Informatics Enables Just-in-Time Responses to Crises

Chul Hyun Park and Erik W. Johnston

Abstract Policy informatics not only gives new approaches to analyzing policy challenges, but also provides guidance for understanding new forms of organizing in the digital era. This chapter aims to investigate how technology accelerates the creation of just-in-time efforts while also lowering the barriers for joining such efforts to an increasingly diverse set of formal and informal actors who can make a meaningful contribution in the context of emergency management. In this chapter, we suggest a novel and extended lens called an 'event-driven' lens for integrating formal and informal responses by reviewing the literature on emergency management, crowdsourcing, open innovation, policy informatics, and digital humanitarianism. The novel lens is called an event-driven lens because crises serve as a focusing event that suddenly bring about not only the activation of formal organizations and their latent networks across the levels of government and the sectors, but also the emergence of many informal actors across the globe and from the affected communities to collectively respond to disasters or crises. Traditionally, emergency preparedness and response are in large part the role and responsibility of formal organizations like emergency management agencies and police and fire departments. Due to concurrent advances in a variety of technologies (information, communication, and artificial intelligence), informal groups of publics from both across the globe and the affected regions now regularly emerge and can play a significant role in the response through crowdsourcing vital information and assisting with the allocation of needed resources and services.

C.H. Park (✉) • E.W. Johnston
Clinton School of Public Service, University of Arkansas,
1200 President Clinton Avenue, Little Rock, AR 72223, USA
e-mail: chpark@clintonschool.uasys.edu; erik.johnston@asu.edu

J.R. Gil-Garcia et al. (eds.), *Policy Analytics, Modelling, and Informatics*, Public Administration and Information Technology 24,
DOI 10.1007/978-3-319-61762-6_15

Abbreviations

AIDR	Artificial Intelligence for Disaster Response
DHS	The U.S. Department of Homeland Security
EMAC	Emergency Management Assistance Compact
ICS	Incident Command System
OSM	OpenStreetMap
SBTF	Standby Task Force
SMS	Short Message Service
UAV	Unmanned Aerial Vehicle

Introduction

Policy informatics not only gives new approaches to analyzing policy challenges, but also provides guidance for understanding new forms of organizing in the digital era. In particular, technology accelerates the creation of just-in-time efforts while also lowering the barriers for joining such efforts to an increasingly diverse set of actors who can make a meaningful contribution. For example, large-scale crises such as natural disasters (earthquake and hurricane) and man-made crises (terrorism and ethnic violence) are events that necessitate responses at multiple scales by an increasingly diverse set of actors. Traditionally, emergency preparedness and response are in large part the role and responsibility of formal organizations like emergency management agencies and police and fire departments under related laws and regulations. The U.S. Federal Emergency Management Agency (FEMA) is responsible for developing emergency management policies and plans, such as the National Response Framework, and for coordinating response to natural disasters at the federal level under the Robert T. Stafford Disaster Relief and Emergency Assistance Act. State and local governments also have their own similar departments or agencies. Today, it is increasingly rare for a single department or agency to address catastrophic disasters and crises due to a lack of capacities and resources to prepare for and respond to every possible type of catastrophe. Thus, when a catastrophe occurs, multiple public agencies across local, state, and federal governments are mobilized and deployed to deal with the disaster. Also, formal nonprofit organizations like the American Red Cross and the Salvation Army and for-profit organizations collaborate with public agencies to offer rescue and relief services to disaster-affected people.

Importantly, because of concurrent advances in information, communication, and computational technologies, communities now regularly emerge where amateurs or concerned publics can play a meaningful role in the response through crowdsourcing vital information and assisting with the allocation of needed resources or services. In recent catastrophic disasters or crises such as the 2010 Haiti earthquake, the 2011 Japan earthquake, tsunami, and nuclear disaster, and the 2015 Nepal earthquake, informal, emergent groups of individuals across the globe and from the affected areas made significant contributions to the effective response. As a novel phenomenon in the digital era, informal actors were able to provide a vital component of the emergency response. In some cases they filled information gaps on disaster conditions and the affected people's needs, voluntarily mobilized, delivered, and allocated relief resources, and helped coordinate formal organizations' tasks and activities in complex, urgent disaster situations. These informal groups are often a kind of virtual community consisting of digital volunteers who are loosely connected through information and communication technologies.

Specifically, these informal groups created a reporting system through which the affected people submitted their requests for rescue or relief and the information of disaster conditions by using various technologies. These informal actors also gathered, verified, and visualized a large amount of disaster data from social media, mainstream media, satellite imagery donated by for-profit companies, and reports from the ground by using open source web platforms, crowdsourced human computation, and artificial intelligence. The information processed by these informal actors increased situational awareness of the current state of disasters, the affected peoples' needs and requests, and which organizations were working with what and where to meet the unmet needs of the affected people. By providing real-time, verified, and reliable disaster information, these informal actors enabled the affected communities (local residents affected by disasters and local community-based nonprofit organizations) to quickly mobilize aid resources and help the communities, formal emergency management agencies and first responders to make timely and effective decisions about rescue missions and relief services. In addition, numerous organizations from the public, private, and nonprofit sectors could efficiently coordinate their various tasks and activities in response to disasters. These informal actors' efforts are becoming now an anticipated and legitimate part of the overall response to catastrophes (Meier 2015).

The Importance of Considering Both Formal and Informal Emergency Responses

Institutionalized formal organizations and their emergency responses are relatively effective in dealing with small-scale or routine emergencies, but formal emergency responses are likely severely delayed and challenged when addressing large-scale, catastrophic disasters and crises. According to Leonard and Howitt (2005), a formal emergency management organization "functions best when it is directed at a well-defined, reasonable consistent or clear prioritized set of purposes" (citied in Buck et al. 2006, p. 5). However, the emergency response systems of formal organizations sometimes operate "poorly for large disasters which often involve a. multiple hazards occurring in close temporal and spatial succession, b. with multiple agent-generated demands, with c. multiple responding agencies, d. attempting to satisfy often conflicting goals that cannot be anticipated and reconciled" (Buck et al. 2006, p. 5). Formal response systems characterized by hierarchical decision making, standard operating procedures, and internal communication channels likely have difficulty responding to or fail to deal with catastrophic disaster situations (Crowley and Chan 2011; Yuan et al. 2013). Formal response actors' difficulty or failure is caused primarily by a lack or absence of information on the current state of disasters (e.g., fatalities, injuries, damages, and the needs of those affected) and on real-time response efforts (i.e. who is working on what, where). Existing emergency management systems often do not have open communication channels and related formal protocols that aggregate or prioritize local intelligence from outside sources and share freely the intelligence with the affected people and informal actors (Yuan et al. 2013). Hence, there are disconnected communications not only within a network of formal organizations, but also between formal organizations and the affected people on the ground. Such communication problems likely result in inefficient coordination (e.g., the duplication of response efforts) among public agencies, first responders in the field, local or international nonprofit organizations deployed to help address a disaster or a crisis, and the affected people on the ground (Kapucu 2006). Indeed, these problems were apparent in recent catastrophic disasters. For example, during Hurricane Katrina that struck the Gulf Coast of the U.S. in 2005, a lack of information on the ground seriously delayed the response of emergency management agencies and nonprofit or for-profit organizations involved. "[D]uring Katrina, federal, state, and local government agencies and private organizations did not know what actions to take in the response, did not have any guidance on how to coordinate and interrelate their activities,…and had no system to track and share information" (Jaeger et al. 2007, p. 593).

Importantly, with advances in information and communication technologies, the role and contributions of informal groups of publics have become more useful in response to recent disasters or crises. Hence, one needs "modern" emergency response systems integrating both formal organizations and informal groups. That is, if the strengths of both formal and informal actors are incorporated, the capacity to deal with disasters or crises would be tremendously increased. Moreover, theoretically and conceptually, a more integrative lens needs to be developed to help explain and understand various responses to catastrophic disasters by taking into account not only formal actors with legal responsibilities, but also informal actors who actively involve emergency response. Thus, this chapter suggests an "event-driven" lens for bridging formal organizations and informal groups of individuals in response to crises. For this purpose, we conducted a comprehensive literature review on emergency management, disaster policy, policy informatics, crowdsourcing, and open innovation.

An Event-Driven Lens for Integrating Formal and Informal Emergency Responses

Recent catastrophic events led to the emergence of informal, online groups of publics that collaborated with formal organizations or worked independently (outside of formal organizations) to respond to the events. That is, one can witness that the actual disaster response systems in the networked age are much more complex and dynamic than the existing emergency management literature and disaster policies based on formal organizations have understood, because a wide range of formal and informal actors work together or independently in response to the catastrophic events. Thus, it is necessary to build a novel and extended lens for integrating both formal and informal responses in the networked age. The novel lens is called an event-driven lens, because crises serve as a focusing event that suddenly bring about not only the activation of formal organizations and their latent networks across the levels of government and the sectors, but also the emergence of many informal actors across the globe and from the affected communities to collectively respond to disasters or crises. Specifically, the event-driven lens takes into account: (1) formal emergency response (i.e. how do formal organizations respond to catastrophic disasters or crises, following predetermined policies, procedures, and related laws? And what are the challenges and limitations of formal organizations in large-scale disasters or crises?); (2) informal emergency response (i.e. what do informal actors perform to voluntarily respond to disasters? And what are their contributions to the effective emergency response?); and (3) how formal actors and informal actors interact with each other in catastrophic disaster or crises (i.e. what are the types of the relationships between formal and informal actors?)

Formal Emergency Response

Formal emergency response is an official system consisting of institutionalized organizations at all levels of government and across the public, nonprofit, and for-profit sectors, their resources and personnel, established policies, procedures, plans, and agreements, and inter-organizational relationships and coordination mechanisms among these organizations. In the formal emergency response system, institutionalized organizations conduct a wide range of activities and tasks to respond to a disaster or a crisis (Haddow et al. 2008). Such activities and tasks include search and rescue missions, emergency medical services, and relief services (foods, water, and temporary shelters).

Formal Organizations

A key feature of formal emergency response is the dependence on (networks of) formal organizations across jurisdictions and the sectors (Schroeder et al. 2001). These formal organizations include public emergency management departments and agencies and other public organizations at the local, state, or federal level (e.g., the Federal Emergency Management Agency, the U.S. Forest Service, the U.S. Coast Guard, and the U.S. Army Corps of Engineering), first responders (e.g., police and fire departments and emergency medical services), institutionalized nonprofit organizations (e.g., the American Red Cross, the Salvation Army, World Vision, and National Voluntary Organizations Active in Disaster), and private corporations (e.g., private utility companies).

Laws, Policies, Procedures, Plans, and Agreements

Prior to the occurrence of a disaster or a crisis, a wide range of established laws, disaster policies, standard operating procedures, emergency preparedness plans, and mutual aid agreements already exist for an effective formal emergency response. For example, these formal rules, plans and policies include the Robert T. Stafford Disaster Relief and Emergency Assistance Act,[1] the National Incident Management System (NIMS),[2] the National Response Framework (NRF),[3] and the Emergency Management Assistance Compact (EMAC)[4] in the United States (Kapucu and Garayev 2011; Lindsay 2012; DHS 2008, 2013; Waugh 2007). These formal rules, plans and policies describe the key principles and concepts of emergency management, the specific roles and responsibilities of each formal organization involved, and detailed procedures on how resources and personnel are mobilized, deployed and reimbursed.

Inter-Organizational Coordination Mechanisms

Soon after a disaster or a crisis occurs, formal organizations and their latent networks are activated, following predetermined policies, procedures, and agreements. That is, the formal emergency response system becomes a large network(s) of formal organizations across the public, nonprofit, and for-profit sectors. The size of the network of formal organizations involved is mostly determined by the magnitude of a disaster or a crisis and its immediate impacts, including fatalities, injuries, collapsed buildings, and displaced people (McGuire and Silvia 2010; Waugh 2006). There are the two types of inter-organizational coordination mechanisms in a network(s) of formal organizations: the Incident Command System (ICS) and collaborative decision making processes. The ICS that emphasizes unified command and control was developed by local forest firefighting agencies in California in the 1970s. Since the inception of an initial version of the ICS called FIRESCOPE (FIrefighting RESources of California Organized for Potential Emergencies) in California, the ICS has spread across the United States (Buck et al. 2006; Cole 2000; Harrald 2006). The ICS is used primarily for on-scene operational activities (Buck et al. 2006; Moynihan 2008). All formal organizations and their personnel involved in on-scene tactical and operational tasks perform their various missions under the authority of an Incident Commander. These operational activities are coordinated through hierarchical decision making structures, division of labor, span of control, and integrated communications among formal actors.

---

[1] https://www.fema.gov/robert-t-stafford-disaster-relief-and-emergency-assistance-act-public-law-93-288-amended.

[2] http://www.fema.gov/national-incident-management-system.

[3] http://www.fema.gov/national-response-framework.

[4] http://www.emacweb.org/.

Unlike the on-scene operations based on unified command and control, off-scene supporting organizations are coordinated through cooperative inter-organizational structures and procedures based on predefined multi-agency agreements and related policies like the Emergency Management Assistance Compact at the state level and the federal Emergency Support Functions (Kapucu and Garayev 2011; DHS 2013; Waugh 2007). Such off-scene inter-organizational structures and procedures are "to coordinate [a variety of supporting] activities above the field level and to prioritize the incident demands for critical or competing resources, thereby assisting the coordination of the operations in the field" (DHS 2008, p. 64). These off-scene arrangements aim to provide timely supports and assistance to first responders on the ground, mostly following predefined standard operating procedures and protocols. A typical example of these arrangements is the Emergency Management Assistance Compact (EMAC). The EMAC is a state to state mutual aid agreement for mobilizing and delivering personnel and equipment to the affected areas (Kapucu and Garayev 2011; Waugh 2007). According to the literature, cooperative inter-organizational coordination is likely influenced not only by established formal rules (i.e. the ICS and collaborative arrangements), but also a variety of the characteristics of inter-organizational relationships including prior history of collaboration, trust, and intergovernmental political dynamics (Haddow et al. 2008; Harrald 2006; Kapucu et al. 2013; Quarantelli 2005; Waugh 2006).

Products and Services

Formal organizations' responding activities create a wide range of products and services. These products and services in response to a disaster or a crisis include hazard warnings and public information on emergency services, transportation services for evacuating the affected people or animals, mass care services (food, water, and temporary shelters), family reunification support services, search and rescue missions, public health and emergency medical services, on-scene security and protection through law enforcement, and situational assessment (DHS 2013). These products and services are the outputs of formal organizations' responding activities.

Informal Emergency Response

For the past four decades, disaster sociologists have researched informal, emergent groups of individuals and their behavior in response to a disaster or a crisis. Most prior studies have focused on emergent collective behavior at the local community level before, during and after a disaster or a crisis (e.g., Drabek and McEntire 2003; Drabek 1985; Helsloot and Ruitenberg 2004; Kreps and Bosworth 1993; Rodriguez 2006; Stallings and Quarantelli 1985). The disaster sociology literature has provided useful knowledge on collective behavior and organizational structures of informal, emergent groups of individuals. However, most prior studies in disaster sociology have investigated relatively small groups of the affected people who helped each other in extreme events. Hence, the existing disaster sociology literature has paid little attention to large-scale collaboration enabled through information, communication, and computational technologies in disaster or crisis situations.

In the fields of crisis informatics, digital humanitarianism, emergency communications, and computer science, many scholars and practitioners recently began to note the contributions and potentials of informal, citizen-driven, volunteer-based groups in a variety of disaster or crisis situations including natural disasters like earthquakes, hurricanes, volcanic eruptions, and tsunamis and manmade disasters like riots, terrorism, and ethnic conflicts (Crowley 2013; Palen et al. 2010; White et al. 2014; Zook et al. 2010). Informal groups of publics are characterized as loosely connected, decentralized, and emergent groups of individuals across the globe and from the affected areas in response to a disaster or a crisis. These informal groups are often crowdsourced communities that perform collective responses by using information and communication technologies and automated data mining tools. Specifically, in recent disaster situations, these informal groups of ordinary people like students, software developers, and bloggers were interconnected through information and communication technologies. These informal groups mostly gathered, processed, and visualized timely, accurate, and reliable disaster information through crowdsourced human computation and artificial intelligence. Also, these informal groups self-organized and coordinated the mobilization and allocation of various relief resources (food, water, temporary shelters, and transportation) through simple web platforms and social networking sites, including Airbnb,[5] Uber,[6] crowdfunding websites like Indiegogo[7] and GoFundMe,[8] Facebook, Twitter, Google Docs, Reddit,[9] and other online platforms.

For instance, in the immediate aftermath of the 2010 Haiti earthquake, students from Tufts University, the Haitian diaspora, and disaster-affected people on the ground remotely collaborated with each other to collate actionable pieces of disaster-related information from mainstream media and social media (Meier 2011, 2015). The disaster-related information consisted mostly of disaster conditions and the affected people's requests for rescue and aid. Such disaster-related information was verified, processed and updated by a large group of informal actors from about 50 countries including Canada, Colombia, Haiti, Switzerland, and the United States

---

[5] https://www.airbnb.com/; Airbnb was used to provide post-disaster accommodations for the affected people in the 2015 Nepal earthquake and the 2015 Paris Terrorist attacks.

[6] https://www.uber.com/; In the aftermath of the 2015 Nepal earthquake, Uber was used for picking up donations for the affected people in India.

[7] https://www.indiegogo.com; In the aftermath of the 2015 Nepal earthquake, a group of Nepalese young volunteers and a nonprofit organization used indiegogo to crowdfund their voluntary responding activities.

[8] https://www.gofundme.com/; During and after the 2015 Nepal earthquake, an informal group of publics called One Stop Portal initiated a crowdfunding campaign though gofundme.

[9] https://www.reddit.com/; In the aftermath of the 2013 Boston Marathon bombing, an informal group of volunteers used reddit to deliver foods (pizzas in particular) to first responders and the affected people.

(Meier and Munro 2010). Also, urgent requests for rescue were directly sent to international first responders in the field. That is, these informal groups of publics helped international and local community-based responding organizations effectively coordinate their missions and tasks on the ground.

The informal emergency response of amateurs and concerned publics is a system consisting of a large group of informal actors including individuals across the globe and the affected people, technologies, and decentralized, open, adaptable organizational structures.

Informal Actors

Numerous individuals like students, software developers, reporters, and GIS professionals around the world and from the affected areas participate in and contribute to their own collective responses. These people have diverse backgrounds regarding age, gender, race and ethnicity, nationality, skill, education, and socioeconomic status (Howe 2009; Meier and Munro 2010). Informal actors can be categorized into a small group of key contributors and a large group of micro-contributors (Howe 2009; Shirky 2008; Tapscott and Williams 2006). In disaster or crisis situations, key contributors often create online platforms as communication channels and collect and process large amounts of disaster-related information from mainstream media and social media. These key contributors serve as builders, processors, and facilitators of informal, emergent groups of publics. In addition, large groups of micro-contributors are likely mobilized by key contributors' open calls for volunteering. Micro-contributors perform small, discrete tasks as much as they can do (e.g., collecting and verifying actionable pieces of disaster-related information and offering couches and beds to those who are stranded). Both key contributors and micro-contributors are essential for an effective informal emergency response.

Information, Communication, and Computational Technologies

Informal groups of individuals in response to a disaster or a crisis mostly rely on information and communication technologies and advanced computing. These technologies are used to mobilize their members across the globe and from the affected regions, to communicate with each other in real-time, and to take collective action by aggregating a variety of intelligence, skills, and resources from the ground and their mobilized members across the globe. In most cases, these informal groups of individuals do not have physical spaces for their collective action. Online chat rooms like Skype and social networking sites like Facebook Groups or Google Groups are their virtual headquarters.

These informal groups of publics utilize multiple technologies for the collection, mining, verification, and visualization of disaster-related data. Specifically, disaster-related data (both text- and image-based data) is collected from multiple sources including reports from the ground (via short message service (SMS), email, and online forms like Google Forms), social networking sites (Facebook, Flickr, Instagram, Twitter, and YouTube), mainstream media (news articles), and satellite imagery donated by for-profit satellite imagery providers.

Moreover, Unmanned Aerial Vehicles (UAVs), commonly called drones, are beginning to be used. The aerial imagery of disaster-affected areas is essential for assessing disaster conditions (collapsed buildings and roads) and creating post-disaster maps. Aerial images captured by drones have several advantages compared to commercial satellite images. "First off, cloud cover is regularly a big challenge for commercial satellites….UAVs fly below the clouds. This is especially critical following typhoons and hurricanes since clouds may linger for days after the devastation…. In addition, it generally takes 48–72 h to task a satellite over an area of interest. In contrast, a locally deployed UAV can capture imagery within hours and even minutes" (Meier 2015, p. 84). When Typhoon Haiyan struck the Philippines and caused over 6000 fatalities in 2013, an informal group of publics for voluntary online mapping called the Humanitarian OpenStreetMap Team[10] used UAV imagery to "quickly identify destroyed buildings and trace up-to-date roadmaps of the hardest hit areas, thus providing humanitarian organizations with critical information on which roads could still be used to provide urgent aid" (Meier 2015).

Notably, informal groups of individuals utilize both advanced computational technologies (i.e. machine learning) for data mining and crowdsourced human computation for data verification and analysis. A large amount of disaster-related information (texts and images) is often posted on social media in the immediate aftermath of disasters or crises. For example, over 250,000 disaster-related contents were posted on Twitter right after Typhoon Haiyan in 2013 (Meier 2015). However, such information is not likely to be useful to emergency management agencies and first responders due to information overload (Boulos et al. 2011; Edmunds and Morris 2000). It is because public agencies and departments usually do not have enough human and technical resources to quickly process such a large amount of disaster data. In fact, informal groups of publics have the same issue as formal organizations do. However, to deal with disaster information overload, informal online communities sometimes use an open-source, automated data-mining tool called Artificial Intelligence for Disaster Response (AIDR) developed and donated by a Qatar-based nonprofit computing research institute.[11] The AIDR uses both machine learning and human intelligence. The general architecture of the AIDR consists of collector, trainer, and tagger (Imran et al. 2014). The collector compiles messages from Twitters. It enables users to filter tweets posted during a disaster or a crisis by using keywords and hashtags (e.g., #Nepalquake and #Fukushima). Next, the messages filtered through the collector are passed to the tagger. The tagger performs the classification of each tweet by user-defined topics or categories such as collapsed buildings, casualties, and urgent needs (Imran et al. 2014).

Messages from Twitter are often too complex for machines to accurately classify them (Castillo 2015). To address this issue, the trainer "allows one or more users to

[10] A global community of online volunteer mappers for humanitarian disaster response.

[11] http://qcri.com/.

train the AIDR platform to automatically tag tweets" (Meier 2013). The collection owner creates a trainer page for tweets of interest by identifying categories of interest such as damages or rumors and manually labeling tweets by the identified categories. This training task can be performed by the collection owner him or herself or be crowdsourced to the public (i.e. online volunteers). If the collection owner wants the crowd to help classify the tweets, he or she can invite volunteers by sharing a link to the training page. As humans (i.e. the collection owner or a group of volunteers) manually label a small set of the tweets, the AIDR learns how to classify the tweets. Once enough tweets are labeled by humans (at least 20 tweets), the tagger automatically begin to apply the human-labeled classifier to incoming tweets by assigning one of the user-defined categories to each tweet (Imran et al. 2014). Then the tagger displays the automatically classified tweets on an online map or another visualization platform.

In the immediate aftermath of Typhoon Haiyan in 2013, an online volunteer community for crisis mapping called the Standby Task Force (SBTF) utilized the AIDR. The SBTF first collected tweets by using keywords and hashtags related to Typhoon Haiyan. Then the SBTF uploaded the collected tweets to a micro-tasking platform called the Tweet Clicker (i.e. a kind of the trainer for the AIDR). By using this platform, online volunteers of the SBTF manually labeled the tweets by the identified categories. Such human intervention provided training examples for the AIDR to learn how to classify the tweets. Ultimately, the AIDR enabled the SBTF to reduce over 250,000 tweets to approximately 55,000 disaster-related tweets that formal organizations could use for their response activities.

Lastly, informal groups of individuals sometimes utilize several data verification tools. For instance, Swift River, which was developed by Ushahidi, a nonprofit, open-source software company, organizes and filters incoming reports from the ground and social media. "Specific report sources are tracked through unique IDs (phone numbers or e-mail addresses) and thus can be ranked according to their record of veracity (Heinzelman and Waters 2010, p. 12).

Informal Organizational Characteristics and Structures

These informal groups are a new form of organization in the networked age (Capelo et al. 2012; Roberts 2011; White et al. 2014). From a systems theory viewpoint, these informal groups are mostly open systems. It means that anyone can participate in and contribute to these informal groups. Moreover, these informal groups have decentralized and horizontal organizational structures (Capelo et al. 2012; White et al. 2014). Such organizational structures enable the informal groups of people to quickly and constantly change their organizational structures and collective action processes to respond to unexpected challenges and problems in an agile, adaptive manner. Furthermore, the members of these informal groups likely share a sense of community and willingly provide mutual aid for learning and problem-solving.

These informal groups are often a kind of virtual community. In other words, a lot of individuals across the globe and the affected people are loosely connected and work together through a variety of information and communication technologies (e.g. short message service (SMS), email, social media, mobile applications, etc.). In many cases, these groups' collective action is coordinated through online crowdsourcing platforms. A large amount of labor is broken into small pieces and is distributed to a large group of individuals around the world and the affected people on the ground.

Resources and Services

Informal groups of people create a variety of products and services to help respond to a disaster or a crisis. In most cases, these informal groups of individuals around the world collectively create the following products and services: online or offline post-disaster base maps of the affected areas; timely, accurate disaster-related information on the current state of a disaster or a crisis, relief resources available, and the affected people's needs and requests; and real-time humanitarian '3W' information regarding **W**hich formal organizations are doing **W**hat, **W**here to meet the affected people's unmet needs; and the self-organized mobilization and delivery of a variety of aid resources by matching those who donate resources to those in need.

Interactions and Relationships Between Formal and Informal Actors

In some disaster or crisis situations, informal groups of individuals actively collaborate with formal organizations such as public emergency management agencies and nonprofit organizations. But in other situations, formal and informal actors perform their own responding activities independently. Interactions and relationships between formal organizations and informal groups can be categorized into four types according to whether formal and informal actors are aware of each other and whether there is alignment in formal and informal emergency responses (Kathuria et al. 2007; Thellufsen et al. 2009) (See Fig. 1).

In the first type of relationship, formal and informal actors recognize each other's existence, resources, and responding activities and these actors complement each other's efforts in a coordinated manner. In such a situation, the efficiency and effectiveness of overall emergency response systems are likely to be maximized. When formal and informal actors develop complementary and synergic relationships, these actors tend to achieve just-in-time performance. Roe and Schulman categorize crisis management in the context of high reliability organizations (particularly, electricity infrastructure) on the basis of instability[12] and options variety[13] (Roe and

[12] "Instability is the extent to which [a high reliability organization]…faces rapid, uncontrollable changes or unpredictable conditions that threaten the grid and service reliability of electricity supply, i.e., that threaten the task of balancing load and generation" (Schulman et al. 2004, p. 19).

[13] "Options variety is the amount of…[a high reliability organization] resources…to respond to events" (Schulman et al. 2004, p. 19).

		Awareness	
		+	-
Alignment	+	**Type 1** **(Complementary and synergic)** - **2010 Haiti earthquake**: The Tufts team, Mission 4636, OpenStreetMap, and international first responders	**Type 2** **(Additive)** - **2011 Japan earthquake**: Action-oriented local groups and the Japanese local government - **2012 Hurricane Sandy:** Online pet advocates and related formal agencies
	-	**Type 3** **(Ignoring and uncooperative)** - **2011 Japan earthquake**:The Safecast team and the Japanese government	**Type 4** **(Inconsistent and confusing)** - Within the first 24 to 72 hours of catastrophic disasters or crises

Fig. 1 Relationships between formal organizations and informal online groups

Schulman 2008; Schulman et al. 2004). When both options variety and system instability are high, just-in-time performance is likely to be dominant. "This performance condition demands 'real-time' flexibility, that is, the ability to utilise and develop different options and strategies quickly" through rich, dynamic communications between related stakeholders (Schulman et al. 2004, p. 20).

The 2010 Haiti earthquake is an exemplary case of complementary and synergic relationships between formal and informal actors. Soon after a devastating earthquake struck Haiti in 2010, informal groups of individuals emerged globally on a large scale. Students from Tufts University in the United States collected, verified, and visualized actionable pieces of disaster information from social media and mainstream media by using Google Docs and Ushahidi (free and open-source software for crisis mapping) in near real-time (Heinzelman and Waters 2010). A global volunteer community for creating a free editable map called OpenStreetMap (OSM)[14] collectively created an online map of post-disaster Haiti (Crowley and Chan 2011; Roche et al. 2011). Over two thousand volunteers from both across the globe and the affected regions participated and contributed to a SMS-based reporting system called the Mission 4636 project that allowed the affected people on the ground to submit their disaster conditions and requests for rescue or aid. In this project, such numerous volunteers translated reports from and to Haitian Creole, French, and English and verified these reports (Munro 2013). Three informal groups of volunteers (Tufts students, the OSM community, and the Mission 4636 volunteers) were connected to one another through information and communication tech-

[14] https://www.openstreetmap.org/.

nologies such as online chat rooms and collaborative crisis-mapping platforms and shared their disaster and geographic information.

Moreover, all information created by informal groups of volunteers was shared with formal emergency management agencies (e.g., the U.N. Office for Coordination of Humanitarian Affairs and the International Federation of Red Cross and Red Crescent Societies) and international first responders (e.g., the U.S. Marine Corps and the U.S. Coast Guard) on the ground in real-time. Particularly, the international first responders used the information to coordinate their search-and-rescue missions and the allocation and delivery of relief resources. After these first responders completed their response operations, they gave the informal groups of volunteers an update on the current status of response operations. Therefore, there were dynamic collaboration and communications between formal and informal actors during the Haiti earthquake response period.

In the second type of relationship, formal and informal actors are not aware of each other, but there is 'unintentional' alignment in response efforts between formal and informal actors. In such conditions, both actors' responding efforts are simply additive, but not able to be coordinated in a synergistic manner due to a lack of inter-organizational awareness. Particularly, such relationships between formal and informal actors may appear in catastrophic disaster situations. According to Quarantelli (2005), a catastrophe is systemically different from a disaster. In a catastrophic event compared to a disaster, "[m]ost or all of the community built structure is heavily impacted.... Local [emergency management] officials are unable to undertake their usual work role.... Most, if not all, of the everyday community functions are sharply and concurrently interrupted." Such catastrophe likely leads to the emergence of numerous informal actors in situations where formal organizations do not have sufficient resources and capacities for dealing with the catastrophe. For example, in the aftermath of the 2011 Japan earthquake, a lot of action-oriented local groups of people voluntarily emerged and self-organized the mobilization and delivery of aid resources within their communities by using simple webpages (Slater et al. 2012). Moreover, during and after the 2012 Hurricane Sandy, loosely organized online pet advocates launched a lost-and-found-pets page on Facebook. By using this page, the online pet advocates aimed to reunite lost pets with their owners. For this purpose, these advocates shared visual information about lost or found pets on Facebook, circulated paper flyers in the areas where the affected people could not get access to the Internet, and self-organized pet transports (White et al. 2014). Importantly, these local groups are often disconnected from formal emergency management organizations (Slater et al. 2012). In other words, formal and informal actors may have different aims and target groups and provide different relief services independently.

In the third type of relationship, formal and informal actors recognize each other's existence and responding activities, but both actors do not interact and collaborate with each other. Rather, they criticize and ignore their counterparts' responding activities. This situation likely occurs when a disaster or a crisis becomes a politically sensitive problem or issue. For example, in the immediate aftermath of the 2011 Japan earthquake and nuclear crisis, over one hundred volunteers from both across the globe called Safecast worked together to develop a low-cost Geiger counter (i.e. bGeige Nano) and measure and publish radiation data on their webpage.[15] Thus, any individual or organization could download and use the data for free. The Safecast team also crowdsourced the collection of radiation data to the public (Hemmi and Graham 2014). Anyone who had a Geiger counter could measure radiation levels and upload the data to the webpage. On the other side, the Japanese government (particularly, the Ministry of Education, Culture, Sports, Science and Technology) monitored and published radiation data online. Although the Japanese government and the Safecast team were aware of each other's efforts, these formal and informal actors did not collaborate with each other. Particularly, the Safecast team did not want to work with the Japanese government to be independent of a political debate on the nuclear crisis.

Lastly, in the fourth type of relationship, formal and informal actors do not recognize each other's existence and response efforts. Also, both actors' response efforts are inconsistent and confusing due to false, outdated, competing information on disaster conditions, the affected people's needs, and relief resources. This type of the relationships likely appears in the immediate aftermath of catastrophic disasters or crises such as the 2010 Haiti earthquake and the 2015 Nepal earthquake. In such catastrophic situations, many formal organizations are likely mobilized across the levels of government and the sectors. Moreover, numerous informal groups often emerge to deal with such catastrophes. For example, during and after the 2015 Nepal earthquake, over 500 formal and informal actors were involved in response to the catastrophic natural disaster. Thus, within the first 24–72 h of catastrophes, formal and informal actors are unlikely to recognize which formal organizations and informal groups are doing what and where. Such a lack of situational awareness often leads to inefficient and ineffective coordination of various responding activities between formal and informal actors.

Conclusion and Implications

Traditionally, emergency response is in large part the role and responsibility of public organizations like emergency management departments and agencies and police and fire departments and their collaborative partners such as nonprofit and for-profit organizations. This chapter focused on not only formal organizations from the public, nonprofit, and for-profit sectors, but also informal groups of publics that emerged spontaneously to respond to recent disasters and crises. The emergence and contributions of these informal actors are a novel phenomenon in the digital age because advances in information, communication, and computational technologies enable mass collaboration among numerous informal actors around the world and from the affected regions in overcoming space and time limits (Tapscott and Williams 2006).

[15] http://blog.safecast.org/.

Despite potential contributions of informal actors, we emphasize that informal actors do not replace formal actors, but supplement formal actors to increase the capacities of the overall emergency response systems for addressing disasters or crises. We note informal actors may have several limitations: data inaccuracy; privacy and security issues; and volunteers' burn out. One of the key criticisms of informal actors is the collection and circulation of inaccurate information. For example, soon after the 2013 Boston Marathon bombing, the general public participated in amateur investigations (Tapia et al. 2014). They launched online forums on reddit.com[16] and 4chan.org[17] to collectively identify suspects. These amateur detectives posted related images and videos to the online forums and conducted voluntary investigations. Unfortunately, the amateur detectives not only produced incorrect information about suspects, but such incorrect information was also circulated rapidly on the Internet, thus leading to serious privacy issues. Moreover, informal emergency response is mostly dependent on the contributions of online volunteers. However, these "volunteers are a fragile and finite resource, frequently subject to burnout" (Korset 2013, p. 138).

Most importantly, we argued that one needs a new and extended lens for integrating formal and informal emergency responses called an event-driven lens. The event-driven lens first takes into account formal emergency response characterized by institutionalized organizations, formal rules and procedures, and hierarchical organizational structures. The event-driven lens also considers informal emergency response based on the voluntary contributions of publics, decentralized, open, adaptive organizational structures, and technologies. Moreover, the event-driven lens concentrates on the interactions and relationships between formal and informal actors in response to a disaster or a crisis. It is argued that the event-driven lens is more useful for understanding and explaining complex and dynamic emergency response systems in the networked age than an existing framework based primarily on formal organizations and their roles and responsibilities.

Lastly, from a practical point of view, we argue that formal organizations need to develop technical and management capacities for collaborating with informal groups of individuals. These informal groups mostly create disaster-related information by collecting, processing, and visualizing content from social media and mainstream media in near real-time. Formal organizations are required to build technical capacities for integrating the information created by informal actors into official emergency information management systems to improve situational awareness and to effectively coordinate on-scene operations and off-scene supports. Moreover, informal groups of individuals are a new form of organization characterized by openness, decentralized organizational structures, and the use of technologies. To collaborate with these informal groups, formal organizations need to not only understand such organizational characteristics of informal groups, but also build and facilitate dynamic communications and robust partnerships with informal groups.

---

[16] https://www.reddit.com/.

[17] http://www.4chan.org/.

References

Boulos MNK, Resch B, Crowley DN, Breslin JG, Sohn G, Burtner R et al (2011) Crowdsourcing, citizen sensing and sensor web technologies for public and environmental health surveillance and crisis management: trends, OGC standards and application examples. Int J Health Geogr 10(1):67

Buck DA, Trainor JE, Aguirre BE (2006) A critical evaluation of the incident command system and NIMS. J Homeland Secur Emerg Manag 3(3):1–27

Capelo L, Chang N, Verity A (2012) Guidance for collaborating with volunteer and technical communities. Retrieved September 1, 2015, from http://digitalhumanitarians.com/content/guidance-collaborating-volunteer-technical-communities

Castillo C (2015) AIDR overview. Retrieved August 10, 2016, from https://github.com/qcri-social/AIDR/wiki/AIDR Overview

Cole D (2000) The incident command system: a 25-year evaluation by California practitioner. Retrieved February 10, 2016, from http://www.alnap.org/resource/9484

Crowley J (2013) Connecting grassroots and government for disaster response. Commons Lab of the Woodrow Wilson International Center for Scholars, Washington

Crowley J, Chan J (2011) Disaster relief 2.0: the future of information sharing in humanitarian emergencies. Retrieved June 24, 2014, from http://hhi.harvard.edu/publications/disaster-relief-20-future-information-sharinghumanitarian-emergencie

Drabek TE (1985) Managing the emergency response. Public Adm Rev 45(Special Issue):85–92

Drabek TE, McEntire DA (2003) Emergent phenomena and the sociology of disaster: lessons, trends and opportunities from the research literature. Disaster Prev Manag 12(2):97–112

Edmunds A, Morris A (2000) The problem of information overload in business organisations: a review of the literature. Int J Inf Manag 20(1):17–28

Haddow GD, Bullock JA, Coppola DP (2008) Introduction to emergency management, 3rd edn. Butterworth-Heinemann, San Diego

Harrald JR (2006) Agility and discipline: critical success factors for disaster response. Ann Am Acad Pol Soc Sci 604(1):256–272

Heinzelman J, Waters C (2010) Crowdsourcing crisis information in disaster-affected Haiti. The United States Institute of Peace, Washington

Helsloot I, Ruitenberg A (2004) Citizen response to disasters: a survey of literature and some practical implications. J Conting Crisis Manag 12(3):98–111

Hemmi A, Graham I (2014) Hacker science versus closed science: Building environmental monitoring infrastructure. Inf Commun Soc 17(7):830–842

Howe J (2009) Crowdsourcing: why the power of the crowd is driving the future of business. Three Rivers Press, New York

Imran M, Castillo C, Lucas J, Meier P, Vieweg S (2014) AIDR: Artificial intelligence for disaster response. In: The international world wide web conference committee, Seoul, Korea

Jaeger PT, Shneiderman B, Fleischmann KR, Preece J, Qu Y, Fei Wu P (2007) Community response grids: e-government, social networks, and effective emergency management. Telecommun Policy 31(10–11):592–604

Kapucu N (2006) Interagency communication networks during emergencies: boundary spanners in multiagency coordination. Am Rev Public Admin 36(2):207–225

Kapucu N, Garayev V (2011) Collaborative decision-making in emergency and disaster management. Int J Public Admin 34(6):366–375

Kapucu N, Garayev V, Wang X (2013) Sustaining networks in emergency management. Public Perform Manag Rev 37(1):104–133

Kathuria R, Joshi MP, Porth SJ (2007) Organizational alignment and performance: past, present and future. Manag Decis 45(3):503–517

Korset R (2013) World disasters report 2013: focus on technology and the future of humanitarian action. Geneva, Switzerland. Retrieved from http://www.ifrc.org/PageFiles/134658/WDR2013 complete.pdf

Kreps GA, Bosworth SL (1993) Disaster, organizing, and role enactment: a structural approach. Am J Sociol 99(2):428–463
Leonard D, Howitt A (2005) A command system for all agencies? Crisis Response J 1(2):40–42
Lindsay BR (2012) Federal emergency management: a brief introduction. Retrieved June 15, 2015, from https://www.everycrsreport.com/reports/R42845.html
McGuire M, Silvia C (2010) The effect of problem severity, managerial and organizational capacity, and agency structure on intergovernmental collaboration: evidence from local emergency. Public Adm Rev 70(2):279
Meier P (2011) New information technologies and their impact on the humanitarian sector. Int Re Red Cross 93(884):1239–1263
Meier P (2013) AIDR: Artificial intelligence for disaster response. Retrieved August 9, 2016, from https://irevolutions.org/2013/10/01/aidr-artificial-intelligence-for-disaster-response/
Meier P (2015) Digital humanitarians: how big data is changing the face of humanitarian response. CRC Press, New York
Meier P, Munro R (2010) The unprecedented role of SMS in disaster response : learning from Haiti. SAIS Rev Int Aff 30(2):91–103
Moynihan DP (2008) Combining structural forms in the search for policy tools : incident command systems in U.S. crisis management. Governance 21(2):205–229
Munro R (2013) Crowdsourcing and the crisis-affected community: lessons learned and looking forward from Mission 4636. Inf Retr 16(2):210–266
Palen L, Anderson KM, Mark G, Martin J, Sicker D, Palmer M, Grunwald D (2010) A vision for technology-mediated support for public participation and assistance in mass emergencies and disasters. In Proceedings of the 2010 ACM-BCS Visions of computer science conference. The British Informatics Society Ltd., Edinburgh, pp 1–12
Quarantelli EL (2005) Catastrophes are different from disasters: some implications for crisis planning and managing drawn from Katrina. Retrieved February 15, 2015, from https://www.unitedsikhs.org/katrina/catastrophes_are_different_from_disasters.pdf
Roberts NC (2011) Beyond smokestacks and silos: open-source, web-enabled coordination in organizations and networks. Public Adm Rev 71(5):694
Roche S, Propeck-Zimmermann E, Mericskay B (2011) GeoWeb and crisis management: issues and perspectives of volunteered geographic information. GeoJournal 78(1):21–40
Rodriguez H (2006) Rising to the challenges of a catastrophe: the emergent and prosocial behavior following Hurricane Katrina. Ann Am Acad Pol Soc Sci 604(1):82–101
Roe E, Schulman P (2008) High reliability management: operating on the edge. Stanford University Press, Stanford
Schroeder A, Wamsley G, Ward R (2001) The evolution of emergency management in America: from a painful past to a promising but uncertain future. In: Farazmand A (ed) Handbook of crisis and emergency management. Marcel Dekker, Inc., New York
Schulman P, Roe E, van Eeten M, de Bruijne M (2004) High reliability and the management of critical infrastructures. J Conting Crisis Manag 12(1):14–28
Shirky C (2008) Here comes everybody: the power of organizing without organizations. The Penguin Press, New York
Slater DH, Nishimura K, Kindstrand L (2012) Social media, information, and political activism in Japan's 3.11 crisis. Asia-Pac J Jap Focus 10(24.1):1–41
Stallings RA, Quarantelli EL (1985) Groups emergency management. Public Admin Rev, 45(Special Issue: Emergencies Management: A Challenge for Pubblic Administration):93–100
Tapia AH, Lalone N, Kim H (2014) Run Amok: group crowd participation in identifying the bomb and bomber from the Boston Marathon Bombing. In: Proceedings of the 11th international ISCRAM conference. University Park, pp 270–279
Tapscott D, Williams AD (2006) Wikinomics: how mass collaboration changes everything. The Penguin Group, New York
The U.S. Department of Homeland Security (2008) National incident management system. Retrieved January 28, 2016, from https://www.fema.gov/pdf/emergency/nims/NIMS_core.pdf

The U.S. Department of Homeland Security (2013) National response framework. Retrieved January 27, 2016, from https://www.fema.gov/medialibrary-data/20130726-1914-25045-1246/final_national_response_framework_20130501.pdf

Thellufsen C, Rajabifard A, Enemark S, Williamson I (2009) Awareness as a foundation for developing effective spatial data infrastructures. Land Use Policy 26(2):254–261

Waugh WL (2006) The political costs of failure in the Katrina and Rita disasters. Ann Am Acad Pol Soc Sci 604(1):10–25

Waugh WL (2007) EMAC, Katrina, and the governors of Louisiana and Mississippi. Public Adm Rev 67(Suppl 1):107–113

White JI, Palen L, Anderson KM (2014) Digital mobilization in disaster response: the work and self-organization of on-line pet advocates in response to Hurricane Sandy. In: The 17th ACM conference on computer supported cooperation work and social computing, Baltimore

Yuan W, Guan D, Huh E, Lee S (2013) Harness human sensor networks for situational awareness in disaster reliefs : a survey. IETE Tech Rev 30(3):240

Zook M, Graham M, Shelton T, Gorman S (2010) Volunteered geographic information and crowdsourcing disaster relief: a case study of the Haitian earthquake. World Med Health Policy 2(2):6–32

Chul Hyun Park is an Assistant Professor at the University of Arkansas Clinton School of Public Service. He received a Ph.D. in Public Administration and Policy from Arizona State University and a Master of Public Policy from Georgia State University. His research interests include emergency management and disaster policy, information technology for collaborative decision making, and policy analysis and program evaluation. Currently, he focuses on how participatory online groups formed by the general public collaborate with public organizations to address complex social problems, using information, communication, and computational technologies.

Erik W. Johnston is an Associate Professor in the School for the Future of Innovation in Society and Director of Policy Informatics at the Decision Theater, both at Arizona State University. His research focuses on policy informatics, the study of how computational and communication technology is leveraged to specifically understand and address complex public policy and administration problems and realize innovations in governance processes and institutions. With undergraduate degrees in Computer Science and Psychology as well as a Master in Business Administration and a Master of Science in Information Technology from the University of Denver, Johnston holds a Ph.D. in Information from the University of Michigan with a certificate in complex systems.

Social-Media-Based Policy Informatics: Cyber-Surveillance for Homeland Security and Public Health Informatics

Wingyan Chung and Daniel Zeng

Abstract Nowadays, policy makers face complex challenges such as cybersecurity, infectious disease control, and political rumors, which can quickly elicit large-scale crises in a world connected by social media. However, scarce work is found in using social media analytics to support high-impact policy areas that span across nations and population groups. In this chapter, we describe a framework for social-media-based policy informatics and its application to addressing policy issues concerned by governments and general public. Based on the framework, we conducted two case studies: The first study concerns with the U.S. immigration and border security, which is increasingly motivated by cyber means and affects the U.S. economy, national security, and foreign policies. The second study concerns with the West African Ebola disease outbreak, which has caused over 11,305 people to die and is the largest and most complex in the history of the disease. We present empirical findings obtained from the data analyses and discuss the implication for public policy decision making. The research should contribute to developing a new social-media-based framework for policy informatics and to demonstrating its use in high-impact policy issues.

Abbreviations

BBC	British Broadcasting Company
BI	Business intelligence

W. Chung (✉)
Institute for Simulation and Training, University of Central Florida,
3100 Technology Pkwy, Orlando, FL 32826, USA
e-mail: wchung@ucf.edu

D. Zeng
Department of Management Information Systems, Eller College of Management, The University of Arizona, 1130 E. Helen Street, Tucson, AZ 85721, USA

State Key Laboratory of Management and Control for Complex Systems, Institute of Automation, Chinese Academy of Sciences, Beijing 100190, China
e-mail: zeng@email.arizona.edu

J.R. Gil-Garcia et al. (eds.), *Policy Analytics, Modelling, and Informatics*, Public Administration and Information Technology 24,
DOI 10.1007/978-3-319-61762-6_16

D.C.	District of Columbia
eMood	Ebola disease outbreak social media mood system
iMood	The U.S. immigration and border security mood system
MIS	Management information systems
SM	Social media
SMA	Social media analytics
U.C.F.	University of Central Florida
U.S.	United States

Introduction

Policy makers face tremendous challenges nowadays, as issues in homeland security, infectious disease control, and political rumors can quickly elicit large-scale crises in today's connected world. Traditionally, policy makers relied on human-contact and media outreach to collect information about public sentiment and concerns. Information technologies have significantly increased the efficiency of this data collection and dissemination process. In particular, social media analytics (SMA) holds the promise of supporting effective policy decision making and analysis. SMA entails a rich set of tools and technologies applied to analyzing web-based social content, such as social media posts, blogs, and online reviews. Traditional domains where SMA is used include online advertising, product review analysis, and social profiling. Applications of SMA to understanding and responding to complex problems faced by governments are still emerging. These applications require advanced development and use of information technologies, data collections, analysis, modeling, and simulation to address specific needs for public safety and policy concerns.

Despite emerging research in SMA (e.g., Schreck and Keim 2013; Song et al. 2007; Zeng et al. 2010) and in policy informatics (e.g., Hagen et al. 2015; Lazer et al. 2009; SUNY Albany 2013), little work is found in sentiment and network analyses among users of social media on high-impact public policy areas. In particular, scarce work is found in the context of high-impact policy areas that span across nations and population groups.

In this chapter, we describe a framework for social-media-based policy informatics and its application to addressing policy issues concerned by governments and general public. We present two case studies of applying the framework to addressing two important policy areas: homeland security and public health. The first study concerns with the U.S. immigration and border security. The second study concerns with the West African Ebola disease outbreak. In each case study, we developed a proof-of-concept system that collected large amounts of social media content, tracked the development of SM discussion, analyzed and visualized the SM content to reveal public concerns, extracted user emotion and sentiment, and revealed network relationships of these users. We present empirical findings obtained from the data analyses and discuss the implication on public policy decision making. These findings should be useful for answering the following research questions.

1. What is the sentiment and emotion of SM users on the selected policy issues?
2. How may relationships of SM users be characterized in the policy discussion community?
3. Among the SM users who discuss the policy issues, who are the opinion leaders? Who are the authorities and hubs? Who are the most active participants in the discussion?
4. Are there significant observations in user emotion, sentiment, and network characteristics shown in the SM discussion?

The rest of the chapter is structured as follows. Section "Literature Review" surveys existing research in policy informatics and social media analytics. Section "A Framework for Social-Media-Based Policy Informatics" presents a framework for social-media-based policy informatics. Sections "Case Study 1: The U.S. Border Security and Immigration" and "Case Study 2: The West African Ebola Outbreak" describe two case studies of using the framework to address two high-impact policy domains. Section "Conclusion" summarizes the findings, discusses implications, and suggests future directions.

Literature Review

To address complex problems related to public policy making, the field "policy informatics" was developed to find effective ways to use information and computation to understand and tackle complex problems of society (Dawes and Janssen 2013). Understanding research projects in policy informatics and relevant techniques and applications of social media analytics (SMA) can inform about the tools available and potential use of them. In this literature review, we attempt to answer several questions to help us identify research needs and potential gaps: What types of research does policy informatics communities conduct? How do information technologies help in policy informatics research? What does social media analytics (SMA) contribute to policy informatics? How are tools and methods in SMA used to support sentiment analysis and assessment? The criteria for selecting literature include relevance to the topics of policy informatics and SMA, recency of publication, and application to government administration.

Policy informatics is an analytical approach that comprises concepts, methods, and processes for understanding complex public policy and management problems (Dawes and Janssen 2013). The approach is increasingly explored in the digital government research community. For instance, the Digital Government Society has embraced social media, Internet, big data, and related technologies in its conference held in 2016 (http://dgo2016.dgsociety.org/call-papers). The PoliInformatics project at the University of Washington (http://poliinformatics.org) aims to encourage scholars representing diverse disciplines such as computer science, information science, electrical engineering, statistics and computational linguistics to collaborate and apply their skills to the analysis of U.S. government activity. Policy issues that have widespread impact may benefit from policy informatics research. Examples include immigration and border security policy, voting and citizen participation, and health care.

Projects on Policy Informatics

The U.S. immigration policies have been a focus of debates over many years (Gans et al. 2012) and have attracted attention from policy informatics researchers. These debates touch on a wide variety of issues including legal immigration selection system, gender and refugee status, legalization for unauthorized immigrants, birthright citizenship, immigrant voting, nativism, border control, immigrant detention and deportation, and so on. The Center for Border Security and Immigration at the University of Arizona develops innovative approaches such as avatar-enabled custom check-points to address some of these issues (http://www.borders.arizona.edu/). Meanwhile, the Center for Policy Informatics at Arizona State University (https://cpi.asu.edu/) is developing a trans-disciplinary study of how computation, communication, and information technology are leveraged to specifically understand and address complex public policy and administration problems, such as border crimes and human trafficking. The Center for Technology in Government at the State University of New York at Albany is involved in cultivating a new community of inquiry and practice within the public policy research and management arena (SUNY Albany 2013).

Engaging citizens in participatory policy making has been an important application of policy informatics and understanding the society (Lazer et al. 2009). Krishnamurthy et al. examined the dissemination of empathy in human-centered electronic participatory platforms (Krishnamurthy et al. 2013) and provided four examples of electronic participatory platforms that encourage participants to share their views and connect with others. Hagen et al. studied electronic petitioning and introduced the use of textual analysis tools to extract named entities and topics from "We the People" petition texts (Hagen et al. 2015). They found that informativeness, named location, and several topics are significantly correlated with the log of the signature counts in e-petitioning. Chung and Lewis used business intelligence tools to predict image perception of a county government (Chung and Hershey 2012) and demonstrated robust performance and high accuracy of their approach in supporting branding government services and predicting public opinions. Marathe et al. developed an integrated program to represent and reason about large co-evolving social, technological, information and organization networks (Marathe et al. 2009). The integrated environment provides analysts and decision makers Web-based access to models and synthetic networks for policy planning and response. Lampe, LaRose, Steinfield, and DeMaagd (Lampe et al. 2011) developed a crowd-sourcing system in the "AdvanceMichigan" project that aimed to use social media to engage with stakeholders in policy decision making. Unfortunately, the project failed because the audiences did not know how to use social media, the software was hard to use, stakeholders had low motivation to participate, the timeframe was too short, and the tasks were not suitable to be done in social media. The work highlights the need to develop new frameworks and technologies that use social media to support policy informatics research.

Social Media Analytics

Social media platforms provide a trove of public data suitable for use in policy informatics research due to its social nature and large-scale organizational capability (Miller 2011). Social media analytics (SMA) is concerned with "developing and evaluating informatics tools and frameworks to collect, monitor, analyze, summarize, and visualize social media data, usually driven by specific requirements from a target application" (Zeng et al. 2010). Traditional domains where SMA is applicable include business intelligence, online product review analysis, and security informatics (Chen et al. 2012; Chung 2014; Chung and Tseng 2012). Sentiment analysis and emotion extraction are important tools for understanding large amounts of social media data (Fan and Gordan 2014; Feldman 2013). For example, researchers have developed a specialized lexicon to support emotion extraction (Mohammad and Turney 2013). The lexicon contains over 13,000 entries (each entry being a combination of a term and its emotional category), in which a term may belong to one or more of the eight emotion categories (anger, anticipation, disgust, fear, joy, sadness, surprise, trust) (Plutchik 1980).

Network analysis is another category of SMA application, in which users within a large community are analyzed to reveal their levels of centrality and influence. Traditionally, social network analysis examines network dynamics using small datasets (e.g., representing networks with fewer than 10 persons) (Wasserman and Faust 1994). Advances in information technologies enable new capabilities of network analysis, such as identifying opinion leaders in discussion forums (Song et al. 2007), finding expertise in networks (Zhang et al. 2007), characterizing social influence in networks (Liang et al. 2015), analyzing business stakeholder networks (Chung et al. 2009), and visualizing networked relationships of infectious disease transmission (Zeng et al. 2011). As the amount of data grows, visual analysis of social media data has become an important capability to support understanding. Schreck and Keim (Schreck and Keim 2013) explore how visual analytics can support trend analysis in disease informatics, marketing, opinion analysis, demographics, and civil protection from social media such as Twitter, Flickr, and news blogs.

Based on our review of existing works described above (e.g., Liang et al. 2015; Schreck and Keim 2013; Song et al. 2007; Zeng et al. 2010), we find that social media are increasingly important information sources for policy makers and social scientists to understand public sentiment and to identify important stakeholders.

The emerging field of policy informatics should benefit from various concepts, methods, and processes established and also the technologies and applications developed in social media analytics. Among these technologies and applications, little work is found in discovering sentiment from and network relationship among users of social media on high-impact public policy areas. While capabilities to analyze and visualize social media data are increasingly important, scarce work is found in the context of high-impact policy areas that span across nations and population groups.

A Framework for Social-Media-Based Policy Informatics

This section describes a framework for social-media-based policy informatics and explains design rationale and components of the framework. The characteristics of data collection and the systems developed based on the framework are explained.

Rationale of Framework Development

The challenges of using SMA in policy informatics include a lack of treatment of enriched data or metadata found in social media, context-dependent user profiling, a lack of data structure, and rapidly evolving dynamics and growing volumes of data. The networked relationships formed among SM participants and their sentiment need to be extracted and analyzed to provide insights to support decision making. Despite availability of SMA research (e.g., Liang et al. 2015; Schreck and Keim 2013; Song et al. 2007; Zeng et al. 2010), there is little work found on sentiment and network analyses and visualization of public policy SM data.

To address the needs, different approaches can be considered, such as questionnaire survey, field studies, simulation, interview, artifact development, and case studies, among others (Nunamaker et al. 1991). These approaches were categorized into two major paradigms: behavioral science and design science (Hevner et al. 2004). The behavioral science paradigm seeks to develop and verify theories to explain or predict human or organizational behavior. The design science paradigm creates new and innovative artifacts to extend human and organizational capabilities. These paradigms are foundational to knowledge creation. Specific to our task of building SM analytics to address policy issues that involve large amounts of SM data, we used a design science paradigm to develop a framework for social-media-based public policy informatics. Hevner et al. proposed seven guidelines for conducting design science research in their paper (Hevner et al. 2004), which is the fourth most cited articles ever published in *MIS Quarterly* (impact factor = 5.384), a top-tier academic journal in the fields of business and management information systems (http://www.misq.org/skin/frontend/default/misq/pdf/MISQStats/MostCitedArticles2013.pdf). Following these guidelines, we developed the framework to tackle important problems in the field of public policy informatics. The artifacts, which are instantiations of the framework and the relevant methods, include an integrated system for sentiment discovery, emotion extraction, network analysis, and activity visualization. The system could be used to discover both sentiment and network relationship among users, relate them to the influence of the participants, and present the results in a dynamic, user-friendly visualization.

In addition to using the guidelines, we systematically reviewed previous work on social media analytics in the domain of policy informatics. We found these relevant approaches: web-based crowdsourcing, data mining, text mining, and network analysis. An example is the social media crowdsourcing system described in (Lampe et al. 2011) that failed to achieve its goals. Another large-scale experiment found that Facebook messages directly influenced political self-expression, information seeking, and real-world voting behavior of millions of people (Bond et al. 2012). The study primarily examined subjects' behavior without analyzing the message sentiment and the network linkage among users. Kane et al. developed an approach to detect opinion shifts in the U.S. presidential debates (Kane et al. 2014). Their approach groups users with similar opinions together and track time-dependent collective measures of user opinions. Our study advances prior research (e.g., Bond et al. 2012; Kane et al. 2014; Lampe et al. 2011) by introducing network analysis that tracks not only the temporal changes but also complex relationship among social media participants. The policy issues examined in the cases described below demonstrate a wider scope than those chosen in prior research.

Framework Components

As shown in Fig. 1, the framework consists of various steps of data collection, cleaning, filtering, and analysis. Input sources for the framework include social media data, a specialized lexicon for emotion extraction and sentiment analysis, an algorithm to compute emotion indices, and program modules for performing network analysis and visualization. Compared with the approaches and frameworks developed in previous works (e.g., Bond et al. 2012; Kane et al. 2014; Lampe et al. 2011; Schreck and Keim 2013), our framework supports a comprehensive analysis of political social media content and networks and provides a dynamic, user-friendly visualization of the user sentiment, reputable leaders, and their social activities. The framework is adaptable to any policy domains where social media discussions are used.

Data Collection

Social media sites provide large streams of live data for building domain-specific collections. These data contain textual information, temporal information, and structural information. Textual information refers to the text message that an author

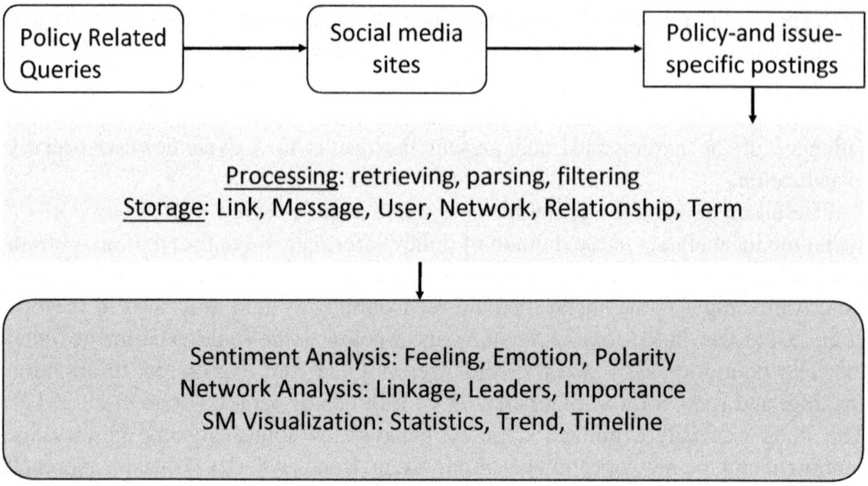

Fig. 1 A framework for social-media-based policy informatics

writes on the social media site (e.g., an opinion on a political issue). Temporal information is the time at which the message was posted on the site. Structural information can refer to the relationship of the message with respect to other messages (e.g., whether a message was re-posted by another author) and to the relationship among authors (e.g., whether an author re-posted another author's message). Analyses are performed to discover sentiment and to extract emotion from posted messages. Because social media sites receive a large number of messages expressed in different sentiment by different authors every day, comprehensive lexical resources must be used to perform the analyses.

Social media site users interact with each other and form networked relationships. The network embeds important information about the level of activity of the users, the centrality of the users, and the density and structure of the network. Policy makers and social scientists will find the information useful for identifying the influential users, the strength of bonding among users, and the key links that contribute to information flows within the network. In addition to static analyses, dynamic analyses of changes in sentiment and network can provide details about the evolution of public opinion and group structures, offering insights into strategies for gaining public approval on political issues. The proposed framework was designed to support dynamic analysis by aggregating temporal data and by presenting the data in visually perceivable formats.

Systems Design and Development

To demonstrate the applicability of the framework, we developed in each case study a proof-of-concept online system that extracts real-time sentiment and emotion from social media messages and discover network relationship among the message authors. The system implements the steps of the framework and collected social media messages posted on Twitter, a major social media website where users post short messages called "tweets" on issues that they are interested. Each tweet consists of no more than 140 alphanumeric characters.

Sentiment Analysis

We developed customized algorithms for computing eight emotional scores for each social media post (e.g., a tweet). As defined in (Plutchik 1980), the eight emotional categories are anger, disgust, fear, sadness, surprise, anticipation, joy, and trust. Then, an aggregated index was computed from the scores for each category. Each index represents aggregated emotional intensity of all tweets posted over a specific period of time. We used the emotion lexicon developed in (Mohammad and Turney 2013) to identify emotional words that appear in the posts. Each lexicon word of the 13,901 word-emotion entries was tagged by over 2216 human users who completed a total of 38,726 tagging assignments. Each word was assigned a sentiment polarity (positive (+1) or negative (−1)) and is categorized into one (or more) emotional category. The score of an emotion category for a tweet is the proportion of the sum of the number of words tagged by the lexicon in that category to the total number of words in the tweet. An aggregated score (e.g., daily average index) of an emotion category for a set of tweets is calculated by averaging the scores of that category for the set of tweets.

Network Analysis

Social media users interact with each other by sending messages or replying to their messages (e.g., tweets). These interactions enable researchers and policy makers to analyze user activities and roles in the community. To support the analysis, we constructed a user interaction network by using the interactions and roles of the users. A node in the network represents a user; a link represents an interaction between two users. The system identifies a link between User A and User B when User A sends a tweet targeted to User B, or re-tweets another tweet written by User B, or modifies and then sends out a tweet written by User B.

The user interaction network changes over time as user relationships change. We define an interaction window to be a specific time frame during which user activities are considered in building the network. This time frame is a sliding window of 36 days that end on the day on which new data are collected. While other lengths are possible, we chose the 36-day length to reflect a typical time span of user activity spike based on empirical observation. This length could be adjusted to adapt to specific context of a chosen domain. The system automatically constructs a new user interaction network for the most recent interaction window.

We used several metrics to identify importance of users in a network and the overall characteristics of the network (Jackson 2008; Wasserman and Faust 1994). *Degree* measures the number of interactions that a user has in a network. *Density* measures the proportion of the number of links in a network to the maximum possible number of links in the network. To assess the influence of a user, we use *Betweenness Centrality* that measures how well-situated a node is in terms of the paths that it lies on. While other measures of centrality are available (Freeman 1977; Jackson 2008), *Betweenness Centrality* identifies the extent to which a user serves as a bridge in immigration reform discussion, which heavily relies on the ability to find compromises and common grounds in participants of highly polarized views.

Social Media Visualization

As data are collected and analyzed to reveal sentiment and network relationships among participants, real-time visualization of the analyzed results can provide timely overview of the various concerns and emotion of the social media community, thereby providing decision support to policy makers. Our framework includes a user-friendly graphical interface of the visualization that seeks to facilitate user understanding of these concerns and emotion. The visualization contains various bar charts to show changes in sentiment and emotion over time, a dynamic table of key statistics such as the number of messages posted and the number of participants, and various ranked lists of participants who are considered to be leaders in the community. The visualization serves to turn the complex analysis results into pictures and charts that can be understood easily by humans.

Case Study

We conducted two case studies on important policy issues to demonstrate the capability of the framework to support policy informatics tasks. These tasks include collecting social media (SM) posts on public policy issues, discovering sentiment from user discussions, extracting user emotion, and identifying network relationships among SM users. The case studies should inform about the sentiment and emotion of SM users on the selected policy issues, characterize relationships of SM users in the policy discussion community, identify opinion leaders and active participants in the community, and reveal significant observations in user emotion, sentiment, and network characteristics.

Case Study 1: The U.S. Border Security and Immigration

The first case study focuses on policy issues of the U.S. border security and immigration, which is increasingly motivated by cyber means such as social media, online movement, and blogs. The background, system implementation, findings, and implications are described below.

Background

Public policies on border security and immigration are major concerns in the United States in recent years. As over eleven million undocumented immigrants reside in the United States, the U.S. Government and Congress are under tremendous pressure of developing timely measures to secure the national border and to reform immigration policies. These measures will help to fight against terrorists, drug traffickers, and illegal gang members who constantly seek to infiltrate the U.S. homeland. Diverse views are expressed on social media and on news, ranging from offering legal status to millions of undocumented immigrants in the U.S. (Dwyer 2014) to mass deportation of all people who entered or stayed in the country unlawfully (Bennett 2015). The former U.S. Secretary of Defense, Leon Panetta, and leading experts in border security concurred that comprehensive immigration reform is necessary for the nation's safety and economic viability (Molnar 2013). Prominent Congress members had proposed a comprehensive immigration reform bill to tackle the long-standing problems in immigration and border security. Immigration and border security was a hotly-debated issue in the 2016 U.S. presidential election. Social media are used actively by diverse groups and opinion leaders to influence public sentiment. Social media analytics offers important opportunities for policy informatics research in the domain of border security and immigration.

System Implementation

The iMood System

Using the framework described in section "A Framework for Social-Media-Based Policy Informatics", we developed a proof-of-concept system named "iMood" (Immigration and Border Security Mood) that extracts real-time sentiment and emotion from immigration reform tweets and discover network relationship among these tweet authors. iMood implements the steps of the framework in the context of U.S. immigration reform discussion on Twitter.com.

Data Collection for iMood

To build the data collection for the iMood system, we focused on collecting tweets that related to the U.S. immigration and border security discussion. To do so, we constructed carefully a list of queries by reviewing a list of literature published on the topic (Bush et al. 2009; Gans et al. 2012; LeMay 2004; U.S. Commission on Immigration Reform 1994). Filtering and collection testing helped to reduce the original 14 queries to 8 domain-specific queries: "comprehensive immigration reform," "illegal alien," "illegal immigration," "immigration debate," "immigration policy," "immigration reform," "US border security," and "US immigration." Following Twitter's guidelines of data collection, the iMood system continually collects on average 4650 tweets per day using the selected queries and then performs analysis on the data. Through the steps of searching, retrieving, parsing, filtering, and storing, the process converts unstructured social media data into a collection of structured social media data that are organized into message, user, message-term association, link, network, and user-network association. This study uses the data collected between May 21, 2013 and September 1, 2015.

Findings

Sample Social Media Messages

Using the algorithms described in section "Systems Design and Development", the iMood system computes the indices and displays charts that summarize the daily emotional indices and sentiment polarity of all collected tweets. It also identified network relationships based on user interactions, such as retweets and modified tweets shown below:

Retweet Example:

Today is Camino Americano, a rally in Washington D.C. demanding comprehensive immigration reform with a path to... http://t.co/sP3wfRW8YX. October 8, 2013.

Modified Tweet Example:

Spite House priorities. MT @FAIRImmigration US tax $ used to protect, police illegal alien rally #noamnesty http://t.co/BTxbimXI4o, October 8, 2013.

Important Events

Several important events happened during the period and were identified by iMood through the number of tweets and evolution of the tweet volume.

The iMarch movement drew over 10,000 tweets in May 2013. The event was developing during that week and reached a climax on May 22, 2013, when more than

7000 tweets were posted expressing support for the movement. Many of the tweets take this form: "*I'm from ... and #imarch for immigration reform because ...* ."

Another major surge on tweet posting happened on June 7, 2013, when several debates on immigration and border security took place. On that day, the White House denounced a Republican-controlled House vote to restart the deportation of hundreds of thousands of immigrants brought illegally to the United States as children. Meanwhile, the U.S. Senate began a formal debate on its version of a comprehensive immigration-reform measure to grant a pathway to citizenship for the estimated 11 million immigrants living in the country illegally (http://www.upi.com/Top_News/US/2013/06/07/Senate-to-begin-immigration-reform-debate/UPI-19961370586600/).

As debates about immigration heated up during the mid- and late-June 2013 as well as the U.S. Presidential Election and its primary elections between 2015 and 2016, more tweets were posted. Opinions on both sides tried to seize the agenda on Twitter. Examples of these tweets include (see Fig. 2):

- *realDonaldTrump: "Fact – Amnesty lowers wages and invites more lawlessness. Obama has unilaterally cancelled any chance of immigration reform."*
- *How Immigration Reform Would Help the Economy—New York Times (blog) http://t.co/R91OqO9ipF*
- *Jose Vargas: "For decades, I have cringed whenever someone called me 'illegal,' as if I'm an insect on someone's back. http://t.co/WQVi5WuQWc"*
- *It's not clear that immigration reform will solve this rampant wasteful border-security spending: http://t.co/fbZv6u24FR*
- *realDonaldTrump: "I told you so. Our country totally lost control of illegal immigration, even with criminals. http://t.co/IZgZqr6BgB"*
- *This bill is amnesty. http://t.co/5X5CdqfaQu #tcot #teaparty #gop*

The daily number of tweets reached record highs of 38,761 and 37,223 on November 20, 2014 and November 21, 2014 respectively shortly after President Obama announced his executive action to provide a path to U.S. citizenship to about four million people who resided in the country without legal documentation. Other notable high points of daily tweet volume was 16,082 on October 8, when a mass protest for comprehensive immigration reform took place in Washington, D.C., resulting in the arrest of 200 people including at least 8 Democratic members of the U.S. Congress (Constable and Morello 2013). The daily tweet volume reached 14,451 on June 27, when the U.S. Senate decided on a 68-32 voting to pass the comprehensive immigration reform bill (Silverleib 2013). While these events alone could attribute to the upsurge, other related events such as the U.S. Presidential election and decisions of the Supreme Court on Defense of Marriage Act triggered the increases as well.

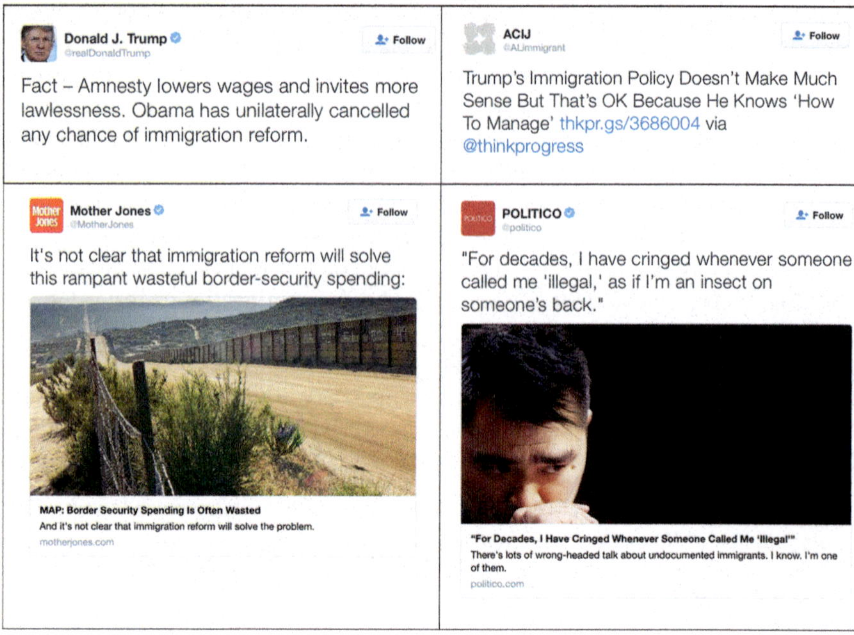

Fig. 2 Sample Tweets on U.S. immigration and border security

Identification of Important Users

A number of leaders were identified in the Twitter discussion on U.S. immigration and border security by using various metrics. These metrics include the influence score (see section "Network Analysis"), number of times the user's tweets are retweeted, number of tweets posted by the user, number of tweets that represent opinions posted by the user, number of targeted tweets received by the user, number of followers of the user, and number of friends of the user.

Among the list of influential users who got highest betweenness centrality scores, Donald Trump is ranked top with an influence score more than triple of the second most influential user, Jose Antonio Vargas. Despite having close rankings, the two leaders hold drastically different views on border security and immigration: Trump, who won the U.S. Presidential Election in November 2016, advocated mass deportation of illegal immigrants, while Vargas, a Pulitzer-winning journalist and an undocumented immigrant, advocated comprehensive immigration reform. Other leaders with high influence scores include Katie McHugh, Ann Coulter, Breitbart News, and CIR Now. iMood updates the influence scores four times per day and adjust user rankings as new data come in. In the sampled results, the users whose tweets were retweeted most frequently by other users are Barack Obama, Donald Trump, Speaker John Boehner, FWD.us, and Charles Garcia. These users are often major politicians or popular press (see "Most Retweeted Users"). Other tweet leaders include mjselker (who strongly opposes immigration reform), Ricardo Parra,

Miles Pierce, and DRM Action Coalition (which supports undocumented youths in the U.S.). Users who expressed the most opinions include Miles Pierce, Benouis Law Office, Linda P, Immigration Law News, and US Senators. Many of these users are related to immigration laws.

Implications

The aforementioned findings demonstrate important implications and trends regarding U.S. borders security and immigration. *First*, tweet volume highly indicates the coverage of news release. For example, the results indicate high tweet daily volumes when approaching and shortly after the dates when President Obama announced his executive action on providing a path to U.S. citizenship to about four million undocumented immigrants (on November 19, 2014). In contrast, these counts drop significantly during July-September 2013 when momentum of passing the comprehensive reform bill at the House stalled. *Second*, the lists of users identified as top leaders change significantly when new tweets were posted and new connections were made with other top leaders and influential users. For example, tweets posted by such users as Donald Trump, Senator Ted Cruz and Barack Obama often spurred waves of new tweets, retweets, and modified tweets, changing the ranks of these users in the top leader lists. *Third*, the use of different influence metrics produced significant different ranked lists of users, prompting for different interpretation of the ranks. In general, the list produced by using betweenness centrality indicates the top users who connect the crowds through civil groups, mass media, and political affiliation. Other lists provide different perspectives of leaderships, each identified by the way users behave on the social media community. These different perspectives allow analysts and decision makers to choose the suitable way to identify influential users.

Case Study 2: The West African Ebola Outbreak

Background

Social media, a growing online platform for sharing opinions and ideas, provides many opportunities for decision makers to understand public emotion. For example, healthcare professionals and public-health policy makers may identify from social media the spread of an infectious disease in advance of an epidemic, helping them to react in a timely manner.

Since the Ebola outbreak in West Africa was first reported in March 2014, more than 11,305 people have died from the disease (BBC 2015). The outbreak is the largest and most complex in the history of the disease, which was first discovered in

1976 (CDC 2015; WHO 2015). There have been more cases and deaths in this outbreak than all previous outbreaks combined. The disease was transmitted first from animals to humans, and then from humans to humans across countries. As of January 2016, the World Health Organization reports that no licensed Ebola vaccines are available (WHO 2015). Fear about the disease spreads widely on social media, raising public panic and concerns about this outbreak.

Systems Implementation

The eMood System

Using the framework described in section "A Framework for Social-Media-Based Policy Informatics", we developed a system for collecting and analyzing emotion expressed in infectious disease social media. The system named "eMood" focuses on Ebola disease outbreak social media mood. The eMood system collects tweet postings automatically and periodically between October 2014 and September 2015. The dataset consists of 2,980,918 tweets posted by 1,460,389 unique users.

Data Collection for eMood

To collect the tweets related to Ebola discussion, we constructed carefully a list of queries by reviewing recent literature published on the topic (BBC 2015; CDC 2015; WHO 2015). Filtering and collection testing helped to reduce the original 12 queries to 6 domain-specific queries: "ebola," "ebola patient," "ebola outbreak," "ebola virus," "ebola vaccine," and "ebola epidemic." In this study, we focus on the spike of Ebola cases in January 2015 (BBC 2015). We selected a subset of the data consisting of 255,118 tweets posted by 210,900 users between January 1st and January 31st of 2015. That month also witnessed the highest number of tweets and the most user activities in the first four months of 2015.

Findings

We used the methods described in section "Systems Design and Development" to analyze sentiment and network relationships of the social media community. The findings of our empirical study provide several key implications for understanding sentiment, emotion, online community leadership, and network structure and relationships.

Sample Social Media Messages

Using the methods described in section "Systems Design and Development", the eMood system computes the indices and displays charts that summarize the daily emotional indices and sentiment polarity of all collected tweets. It also identified network relationships based on user interactions, such as retweets and modified tweets shown below:

Retweet Example:

SYMONDSxSAYS: "They think it prevents Ebola too... RT @Lib_Librarian: No, eating your fruits and veggies won't prevent measles."

Modified Tweet Example:

Henry E. Chang "MT @AAI_AIDSwatch: @AIDSHealthcare New public health preparedness video on #Ebola: Suitable for screening on TV https://t.co/hmhD9p3v4q"

Mindoca "While everyone was freaking re: #ebola... MT @NewsHour Infected @Amtrak rider may have exposed travelers to #measles http://t.co/BDZ9DKjBUY"

Opinion Tweet Example:

Lim Wey Wen: "Something to think about. --> Ebola - Is Culture the Real Killer? http://t.co/m2ACJA42LM #westafrica via @allafrica"

Sentiment and Network Analyses

We identified influential users based on sentiment and network analyses. The first method identified several authorities in infectious diseases. World Health Organization (ranked #1) and Center for Infectious Disease Research and Policy at University of Minnesota (ranked #4) are two major organizations that conduct various activities to help Ebola patients. Médecins Sans Frontières (ranked #7) is an independent organization providing medical care for Ebola patients. Experts and authorities in Ebola are identified, including Ilona Kickbursch (ranked #2) and Ian M. Mackay (ranked #3), who are respectively renowned public health expert and virologist. The remaining users on the list are activists, medical news correspondents, and major news sources related to Ebola such as BBC Africa.

Profiles of Major Users

Figure 3 shows profiles of several top-ranked users. The CIDRAP examines infectious diseases as well as related policy. The medical experts, Ilona Kickbusch and Ian Mackay, provide opinions in health issues, policy, and diplomacy. The MSF_UK (also known as "Doctors without borders") provides medical care and

Fig. 3 Major top-ranked users

assistance to people based on need, irrespective of race, religion, gender or political affiliation. These users are well connected on social media and serve as major influential agents in the community.

Sample tweets by Ilona Kickbusch:

- *"#Ebola and the security sector! Join us on 5 February @IHEID in #Geneva http://t.co/Ni2ooVjhwy"*
- *"Feb issue of @LancetGH covers a variety of #globalhealth topics -#womenshealth #HPV #Ebola #pneumonia #childhealth http://t.co/epkbzZLyD6"*

Sample tweets by CIDRAP:

- *"Guinea sees bump in Ebola cases, group reports vaccine trial results via @ CIDRAP http://t.co/f0AYhZH5Up"*
- *"Oxfam calls for Ebola recovery 'Marshall Plan' -- also, outbreak tops 22,000 cases http://t.co/kiYIrse1Io"*

Sample tweets by MSF_uk:

- *"Check out the brilliant piece written by @PrettyNiceCo about @JuliaGash's donation to @MSF_uk for #Ebola fight. http://t.co/AvcsDRw9lD"*
- *"BBC News - #Ebola: Guinea's battle to change culture http://t.co/vSFPJsoADC"*

Sample tweets by Ian M. Mackay:

- *"#Malishouldnowbe#Ebolavirusdiseasefree(from18JAN)CONGRATULATIONS Mali for kicking Ebola out!42-days since last case tested NEG"*
- *"Why #VirolDU is now using more graphs with _just_ the confirmed #Ebola case numbers. http://t.co/7T08EX5cUI http://t.co/EvjyDn9qhp"*

Implications

The results demonstrate that the eMood system and the emotion analysis methods can be used for social-media-based public health informatics. Health professionals, public health policy makers, and health organizations administrators should benefit from the identification of influential opinion leaders, who can inform situational awareness during a disease outbreak and can provide necessary social support to other SM users. New policies in social media monitoring and analytics should be developed to enhance decision making capability.

Conclusion

The use of online social media has made it easy and convenient to collect data about public concern. However, the large volume and variety of expressions on social media have challenged traditional policy analysis and public sentiment assessment.

Summary of Findings

In this chapter, we describe a framework for social-media-based policy informatics. Based on the framework, we developed online systems that collect related social media messages and extracts user sentiment and network on Twitter. We presented case studies on two important policy topics: the U.S. border security and immigration and the West African Ebola disease outbreak. Two systems, iMood and eMood, were developed based on the framework to collect and analyze social media posts on the policy topics. Our findings indicate the systems are useful in tracking the sentiment and activities of users in the communities. Influential users were identified in each community to indicate opinion leaders and to highlights issues of concerns from these users. Changes in sentiment in different time frames help policy makers to understand shifts in public perception of policy issues.

Implications

The results have implications for public policy decision-making, online social community study, and policy trends. The release of news often coincide with changes in public sentiment and the number of tweets posted, as reflected in the drastic changes in tweet count and emotion and sentiment scores recorded on the systems in different times. The drastic changes in user network produce evolving lists of top leaders. The changes in network structure also indicate a lack of a single core set of leaders within the network, but rather an evolving group of members who took turn to lead the discussion at different times. New public policies could be developed to monitor social media posts, in addition to other information sources. These policies could support cyber-surveillance in the domains of national security and public health. Advanced scientific and technological developments could be undertaken by academia, industry, and governments to understand the sentiment and relationship of people for the betterment of society. New occupations could be created for using, maintaining, and developing social media analytics to monitor public sentiment and to identify influential actors and trends.

Future Work

The research can be extended in several ways. *First*, additional data with a longer time-horizon (e.g., 4–10 years) could be collected to support discovery of trends and fluctuation in public sentiment, network structure, and emotion changes. It will provide new insights on cyclical and seasonal movements of participants' behavior, which can be correlated with other social and economic factors to provide deeper explanation of social movement. *Second*, development of new influence metrics and new influence prediction methods can help enhance understanding of leaders, activists, and online social communities. The results can be generalized to any application where a group of participants engage in some movement (e.g., viral marketing, spread of rumors, online support network). *Third*, designing new measures (or indices) for gauging public sentiment can help reveal community mood more effectively. Extending beyond simple keyword identification and incorporating network characteristics of the authors of messages can provide more insight into the context and mood of the discussion. Tracing the sources of SM user sentiment can possibly provide further insights to policy makers. *Fourth*, modeling the spreading of messages in a social media community can help to simulate the evolution of community activities. The resulting models could help to explain such phenomena as virality and community change. *Fifth*, the aforementioned techniques can be applied to a broader community, including policy makers, managers, and BI analysts, to examine new research questions and trends.

References

BBC (2015) Ebola: mapping the outbreak, British Broadcasting Company. http://www.bbc.com/news/world-africa-28755033

Bennett B (2015) Immigration officials plan stepped-up deportations for new year. LA Times. http://www.latimes.com/nation/la-na-border-deportations-20151224-story.html

Bond RM, Fariss CJ, Jones JJ, Kramer ADI, Marlow C, Settle JE (2012) A 61-million-person experiment in social influence and political mobilization. Nature 489(7415):295–298

Bush J, Mclarty TF III, Alden EH (2009) U.S. immigration policy. Council on Foreign Relations, New York

CDC (2015) Ebola: current outbreaks, centers for disease control and prevention. http://www.cdc.gov/vhf/ebola/outbreaks/index.html

Chen H, Chiang R, Storey V (2012) Business intelligence and analytics. MIS Quart 36(4):1165–1188

Chung W (2014) BizPro: extracting and categorizing business intelligence factors from textual news articles. Int J Inf Manag 34(2):272–284

Chung W, Chen H, Reid E (2009) Business stakeholder analyzer: an experiment of classifying stakeholders on the web. J Am Soc Inf Sci Technol 60(1):59–74

Chung W, Hershey L (2012) Predicting image perception using business intelligence tools: a case study of a county government. In: BI congress 3: driving innovation through big data analytics. Association for Information Systems, Orlando, FL

Chung W, Tseng T-L (2012) Discovering business intelligence from online product reviews: a rule-induction framework. Exp Syst Appl 39(15):11870–11879

Constable P, Morello C (2013) Marchers urge Congress to pass immigration reform; several congressmen arrested. The Washington Post. http://www.washingtonpost.com/local/marchers-urge-congress-to-pass-immigration-reform/2013/10/08/9714e5c0-2f8f-11e3-9ccc-2252bdb14df5_story.html

Dawes SS, Janssen M (2013) Policy informatics: addressing complex problems with rich data, computational tools, and stakeholder engagement. In: Proceedings of the 14th annual international conference on digital government research. ACM, Quebec, 2013, pp 251–253

Dwyer D (2014) President Obama offers legal status to millions of undocumented immigrants. ABC News. http://abcnews.go.com/Politics/president-obama-offer-legal-status-millions-undocumented-immigrants/story?id=27063573

Fan W, Gordan MD (2014) The power of social media analytics. Commun ACM 57(6):74–81

Feldman R (2013) Techniques and applications for sentiment analysis. Commun ACM 56(4):82–89

Freeman LC (1977) A set of measures of centrality based on betweenness. Sociometry 40(1):35–41

Gans J, Replogle EM, Tichenor DJ (eds) (2012) Debates on U.S. Immigration. Sage Publications

Hagen L, Harrison TM, Uzumer O, Fake T, LaManna D, Kotfila C (2015) Introducing textual analysis tools for policy informatics: a case study of E-petitions. In: 16th annual international conference on digital government research, dg.o 2015, May 27, 2015-May 30, 2015, Association for Computing Machinery, pp 10–19

Hevner AR, March ST, Park J, Ram S (2004) Design science in information systems research. Manag Inf Syst Q 28(1):75–105

Jackson MO (2008) Social and economic networks. Princeton University Press, Princeton

Kane GC, Alavi M, Labianca GJ, Borgatti SP (2014) What's different about social media networks? A framework and research agenda. MIS Quart 38(1):275–304

Krishnamurthy R, Desouza KC, Johnston EW, Bhagwatwar AA (2013) glimpse into policy informatics: the case of participatory platforms that generate synthetic empathy. Commun Assoc Inf Syst 33(1)

Lampe C, LaRose R, Steinfield C, DeMaagd K (2011) Inherent barriers to the use of social media for public policy informatics. Innovation J Public Sect Innovation J 16(1):Article 6

Lazer D, Pentland A, Adamic L, Aral S, Barabasi A-L, Brewer D, Christakis N, Contractor N, Fowler J, Gutmann M, Jebara T, King G, Macy M, Roy D, Alstyne MV (2009) Computational social science. Science 323(5915):721–723

LeMay MC (2004) U.S. immigration: a reference handbook. ABC-CLIO, Santa Barbara

Liang Y, Zheng X, Zeng DD, Zhou X, Leischow SJ, Chung W (2015) Characterizing social interaction in tobacco-oriented social networks: an empirical analysis. Nat Sci Rep 5

Marathe MV, Barrett CL, Eubank S (2009) Policy informatics for co-evolving socio-technical networks. In: 2009 fourth international conference on critical infrastructures (CRIS), 27–30 April 2009. IEEE, 1 pp

Miller G (2011) Social scientists wade into the Tweet stream. Science 333:1814–1815

Mohammad SM, Turney PD (2013) Crowdsourcing a word-emotion association Lexicon. Comput Intell 29(3):436–465

Molnar P (2013) Panetta Lecture Series: Border security experts say immigration reform is vital. Monterey Herald. http://www.montereyherald.com/local/ci_22982860/panetta-lecture-series-border-security-experts-say-immigration

Nunamaker JF, Chen M, Purdin T (1991) Systems development in information systems research. J Manag Inf Syst 7(3):89–106

Plutchik R (1980) A general psychoevolutionary theory of emotion. Emotion Theory Res Exp 1(3):3–33

Schreck T, Keim D (2013) Visual analysis of social media data. IEEE Comput 46(5):68–75

Silverleib A (2013) Senate passes sweeping immigration bill. Cable News Networks. http://www.cnn.com/2013/06/27/politics/immigration/

Song X, Chi Y, Hino K, Tseng B (2007) Identifying opinion leaders in the blogosphere. In: Proceedings of the sixteenth ACM conference on conference on information and knowledge management. ACM 971–974

SUNY Albany (2013) Building a Policy Informatics Network, Center for Technology in Government, Albany. http://www.ctg.albany.edu/projects/egovpolinet

U.S. Commission on Immigration Reform (1994) U.S. immigration policy - restoring credibility: a report to congress. U.S. Commission on Immigration Reform, Washington (1825 Connecticut Ave., NW, Suite 511, Washington 2009)

Wasserman S, Faust K (1994) Social network analysis: methods and applications. Cambridge University Press, Cambridge

WHO (2015) Ebola virus disease. World Health Organization. http://www.who.int/mediacentre/factsheets/fs103/en/

Zeng D, Chen H, Castillo-Chavez C, Lober WB, Thurmond M (2011) Infectious disease informatics and biosurveillance. Springer, Berlin

Zeng D, Chen H, Lusch R, Li S-H (2010) Social media analytics and intelligence. IEEE Intell Syst 25(6):13–16

Zhang J, Ackerman M.S, Adamic L (2007) Expertise networks in online communities: structure and algorithms. In: Proceedings of the 16th international conference on World Wide Web, (Banff, Alberta, Canada, 2007). ACM, New York, pp 221–230

Wingyan Chung, Ph.D. is an associate professor in the Institute for Simulation and Training (IST) at University of Central Florida (UCF). Dr. Chung's scholarly interests and expertise include social media analytics, business intelligence, cybersecurity, data and text mining, Web analytics, knowledge management, information visualization, and human-computer interaction. He has published over 90 refereed articles in high-impact academic journals and is ranked among the Top 20 academic authors globally in Business Intelligence and Analytics.

Daniel Zeng, Ph.D. is Gentile Family Professor in the Department of Management Information Systems at the University of Arizona. Dr. Zeng's research interests include intelligence and security informatics, spatial-temporal data analysis, infectious disease informatics, social computing, recommender systems, software agents, and applied operations research and game theory. He has published more than 300 peer-reviewed articles and received more than $28 million research support. An *IEEE* fellow, he currently serves as the Editor in Chief of *IEEE Intelligent Systems* and is President of *IEEE ITS Society* (2016–2017).

Model-Based Policy Analysis to Mitigate Post-Traumatic Stress Disorder

Navid Ghaffarzadegan, Richard C. Larson, Henry Fingerhut, Mohammad S. Jalali, Alireza Ebrahimvandi, Anne Quaadgras, and Thomas Kochan

Abstract A wide range of modeling methods have been used to inform health policies. In this chapter, we describe three models for understanding the complexities of post-traumatic stress disorder (PTSD), a major mental disorder. The models are: (1) a qualitative model describing the social and psychological complexities of PTSD treatment; (2) a system dynamics model of a population of PTSD patients in the military and the Department of Veterans Affairs (VA); and (3) a Monte Carlo simulation model of PTSD prevalence and clinical demand over time among the OEF/OIF population. These models have two characteristics in common. First, they take systems approaches. In all models, we set a large boundary and look at the whole

N. Ghaffarzadegan (✉) • A. Ebrahimvandi
Department of Industrial and Systems Engineering, Virginia Tech,
250 Durham Hall (MC 0118), 1145 Perry Street, Blacksburg, VA 24061, USA
e-mail: navidg@vt.edu; alvandi@vt.edu

R.C. Larson
Institute for Data, Systems, and Society, Massachusetts Institute of Technology,
77 Massachusetts Ave., E17-365, Cambridge, MA 02139, USA
e-mail: rclarson@mit.edu

H. Fingerhut
Institute for Data, Systems, and Society, Massachusetts Institute of Technology,
50 Ames Street, E18-407A, Cambridge, MA 02142, USA
e-mail: henryaf@mit.edu

M.S. Jalali
Sloan School of Management, Massachusetts Institute of Technology,
245 First Street, E94-1568D, Cambridge, MA 02142, USA
e-mail: jalali@mit.edu

A. Quaadgras
Sloan School of Management, Massachusetts Institute of Technology,
100 Main Street, E62-490, Cambridge, MA 02142, USA
e-mail: aquaad@mit.edu

T. Kochan
Sloan School of Management, Massachusetts Institute of Technology,
100 Main Street, E62-334, Cambridge, MA 02142, USA
e-mail: tkochan@mit.edu

J.R. Gil-Garcia et al. (eds.), *Policy Analytics, Modelling, and Informatics*, Public Administration and Information Technology 24,
DOI 10.1007/978-3-319-61762-6_17

system, incorporating both military personnel and veterans. Second, the models are informed by a wide range of qualitative and quantitative data. Model I is rooted in qualitative data, and models II and III are calibrated to several data sources. These models are used to analyze the effects of different policy alternatives, such as more screening, more resiliency, and better recruitment procedures, on PTSD prevalence. They also provide analysis of healthcare costs in the military and the VA for each policy. Overall, the developed models offer examples of modeling techniques that incorporate a wide range of data sources and inform policy makers in developing programs for mitigating PTSD, a major premise of policy informatics.

List of Abbreviations

DOD	Department of defense
OEF	Operation enduring freedom
OIF	Operation Iraqi freedom
PTSD	Post-traumatic stress disorder
VA	Veterans affairs

Introduction

Policy informatics refers to the use of information technologies, data analysis, computational modeling, and simulation techniques to address complex public policy problems (Johnston 2015). The goal is to utilize multiple and extensive data sources and computational methods to inform policy makers and to help improve the design and implementation of policies (Ghaffarzadegan et al. 2015; Kim et al. 2013). A broad definition of policy informatics includes data-driven system dynamics models and other mathematical models developed to study complex systems (Ghaffarzadegan et al. 2015). Example applications include studies of social welfare (Zagonel et al. 2004), environment and energy (Sterman 2014), education systems (Ghaffarzadegan et al. 2014), and healthcare systems (Fallah-Fini et al. 2014; Sabounchi et al. 2014; Teytelman and Larson 2013; Wittenborn et al. 2016). Specifically, in the domain of health care, growing attention has been paid to developing models that can capture population health and prevalence of chronic health conditions or infectious diseases and be used as platforms to test policy alternatives (Finkelstein et al. 2015; Homer and Hirsch 2006; Milstein et al. 2011).

In this chapter, we focus on the application of simulation-based models in a specific mental health policy context, the problem of post-traumatic stress disorder among military personnel and veterans. The materials in this chapter are based on the outputs of modeling different aspects of PTSD in the Post-Traumatic Stress Innovation project that was conducted at the Massachusetts Institute of Technology from 2012 to 2015. Here, we provide brief information about PTSD and the results of the three models. The main contribution of this chapter is to provide three diverse examples of applications of policy informatics tools to study one major societal problem. The models provide examples of how modeling techniques that utilize a wide range of qualitative and quantitative data sources can help analyze a major health issue and inform policy makers. Furthermore, the chapter aims at depicting how, under the umbrella of policy informatics, different modeling techniques can provide different types of insights about a major societal problem with policy implications.

What Is PTSD?

Post-traumatic stress disorder is a mental illness that can occur after a person experiences a traumatic event, such as combat, family violence, sexual assault, a terrorist attack, or serious injury (US Department Of Veterans Affairs 2016a). Individuals suffering from PTSD continue to experience the psychological effects of trauma long after being moved from the original stressor. Symptoms include re-experiencing events, urges to avoid similar stimuli, negative cognition and mood, and increased physical arousal (Solomon et al. 2015). PTSD comorbidities include other psychological effects or mental illnesses that can occur as an effect of trauma, such as depression (Campbell et al. 2007; Ginzburg et al. 2010), substance abuse (Breslau et al. 2003; McFall et al. 1992), guilt and shame (Hendin and Haas 1991; Henning and Frueh 1997), and suicidality (Jakupcak et al. 2009).

PTSD has become a serious public health challenge. About eight million people in the United States suffer from PTSD (US Department Of Veterans Affairs 2016b). The illness is more common among military personnel and Veterans, especially those deployed to combat zones. It is estimated that 11–20% of US military personnel who served in Iraq or Afghanistan have diagnosed or undiagnosed PTSD (US Department Of Veterans Affairs 2016b).

The military PTSD burden incorporates not only the medical challenge of treatment, but also social dimensions that interact with cultural and logistical aspects of the military system. In 2007, the Department of Defense Task Force on Mental Health outlined four goals for improving mental health care, considering both medical and non-medical factors. These goals were to foster a culture of support for psychological health, a full continuum of excellent care in both peacetime and wartime, sufficient and appropriate resources allocated to prevention, early intervention, and treatment, and visible and empowered leaders at all levels to advocate, monitor, plan, coordinate and integrate prevention, early intervention, and treatment (Department of Defense Task Force on Mental Health 2007). Though this effort is an important step in normalizing care-seeking and improving access to mental health services, cultural factors such as stigma and fear of professional repercussions still prevent many individuals from seeking the care they need (Ghaffarzadegan and Larson 2015; Phelan 2005; Vogt 2011).

Moreover, the tools and techniques used for screening and diagnosis are mainly based on self-reported surveys. In order to diagnose PTSD, individuals usually answer a survey consisting of 17 questions, referred to as a post-traumatic stress checklist (Hoge et al. 2014). Each question receives a score of 1 to 5, with a possible total of 17 to 85. While there is not a precise cutoff value, generally people with numbers above 40 or 50 are considered to have PTSD symptoms and are sent for interviews and diagnosis with mental health professionals. Since this is a self-reported survey, it is not difficult to hide symptoms and manipulate responses. This adds to the uncertainties and difficulties of PTSD diagnosis.

Our Project

Mathematical models are well-known to the Department of Defense (DoD). The very birth of operations research, which relies heavily on mathematical models, was DoD-driven during World War II. Among the products of this work were efficient linear programming algorithms to improve wartime logistics; the theory of optimal search, which proved invaluable to US efforts in the North Atlantic to find and destroy enemy submarines; and optimal location theory, which proved most useful for placement of radar stations in Britain to detect incoming enemy aircraft.

However, models of human health and behavior and interventions to improve them were less well-developed at that time. The subsequent seven decades have seen a lot of good work in this area, focused less on hardware-dominated tactical operations and more on human systems. Epidemiology is now a mature field involving many different types of mathematical models of behavior-influenced disease progression and control. In the case of PTSD, mathematical modeling is a relatively new field, and there are just a handful of papers addressing the topic.

The goal of the innovation project has been to incorporate work based on both qualitative and quantitative modeling as a way to capture the potential benefits of a multidisciplinary examination of the burden of PTSD and how it might be addressed in the military health system going forward. Our mathematical modeling work takes a "system" perspective, embedding service members in the military system and then structuring various PTSD-focused models around that, with the type of structure depending on the decisions and policies to be guided and influenced by the model. From the systems point of view, we seek first to frame the problem to understand the overlapping and intertwined subsystems—formal and informal, positive and negative—that influence the treatment of PTSD. Then, from an aggregate level, we seek to project PTSD treatment workloads of the DoD and the VA over the coming years and even decades.

Mathematical models come in many varieties: deterministic or probabilistic; equation-based, or algorithm-based; simulation-based or solution-based; optimizing or descriptive; and so on. Our approach is simulation-based and descriptive in response to the complexity of PTSD. Simulations also come in a number of varieties: Monte Carlo (probabilistic) simulation, system dynamics, and even micro rule-based such as agent-based simulations. Our work utilizes system dynamics and Monte Carlo simulations.

Simulation models have a wide variety of uses. Among others, they support *"what-if" analysis*. A simulation model of a PTSD treatment system can project the multi-year consequences of PTSD workloads and costs under a wide variety of "what-if" scenarios, ranging from those largely outside the control of the PTSD system, such as the intensity of engagements in future wars, to those under the control of the PTSD system, such as the deployment of additional resources and the use of new evidence-based treatments. This can inform projections of budgets and needs for professional manpower and facilities.

Another use is to *improve the understanding of a system*. Sometimes, a model's primary use is problem framing, through which decision makers and other stakeholders—such as PTSD-afflicted service members and veterans, their families, friends, and personal support organizations—can learn from model development and structure about the shared importance of the many intertwined stakeholders in helping to ameliorate the symptoms of PTSD. From a DoD perspective, such a framing model can justify the allocation of government resources to family, friends, and supporting organizations outside the DoD, as these are also seen as critical in a comprehensive treatment program.

We should clarify that no models perfectly depict reality, and they are used to describe reality as best as they can. As George Box and Norman Draper famously wrote, "All models are wrong, but some are useful" (Box and Draper 1987). Also, detailed models do not necessarily imply greater or even equal usefulness as compared to small or simple models. Here a quote often attributed to Albert Einstein is appropriate: "Everything should be made as simple as possible, but not simpler." We have tried to follow these two propositions in our three developed models, which we describe next.

Three PTSD Models

Model I: A Conceptual Systems Model of PTSD

With what we refer to as Model I (developed as part of the innovation project), Ghaffarzadegan and Larson (2015) developed a qualitative representation of the system, seeking to answer a basic question: What are the interrelations among psychological, sociological, and medical factors in PTSD treatment? We think that a qualitative approach to the problem can be the first step in major modeling projects which will lead to better understanding of interconnections within a system and an overall map of the current state of the literature, and can lead to more accurate quantitative modeling efforts in the next stages.

Method The modeling effort at this stage was consistent with the procedures described in qualitative approaches to system dynamics modeling such as in

Luna-Reyes and Andersen (2003) and later in Kim and Andersen (2012). We used textual data: published articles and reports about PTSD in the military and the Department of Veterans Affairs. We carefully coded the discussion and conclusion sections of the papers on factors that influence PTSD treatment, trying to elicit causes and effects. Early in the coding effort, we recognized that causal factors affect PTSD treatment at three different levels (themes): Individual barriers, family and close friends' supports/barriers, and societal forces. This led to categorizing the effects under these three groups. Additional causal links were proposed to clarify causalities. Then the model was presented at different events with experts in both systems modeling and PTSD which further helped us clarify the model and make it more consistent with the domain experts' language. With this model, we uncovered several root causes that contribute to the complexity of treating PTSD.

Outcomes The main outcome of the study was a model to present an overall picture of the system. The model demonstrates how military personnel with PTSD are situated in a complex web of partially overlapping structures, some formal, such as those operated by the Department of Defense (and the VA), and some informal, such as those provided by family and friends. The model represents PTSD treatment as influenced by medical, personal, and social factors. This creates multi-layer dynamics. The individual layer concerns how personal decisions and willingness to seek care affect treatment.

Let us first start with corrective mechanisms. In an ideal situation, we would expect patients to seek care as soon as an individual becomes aware of PTSD symptoms, which leads to PTSD treatment and mitigation of symptoms (loop B1). In the family and friends layer, others may provide social support and help in treatment. With more support from family and friends, a patient can heal in a faster pace (loop B2). In addition, family and friends may encourage patients to seek care, e.g., spouses who not only support their ill spouse but also encourage them to see a doctor (loop B3). These are all idealistic corrective mechanisms which one hopes help cure the illness. Figure 1 shows these mechanisms.

However, the real world is much more complex. There is no guarantee that everyone will receive family and friend support. Furthermore, delaying treatment may lead to increase in illness and symptoms. Let us describe two major vicious cycles which inhibit treatment.

First, the illness can cascade and patients' mental and physical health may get worse if they do not receive timely treatment. Studies indicate that some PTSD patients also develop other psychiatric disorders (Najavits et al. 2009). Such increasing complications render medical interventions progressively less effective. Patients' responses in the form of drug abuse can further complicate health conditions. The entire process ends up in a cascading pattern that eventually converts mild medical illnesses into chronic and life-threatening conditions. This is shown with loop R1 in Fig. 2. Second, cascading illness happens due to isolation from family and friends over time. Figure 2 demonstrates two reinforcing loops of R2a and R2b to depict the tolerance limitation of close friends and family members. At higher levels of PTSD extent and symptoms, when patients significantly affect their family members, they

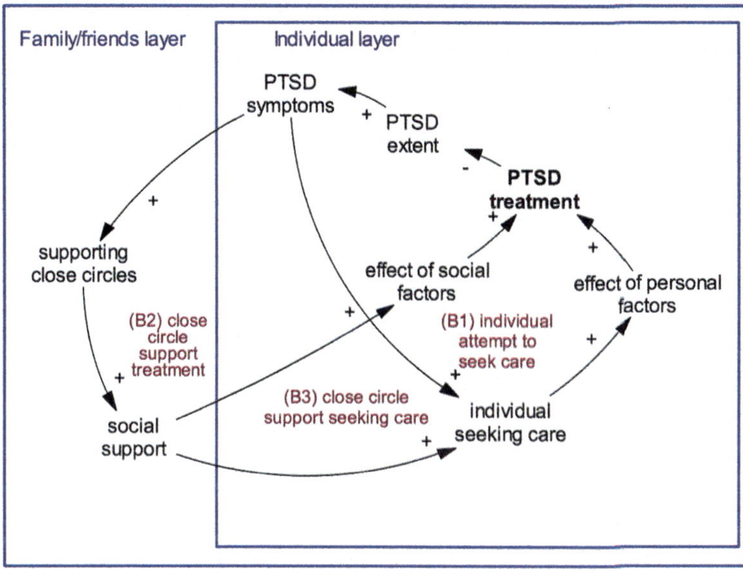

Fig. 1 Three corrective mechanisms for PTSD treatment

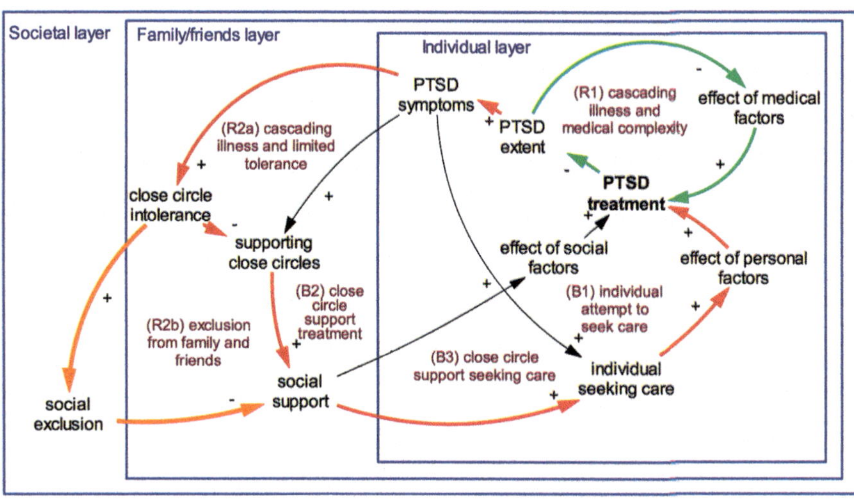

Fig. 2 Cascading medical complexities (loop R1) and isolation (loops R2a and 2b)

lose their social support and therefore get further excluded from the society. A simple example is not being invited to social gatherings, and more serious examples are workplace discrimination, losing a job, or having difficulties renting an apartment.

All of these effects ultimately feed back to the individual layer and adversely affect treatment and willingness to seek care.

These mechanisms are examples of vicious cycles. In this context, a vicious cycle is a feedback loop that, over time, creates cascading negative influences on PTSD sufferers and exacerbates their mental health situation.

We continued the development of the model and uncovered three more vicious cycles for PTSD treatment (see R3–R5 in Fig. 3). In summary, the five cycles are titled:

R1—cascading illness and medical complexity
R2—cascading illness and isolation from family and friends
R3—stigma and social exclusion
R4—self-fulfilling prophecy
R5—the malingerer stigma

Additional information about these cycles can be found in Ghaffarzadegan and Larson (2015).

Overcoming vicious cycles is very difficult, requiring policies and patience over the long term. Without early interventions, these cycles result in an individual's downward spiral into depression, family discord and possible divorce, substance abuse, joblessness, homelessness, and even suicidal ideation or action.

Like a snowball that gets bigger and bigger as it rolls downhill, vicious cycles are difficult to stop as they gain momentum. This analysis points to the need to prevent these situations from developing. Two conceivable policy steps are early effective screening and resiliency-related interventions (e.g., better recruitment procedures and resiliency-related training), where attention should be paid to military personnel and their families. Generally, resilience reflects the ability to maintain a stable equilibrium in highly disruptive events, such as loss of a close relation or life-

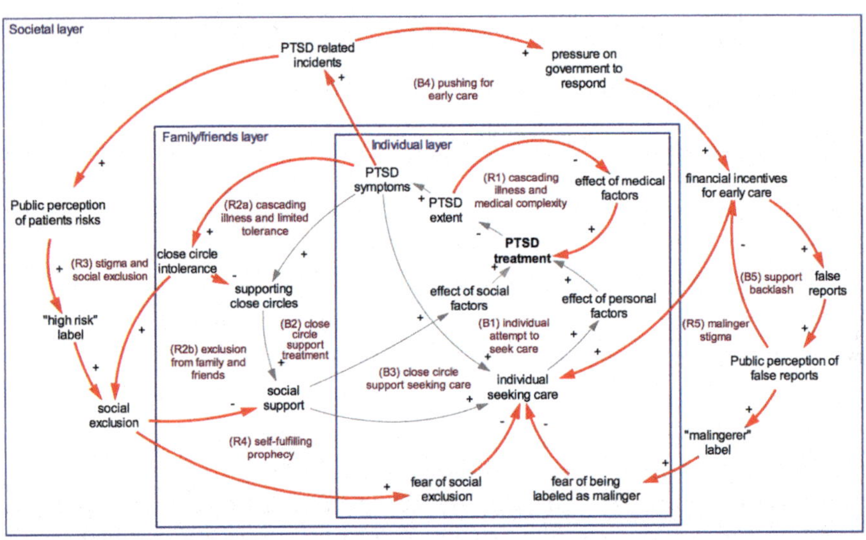

Fig. 3 A conceptual model of PTSD treatment. Adapted from Ghaffarzadegan and Larson (2015)

threatening situation (Bonanno 2004). Analyses on the veterans of Operation Enduring Freedom (OEF) and Operation Iraqi Freedom (OIF) show that unit support and post-deployment social support are some of the effective interventions that may increase resiliency (Pietrzak et al. 2010). Another approach is to help increase stress tolerance capacity through specialized training sessions for military personnel before deployment or immediately after experiencing a traumatic event (Department of the Army 2009).

Model I was a first step in framing the problem and understanding interconnections and complexities surrounding military personnel with PTSD. Next, we needed models to help quantify these effects and compare and contrast the effects of improving resiliency and early treatment. Such quantitative models should include uncertainties in diagnosis, individual health, access to care, and military personnel readiness. The models should also help compare PTSD prevalence and healthcare system costs under different policies and scenarios.

In response to these requirements, two additional models were developed for the project: a system dynamics model of PTSD prevalence and a Monte Carlo simulation model of an individual serviceperson.

Model II: A Population-Level System Dynamics Model of PTSD

With Model II, we moved toward quantifying the effects of different interventions on PTSD prevalence, asking these basic questions: What are the trends in the population of PTSD patients among military personnel and veterans? What policies can help mitigate the effects of PTSD? What are the healthcare cost implications of these policies?

Method To answer these questions, Ghaffarzadegan et al. (2016) developed a system dynamics simulation model of the PTSD population with a broad boundary, where the model incorporates both military personnel and veterans. It encompasses veterans of pre-2000 wars and more recent wars in Iraq and Afghanistan, and can track cases over the entire lives of patients. The overall structure of the model is developed in a way that simply—while as precisely as possible—presents the flow of individuals in the military and post-military and across different stages of developing the illness, diagnosis and treatment. In other words, our simple model captures the core mechanisms of a complex system. The model's equations are fully documented—along with the coding and time series—for further development and assessment (see the supplementary materials in Ghaffarzadegan et al. (2016) for more detail).

The model uses a variety of data sources. The structure is informed by the researchers' prior work, as well as other published articles and reports. The model parameters and time series (2000–2014) come from the DoD, the Institute of Medicine, and the VA. We ran the model for the period 2000–2025, where 2000–2014 was used for model validation and examination of the model's fidelity in replicating the historical data. Then, the model forecasts the period 2015–2025. To create scenarios for forecasts, US involvement in wars and the intensity of future wars (in comparison to OIF) were used as inputs. The outputs are PTSD prevalence, number of PTSD cases diagnosed and undiagnosed in both the military and the VA, and PTSD-related healthcare costs.

Structurally, Model II depicts the flow of people from recruitment into the military, from the military to the post-military stage, and from the post-military stage to death. Figure 4 presents the stock and flow structure of the model, where a stock represents accumulation in the system and a flow is the rate at which the stock is changing. Let us briefly discuss these stocks and flows.

Flow (1) presents the recruitment of individuals, the majority healthy, who enter the 'Healthy Military members' stock; however, a small percentage might already have a history of PTSD and enter the 'Ill-Undiagnosed Military members' stock. As a result of traumatic events, healthy people in the service may develop the illness (Flow (2)) and move to the 'Ill-Undiagnosed Military members' stock. These undiagnosed individuals are either diagnosed with PTSD during their service (Flow (3)), or separated from the military with unknown illness (Flow (6)). Moreover, if 'Ill-Diagnosed Military members' are successfully treated (Flow (4)), they move back to the 'Healthy Military members' stock.

A similar diagnostic process also exists for veterans, where individuals in the 'Ill-Undiagnosed Veterans' stock are diagnosed with PTSD (Flow (8)) and moved to the 'Ill-Diagnosed Veterans' stock—these two stocks are fed by Flows (6) and (7), respectively. The last stock is 'Healthy Veterans,' which includes healthy individuals separated from the military (Flow (5)) and the successfully treated veterans (Flow (9)). It should be noted that all stocks in Fig. 4 have an outflow of death—the death rates are different for each stock, e.g., the death rate of healthy military members and that of healthy veterans are not the same. The death outflows are shown in grey.

For the sake of presentation, Fig. 4 contains a simplified version of the model and only illustrates the stock-and-flow structure. See Ghaffarzadegan et al. (2016) for more detail on the causal relationships among the variables that are key components of any system dynamics model. The model also incorporates the chances of deployment, experiencing trauma, and developing PTSD given that trauma. In the model, individuals do not necessarily reveal PTSD symptoms immediately; the diagnosis may be delayed, in some cases occurring after separation from the military. It should be also noted that since the model includes two subsystems, military and post-military, it helps estimate PTSD-related healthcare costs for both the DoD and the VA.

Outcomes Figure 5 is an example of one of the outputs of Model II. It depicts simulation outcomes of the model for the time period of 2000–2025. The time period can be divided into time frames of the past (2000–2014) and the future (after 2014). For the first time period, the goal was to replicate the data as a validation process, and the model fairly replicates the data.

For the post 2014 period, we assume three different scenarios about US involvement in future wars and estimate PTSD prevalence under these scenarios. In the first and second scenarios which are optimistic, the US deploys 1 and 2% of its troops to

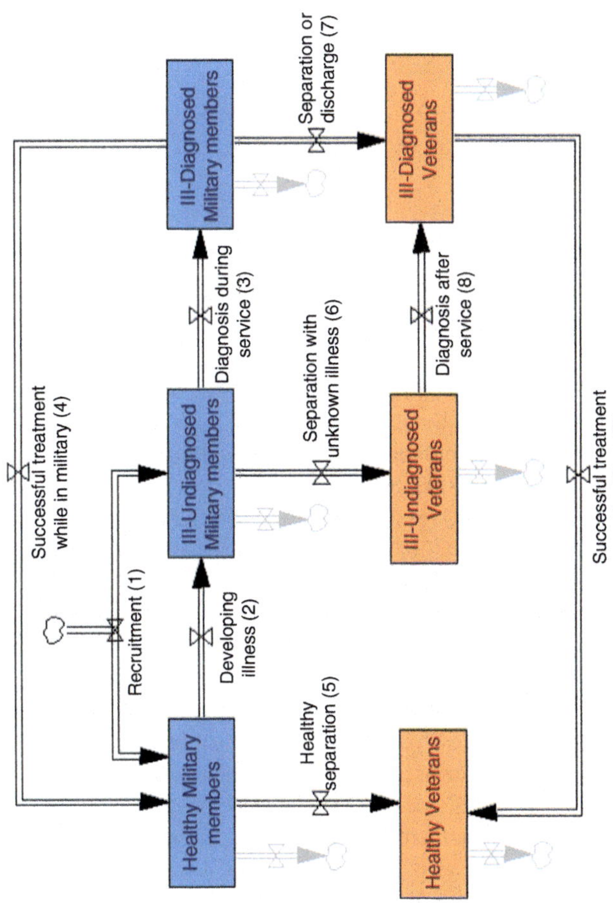

Fig. 4 A simplified version of Model II representing stocks and flows in the system. The *blue boxes* ('stocks') represent the military subsystem, and the *orange* stocks represent the post-military subsystem. Variables (1) through (9) are the main 'flows' in the system. All stocks in the model have an outflow (shown in *grey*) which represents death. For the sake of simplicity in this presentation, causal relationships among the model variables are not illustrated (see Ghaffarzadegan et al. (2016) for more detail)

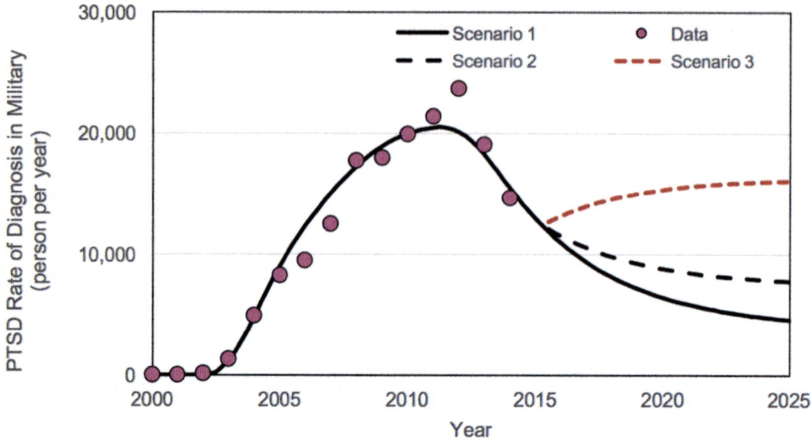

Fig. 5 PTSD diagnosis rate in the military [new cases per year]. Note: The model fits the available historical data for 2000 to 2014 and predicts the trends for 2015 to 2025 for three scenarios: Scenario 1: Minimum deployment to intense/combat zones (1% of military personnel); Scenario 2: 2% deployment to intense/combat zones; and Scenario 3: 5% deployment to intense/combat zones. Adopted from Ghaffarzadegan and Larson (2015)

combat zones. In the third scenario which is less optimistic, the US deploys 5% of the troops to combat zones. As a reference, in the Iraq and Afghanistan wars, deployment was around 8–10%. As the model shows in the optimistic scenarios, PTSD diagnosis per year will be around 5000–8000 new patients per year only in the military. This of course has implications for the population of veterans as well. Based on these estimations the model predicts that (not shown in the Figure), in scenarios 1 and 2, PTSD prevalence among veterans in 2025 will be about 10%.

The model yielded several major results, and we invite interested readers to refer to Ghaffarzadegan et al. (2016) paper. In sum, the model predicted that the population of patients and system costs will be very sensitive to US involvement in future wars, and that screening and treatment policy interventions will have marginal effects in comparison. In fact, more screening increases healthcare costs by increasing demand for health services. Furthermore, they find that it takes a long time, on the order of 40 years, to overcome the psychiatric consequences of a war. This is also consistent with the data on Vietnam-War-era PTSD patients.

Ghaffarzadegan et al. (2016) also provide detailed discussions about direct healthcare costs for the DoD and the VA related to PTSD. In reality, there are also social costs associated with PTSD, but these were not considered in the analysis. In an optimistic scenario (about 1 to 2% deployment to intense/combat zones in the next 10 years), the model's predictions of PTSD healthcare costs for the military in 2025 range from \$130 to \$160 million per year (in 2012 dollars). With greater involvement in future wars (about 5% deployment to intense/combat zones), the costs potentially increase to \$260 million per year. For the VA, the cost estimates are one order higher, with average estimates of \$2.9–\$3.2 billion per year (in 2012 dollars). With greater involvement in future wars (5% deployment to intense/combat zones), this cost can also increase to \$3.6 billion per year.

One major outcome of this modeling effort which has policy informatics implications is its "management flight simulator."

Similar to the airline industry, in that no pilot is sent up in a jumbo jet without enough training in a flight simulator, policy makers should also not make a major decision based only on their knowledge and experience. Management flight simulators provide a virtual "gaming" environment for policy experiments, where policy and decision makers can explore the effects and consequences of their decisions over the long haul.

Our flight simulator helps represent the model outputs in an interactive interface and easy-to-interpret fashion, such that users are not required to use any specific software or be familiar with any modeling language. More importantly, it brings an experiential aspect to learning about the complexity of the system. Similar applications have been developed in other fields, such as People Express in strategic management (Sterman 1992), Climate Interactive (C-ROADS) in climate change (Sterman et al. 2012), and ReThink Health in health policy (McFarland et al. 2016). Our simulator can be tried online at http://jalali.mit.edu/ptsd-simulation. Figure 6 provides a snapshot of the online presentation of the simulator.

Beyond prevalence and cost estimates, as well as providing a flight simulator, one major insight from developing Model II is the importance of considering PTSD as a multi- organizational problem. A systematic approach to PTSD needs to consider the military and post-military stages together, since an effective policy in one stage may create problems for the other. The models should also look at long-term dynamics, considering delays between developing PTSD and showing symptoms. The analysis also shows that a focus on resiliency and decreasing the chances of developing PTSD is potentially one of the most effective policies, which is consis-

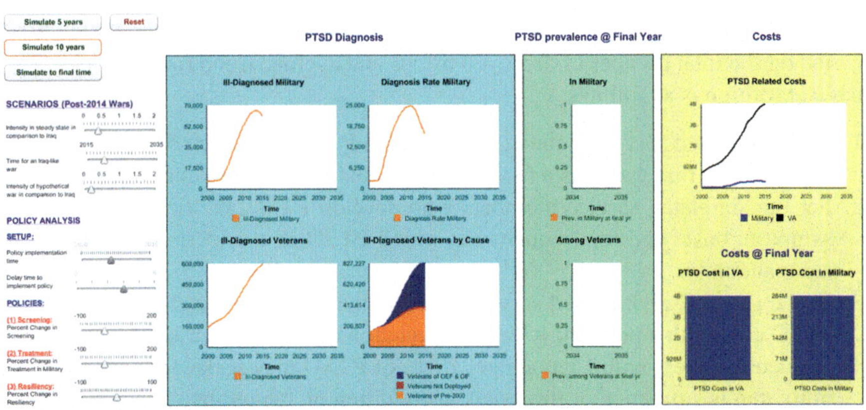

Fig. 6 Snapshot of the online presentation of Model II, available at: http://jalali.mit.edu/ptsd-simulation

tent with Model I's suggestions. As stated before, resiliency is about increasing the ability to maintain stable emotions in disruptive events (Bonanno 2004). In our case, increasing resiliency can be achieved by, e.g., immunizing individuals against PTSD with more support and psychological training sessions (Department of the Army 2009; Pietrzak et al. 2010).

Model III: A Monte Carlo Simulation Model of PTSD

Model III is a Monte Carlo simulation model developed by Fingerhut (2015). It predicts PTSD and associated clinical demand over five decades following OEF and OIF.

This approach creates representative service members who replicate the deployment schedule, PTSD risk, care-seeking behavior, and treatment of actual service members from the two conflicts. After randomly assigning each virtual service member's deployment and trauma exposure, as well as possible PTSD onset, recognition, and treatment events over the period of study, the study aggregates each individual's simulated history to determine population level statistics and trends. This study also provides a series of sample policies designed to replicate possible decisions that policy makers could implement to affect burden posed by PTSD.

The model uses empirically observed distributions of parameters from across the mental health care system (traumatic exposure, onset, recognition, care-seeking, and treatment) to estimate the dynamic trends of a series of individual and population level variables.

Method The model simulates a population of service members deployed to Iraq and Afghanistan. In this model, an individual's use of clinical resources depends on the individual's PTSD symptoms, PTSD recognition status, and treatment characteristics. The model is stochastic and each simulated service member is followed over the length of the study. The model also uses quarterly data for the 2003–2014 period on the total number of deployed troops and average troop rotation rates. The basic algorithm is as follows:

1. Each simulated service member may develop PTSD after each deployment with a probability that varies as a function of combat severity and duration.
2. For the simulated service members who develop PTSD, the timing of the first symptom onset and the frequency at which PTSD is re-experienced are generated randomly.
3. In each period, a simulated service member may be recognized with a possible PTSD case with a probability that is a function of time since trauma.
4. Service members with a recognized cases of PTSD may seek care with a probability that is a function of time since recognition of symptoms.
5. A treatment is successful at remitting the service member's PTSD symptoms with a probability that depends on treatment efficacy and the probability of treatment drop-out.

All the values mentioned here are extracted from empirical studies in the psychology literature. Further details regarding the model structure, parameter derivation, and results are available in Fingerhut (2015).

The model manipulates a time-series form of input and is thus able to provide time-series output. That output takes the form of prevalence estimates from the population perspective, that is, each point in calendar time provides a snapshot of what a real-world population prevalence estimate would look like, given changes in deployment, combat, and other factors over time.

Outcomes The model projects a peak rate of active-case PTSD of about 200,000 by 2016 (17% of the deployed population) which later declines to 150,000 by 2025 (15% of the population). These predictions reflect best-case assumptions about PTSD recognition, care-seeking, and treatment efficacy observed in recent empirical studies. This model further predicts that 29% of OEF/OIF combat veterans will experience PTSD at some point in their lives (Fig. 7).

In terms of care-seeking and treatment, under best-case care-seeking assumptions, Model III predicts that 80% of the ever-PTSD population (the population that at somepoint in their life suffer from PTSD) will seek treatment. The model estimates that 48–63% of the ever-PTSD population (14–18% of the OEF/OIF population) are expected to seek treatment. Under best-case model assumptions, clinical demand peaks at 3% of the OEF/OIF population per year in 2010, decreasing to 0.5% of the OEF/OIF population in 2025.

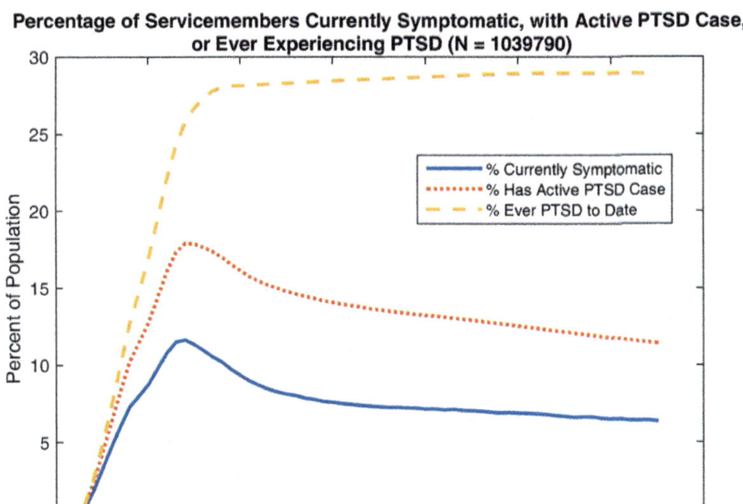

Fig. 7 Baseline (best-case) model predicted percent of service members with currently symptomatic or active case PTSD in each year and service members who have ever experienced PTSD through the indicated year end. Percentage denominator is the number of deployed service members in the indicated year. Figure is adopted from Fingerhut (2015)

Discussion and Conclusions

The three models specific to the innovation project are initial efforts to depict (in a systems context) our knowledge about PTSD treatment system structures. The models embed the all-important psychological and social processes, both formal and informal, underlying the PTSD burden in the populations studied. They provide a good first look at the implications of various policies and managerial actions on future prevalence of PTSD, clinical demand, and costs. For those interested in additional details, each model is fully developed in separate published papers and/or technical reports, as cited in this chapter and in the references.

In this chapter, we aimed to synthesize the models, and provide a systems level perspective on these modeling efforts. The models provide examples of how different modeling techniques that utilize a wide range of qualitative and quantitative data sources can help analyze a major health issue and inform policy makers. We would also like to stress the interconnections between all these modeling efforts. For example, we see how a qualitative model (Model I) leads to a quantitative model (Model II) which raises further questions at the individual level that can be investigated in an agent-level model (Model III). There is no single model that can answer all questions, but we gain insights in the process of modeling, and resolving research questions may lead to further questions that will need more modeling efforts. In a sense, model-based policy informatics as a process is not a linear set of actions but includes the cycle of problem recognition, data gathering, model building, insight generation, and more problem recognition. During the policy informatics process, we are "inform"ed by different data sources and models, and we address additional "policy" challenges.

We also want to stress the importance of models when dealing with huge datasets. Extracting insights from big data requires good models. While larger datasets sound helpful, it is important to not fall into the trap of losing the big picture. And as was discussed in this chapter, models come in different forms and are developed for various purposes.

In this chapter, we also offered an example of a flight simulator. As discussed, Model II was further developed to allow policymakers to easily "play" with the model and test the effects of different policy measures. Ultimately, the goal of the modeling projects was to inform policy makers, but the main challenge is about helping them learn, question, and trust the model outcomes. Given that policy makers often come from a wide range of fields and backgrounds, they might not be familiar with the technical languages of modelers; hence, the modeling projects may not be fully understood or used by the policy makers. Flight simulator platforms are critical tools in this respect which provide a simple and non-technical environment to present the model outcomes clearly and help the policy makers focus on the meaning of the outcomes rather than how to run the models.

Furthermore, the chapter aims to depict how, under the umbrella of policy informatics, different modeling techniques can provide different levels of insight about a major societal problem.

While the discussed models are different in many respects, they share two major methodological themes. First, they take systems approaches. Most previous models of mental health in the military or the VA have a very narrow perspective and focus exclusively on either the military or the VA. A narrow focus on one organization "shifts the burden" to the other organization. Furthermore, most past models take a snapshot of the problem and focus on solving "today's issues." Second, the models presented here were informed by a wide range of qualitative and quantitative data. Model I is rooted in qualitative data, and models II and III are calibrated with multiple data sources.

Going forward, two key questions remain. First, *how might these models be used*? Our suggestion is to view the models as living entities, evolving and improving over time as new knowledge becomes available. This will require professionals in the DoD and VA to take ownership of the models and incorporate model related information in a timely fashion as it becomes available.

Second, *what is the take-away*? The models show the effects of changes ranging from administrative aspects such as multi-year projected budget levels that may constrain system resources, to new scientific knowledge about the efficacy of new treatments for PTSD. Within the models, budget constraints may appear only indirectly in terms of the total numbers of professionals and facilities available for PTSD treatment. Understanding the impact of putting new scientific knowledge to work will require going into the details of the models, feedback loops, flow parameters, response delays, and updating them to be compatible with the new scientific results. New science will, in turn, affect budgets and facilities. Perhaps a new treatment protocol will prove very costly, but demonstrate a very high chance of lifetime cessation of PTSD symptoms after, say, 2 years of the treatment. Such a protocol if discovered, would likely be expensive in the short term but very cost-effective in the long term. All this shows how scientific knowledge of treatment effects could cause major changes in the DoD resource-intensive systems model.

One final thought: Our observation, not only for the DoD but also for the VA and virtually all large service systems (including research universities), is that everyone is so dedicated to their work that on a day-to-day basis they often only "see" the immediate vicinity of their own workplaces. Improvements—some may call them "optimizations"—tend to be local that is, focused only on that small part of the total system in which a group of professionals works. One major value of systems models is that they show how one set of changes affects many aspects of a complex structure. They demonstrate clearly the hazards of local optimization, showing how even attractive local changes have the potential to be detrimental to the total system. In this sense, systems models provide an integrated, unifying framework that can enable key policy makers throughout the system to discuss their problems intelligently and objectively. This attribute may be one of the major arguments in favor of systems models.

Acknowledgments This chapter is based upon MIT work supported, in part, by the Office of the Assistant Secretary of Defense for Health Affairs, under Award No. W81XWH-12-0016. The US Army Medical Research Acquisition Activity, 820 Chandler Street, Fort Detrick MD 21702-5014, is the awarding and administering acquisition office. Opinions, interpretations, conclusions, and recommendations are those of the authors and are not necessarily endorsed by the Department of Defense, MIT, or Virginia Tech.

References

Bonanno GA (2004) Loss, trauma, and human resilience: have we underestimated the human capacity to thrive after extremely aversive events? Am Psychol 59(1):20

Box GE, Draper NR (1987) Empirical model-building and response surfaces, vol 424. Wiley, New York

Breslau N, Davis GC, Schultz LR (2003) Posttraumatic stress disorder and the incidence of nicotine, alcohol, and other drug disorders in persons who have experienced trauma. Arch Gen Psychiatry 60(3):289–294

Campbell DG, Felker BL, Liu C-F, Yano EM, Kirchner JE, Chan D, Rubenstein LV, Chaney EF (2007) Prevalence of depression–PTSD comorbidity: implications for clinical practice guidelines and primary care-based interventions. J Gen Intern Med 22(6):711–718

Department of Defense Task Force on Mental Health (2007) An achievable vision: Report of the Department of Defense Task Force on Mental Health.ES 1–2

Department of the Army (2009) Manual 6-22.5: combat and operational stress control manual for leaders and soldiers. US Department of the Army, Washington, p 111

Fallah-Fini S, Rahmandad H, Huang TT-K, Bures RM, Glass TA (2014) Modeling US adult obesity trends: a system dynamics model for estimating energy imbalance gap. Am J Public Health 104(7):1230–1239

Fingerhut HA (2015) A systems framework and modeling approach to predict and manage the mental health effects of a military conflict. Masters Thesis, Massachusetts Institute of Technology. https://dspace.mit.edu/handle/1721.1/103570. Accessed 4 Aug 2016

Finkelstein SN, Larson RC, Nigmatulina K, Teytelman A (2015) Engineering effective responses to influenza outbreaks. Serv Sci 7(2):119–131

Ghaffarzadegan N, Ebrahimvandi A, Jalali MS (2016) A dynamic model of post-traumatic stress disorder for military personnel and veterans. PLoS One 11(10):e0161405. doi:10.1371/journal.pone.0161405

Ghaffarzadegan N, Hawley J, Desai A (2014) Research workforce diversity: the case of balancing national versus international postdocs in US biomedical research. Syst Res Behav Sci 31(2):301–315

Ghaffarzadegan N, Larson R (2015) Posttraumatic stress disorder: five vicious cycles that inhibit effective treatment. US Army Med Dep J (4–15):8–13

Ghaffarzadegan N, Lyneis J, Richardson GP (2015) Policy informatics with small system dynamics models: how small models can help the public policy process. In: Johnston EW (ed) Governance in the information era: theory and practice of policy informatics. Routledge Press, New York

Ginzburg K, Ein-Dor T, Solomon Z (2010) Comorbidity of posttraumatic stress disorder, anxiety and depression: a 20-year longitudinal study of war veterans. J Affect Disord 123(1):249–257

Hendin H, Haas AP (1991) Suicide and guilt as manifestations of PTSD. Am J Psychiatr 148(5):586–591

Henning KR, Frueh BC (1997) Combat guilt and its relationship to PTSD symptoms. J Clin Psychol 53(8):801–808

Hoge CW, Riviere LA, Wilk JE, Herrell RK, Weathers FW (2014) The prevalence of post-traumatic stress disorder (PTSD) in US combat soldiers: a head-to-head comparison of DSM-5 versus DSM-IV-TR symptom criteria with the PTSD checklist. Lancet Psychiatr 1(4):269–277

Homer JB, Hirsch GB (2006) System dynamics modeling for public health: background and opportunities. Am J Public Health 96(3):452–458

Jakupcak M, Cook J, Imel Z, Fontana A, Rosenheck R, McFall M (2009) Posttraumatic stress disorder as a risk factor for suicidal ideation in Iraq and Afghanistan war veterans. J Trauma Stress 22(4):303–306
Johnston EW (2015) Governance in the information era: theory and practice of policy informatics. Routledge, New York
Kim H, Andersen DF (2012) Building confidence in causal maps generated from purposive text data: mapping transcripts of the Federal Reserve. Syst Dyn Rev 28(4):311–328
Kim H, MacDonald RH, Andersen DF (2013) Simulation and managerial decision making: a double-loop learning framework. Public Adm Rev 73(2):291–300
Luna-Reyes LF, Andersen DL (2003) Collecting and analyzing qualitative data for system dynamics: methods and models. Syst Dyn Rev 19(4):271–296
McFall ME, Mackay PW, Donovan DM (1992) Combat-related posttraumatic stress disorder and severity of substance abuse in Vietnam veterans. J Stud Alcohol 53(4):357–363
McFarland L, Milstein B, Hirsch G, Homer J, Andersen D, Irving R, Reineke E, Niles RD, Cawvey E, Desai A (2016) The NASPAA student simulation competition: reforming the US health care system within a simulated environment. J Public Aff Edu 22(3):363–380
Milstein B, Homer J, Briss P, Burton D, Pechacek T (2011) Why behavioral and environmental interventions are needed to improve health at lower cost. Health Aff 30(5):823–832
Najavits LM, Ryngala D, Back SE, Bolton E, Mueser KT, Brady KT (2009) Treatment of PTSD and comorbid disorders. In: Foa EB, Keane TM, Friedman MJ, Cohen J (eds.) Effective treatments for PTSD: practice guidelines from the International Society for Traumatic Stress Studies. Guilford Press, New York, pp 508–535
Phelan JC (2005) Geneticization of deviant behavior and consequences for stigma: the case of mental illness. J Health Soc Behav 46(4):307–322
Pietrzak RH, Johnson DC, Goldstein MB, Malley JC, Rivers AJ, Morgan CA, Southwick SM (2010) Psychosocial buffers of traumatic stress, depressive symptoms, and psychosocial difficulties in veterans of operations enduring Freedom and Iraqi Freedom: the role of resilience, unit support, and postdeployment social support. J Affect Disord 120(1):188–192
Sabounchi NS, Hovmand PS, Osgood ND, Dyck RF, Jungheim ES (2014) A novel system dynamics model of female obesity and fertility. Am J Public Health 104(7):1240–1246
Solomon Z, Snir A, Fingerhut H, Rosenberg M (2015) Long-term trajectories and recovery from PTSD. In: Bromet EJ (ed) Long-term outcomes in psychopathology research: rethinking the scientific Agenda. Oxford University Press, Oxford, pp 187–204
Sterman J (2014) Interactive web-based simulations for strategy and sustainability: the MIT Sloan Learning Edge management flight simulators, part I. Syst Dyn Rev 30(1–2):89–121
Sterman J, Fiddaman T, Franck T, Jones A, McCauley S, Rice P, Sawin E, Siegel L (2012) Climate interactive: the C-ROADS climate policy model. Syst Dyn Rev 28(3):295–305
Sterman JD (1992) Teaching takes off. OR/MS Today 35(3):40–44
Teytelman A, Larson RC (2013) Multiregional dynamic vaccine allocation during an influenza epidemic. Serv Sci 5(3):197–215
US Department Of Veterans Affairs (2016a) PTSD: National Center for PTSD. http://www.ptsd.va.gov/professional/PTSD-overview/index.asp. Accessed 7 Apr 2016
US Department Of Veterans Affairs (2016b) How Common is PTSD?. http://www.ptsd.va.gov/public/PTSD-overview/basics/how-common-is-ptsd.asp. Accessed 7 Apr 2016
Vogt D (2011) Mental health-related beliefs as a barrier to service use for military personnel and veterans: a review. Psychiatr Serv 62(2):135–142
Wittenborn A, Rahmandad H, Rick J, Hosseinichimeh N (2016) Depression as a systemic syndrome: mapping the feedback loops of major depressive disorder. Psychol Med 46(03):551–562
Zagonel AA, Rohrbaugh J, Richardson GP, Andersen DF (2004) Using simulation models to address "what if" questions about welfare reform. J Policy Anal Manage 23(4):890–901

Navid Ghaffarzadegan is an assistant professor in the department of industrial and systems engineering at Virginia Tech. His research interests include systems sciences, system dynamics, and policy informatics.

Richard C. Larson is a Mitsui Professor of Engineering Systems at MIT, Institute for Data Systems and Society. The majority of his career has focused on operations research as applied to services industries. He is past president of INFORMS, a member of the National Academy of Engineering, an INFORMS Founding Fellow, and a recipient of the INFORMS President's Award, the Lanchester Prize, and the Kimball Medal.

Henry Fingerhut is a PhD student in Engineering Systems at MIT. His research interests include policy decision-making and systems analysis of public policy, particularly in the military mental health domain.

Mohammad S. Jalali (also known as 'MJ') is a research scientist at MIT Sloan School of Management. MJ is interested in simulation and model estimation methodologies, and the applications of dynamic modeling for organizational cybersecurity, complex health and social problems. He is the recipient of the 2015 Dana Meadows Award, the 2015 WINFORMS Student Excellence Award, and the 2014 Lupina Young Researcher Award. He is also a former consultant at the World Bank and a former researcher at the U.S. Department of Energy.

Alireza Ebrahimvandi is a PhD student in the department of industrial and systems engineering at Virginia Tech. His research interests include system dynamics modeling, policy analysis, and simulation. He is also the winner of simulation competition of the Society for Health Systems in 2015.

Anne Quaadgras is Director of the Initiative for Health Systems Innovation, and a Senior Lecturer at MIT Sloan. Her work focuses on health systems transformation, and the role of information technology in supporting that change.

Thomas Kochan is the George Maverick Bunker Professor of Management, a Professor of Work and Employment Research and Engineering Systems, and the CoDirector of the MIT Sloan Institute for Work and Employment Research at the MIT Sloan School of Management. Kochan focuses on the need to update America's work and employment policies, institutions, and practices to catch up with a changing workforce and economy.

Competition and Collusion Between Government and Businesses in Pollution Management: Analysis Based on Stochastic Differential Game

Danlu Jiang and Jing Zhang

Abstract This paper studies the pollution reduction decision process of local government under the influence of social media and PES mechanism using stochastic differential game theory. We built two differential game scenarios simulating the decisions making process between upstream area local government and business. One is Stackleberg game assuming that the business's emission abatement is closely supervised by local government, and the other is cooperative game which is applied when business and local government collude in emission abatement. Comparing simulation results reveals two main findings. One is that transparency make local government perform better in pollution abatement. The other one is that local government tends to collude with business if there is a lack of transparency and supervision. Based on these findings, policy suggestions are presented to overcome this problem.

Introduction

Payments for ecosystem services (PES) received a lot of attention from academics in the recent years. Ecosystem services (ES) are broadly defined as the benefits people obtain from ecosystems in the Millennium Ecosystem Assessment (MEA) (Ecosystem 2005). According to Van Hecken and Bastiaensen, the conceptual basis for PES can be found within neoclassical environmental economics. Essentially, the PES philosophy argues for the internalization of environmental externalities through

---

D. Jiang
School of Business Administration, Chongqing University of Science and Technology, Chongqing, China
e-mail: jiangdl@cqu.edu.cn

J. Zhang (✉)
Graduate School of Management, Clark University, Worcester, MA 01610, USA
e-mail: jizhang@clarku.edu

J.R. Gil-Garcia et al. (eds.), *Policy Analytics, Modelling, and Informatics*, Public Administration and Information Technology 24,
DOI 10.1007/978-3-319-61762-6_18

the creation of markets and quasi-markets of the ES valuables, such as clean water and forestry (Van Hecken and Bastiaensen 2010)· PES are discussed as a mechanism to translate external and non-market values of environment into real financial incentives for local actors who provide ES (Engel et al. 2008).

In China, although not a silver bullet to solve all environmental problems, PES is argued to be one of the most effective tools tailored to address specific sets of environmental pollution. PES schemes in China are carried out through government regulated Eco-Compensation system that uses financial payment mechanisms among conservers, beneficiaries and damagers to internalize externalities, thus correcting the distortion between the private and social interest (Xiong and Wang 2010). They are applied in the domain of legal use of land and forest and wetland recovery in China.

Eco-Compensation can be implemented by employing financial payment, taxation, market and other instruments, which are developed based on the evaluation of the benefits of ecosystem services and the costs associated with conservation, damage control, and lost opportunities, with the goals to maintain ecosystem integrity and promote sustainable co-existence between nature and human beings.

Recently, PES is becoming a popular and important means in solving trans-boundary pollution disputes in river basins. The purpose of PES in river basins is to eliminate the 'free rider' phenomenon, a situation incurred when the pollutions produced by the upstream are passed onto the lower stream with no penalties. By implementing institutional innovation in the form of PES for river basins, ecological resource investors can expect to be rewarded financially. As a result, residents, businesses or local communities have the motivation to engage in ecological protection and conservation, which lead to greater level of ecological capital for the society. At present, PES for trans-boundary water pollution in a particular section is based on the water quality that are monitored periodically, and water quality is monitored across different regions. Local governments in upper stream has to pay for their pollution if the pollution is higher than the standard, or alternatively receive a compensation if the level is lower. This mechanism drives local governments to undertake the responsibility they may have shirked before and make their best efforts in water pollution prevention. This model has been practiced in the Jiangsu, Hebei, Liaoning and Henan provinces (Liu 2011). So far, successful cases of eco-compensation have been found in some small river basins, such as those in Fujian and Zhejiang provinces. According to Liu, the local government plays a key role in monitoring and appraising the effectiveness of eco-compensation mechanisms while central government plays the coordinating role (Liu et al. 2008).

Based on experience and research thus far, we propose that despite of the sound logic behind of the PES scheme design, it won't be successful without openness and transparency in government process management and policy making. This paper tries to explore how transparency make a difference in overcoming difficulties in enforcing environmental regulations. In this paper, we take consideration of the cost of environmental degradation, a parameter that measures the true cost of environment deterioration through damage and depletion of natural resources such as clean air, water and soil; the destruction of ecosystems and the extinction of wildlife; and the damage done to health and wellness of human being. We establish that this parameter could be strongly affected by government transparency. To prove this hypothesis, we built two traditional differential game scenarios that are both constrained by PES mechanism. One is Stackleberg game applied when the business's emission abatement is supervised by local government. The other is cooperative game applied when business and local government collude in emission abatement. Based on the simulation results, this paper reveals two main findings. One is that local government perform better in pollution abatement when there is greater level of transparency. The other one is that local government tends to collude with business when there is a lack of transparency and supervision. The collusion decreases efficiency in emission abatement, which is an equilibrium derived from the model. Therefore, unless there is external force to disclose environmental damages through tools such as open data and social media, it is inevitable that both local governments and businesses choose to collude. We suggest that central government should mandate the disclosure of pollution information and encourage the use of social media by the public to increase transparency and eliminate collusion.

The paper is organized as follows. Second section review literatures on E-government and game theories. This is followed by the description of game formulation. Non-cooperative outcomes are discussed in fourth section and cooperative arrangements, group optimal actions, individual rationality and subgame-consistent imputations are examined in fifth section. A payment distribution mechanism introducing the proposed subgame-consistent solution is derived in sixth section. A numerical example is provided in seventh section and finally, concluding remarks are given in eighth section.

Literature Review

E-Government and Social Media

Environmental regulation has been one of the most intractable problems encountered by public policy. Environmental problems stem from multiple activities and are embedded in interactions across social and natural systems. Because of these interactions and complexity, solutions often lead to unintended consequences. Environmental policy for the most part can broadly be categorized into two distinct types, mandate-based and market-based. The mandate-based approach is more prescriptive and imposes strict limits on the amount of pollutants by setting emission standards for hazardous pollutants to curb environmental pollution. This command and control approach though is criticized for being inflexible with high marginal cost of abatement for individual organizations without the certainty of resulting environmental performance (Stavins 2007). The second type of policy introduces a market based approach to encourage environmental performance by attaching a price to environmental externalities (Jaffe et al. 2002). Compared to the command-and-control approach, the market-based approach has been argued to be more flexible and cost-effective because the reduction tends to gravitate toward the sources where pollution is least costly to reduce, while providing financial incentives to reduce pollution (Stavins 2007). PES is one form of the market-based regulation. The effectiveness of market-based approach, however, relies very much on the accountability and transparency of the policy environment.

In the information age, researchers and practitioners has increasingly turned to the use of ICTs (information and communications technology) and networks of public, private, and non-profit organizations for more innovative solutions (Travis et al. 2010). ICTs, especially E-government tools and social media, are seen by many as a cost-effective and convenient means to promote openness and transparency. Scholars also examine the interrelationships between ICTs, social media, and collaborative e-government in their ability to facilitate transparency. They identified various ways by which members of the public are able to employ the social media to monitor government activities (Bertot et al. 2010, 2012).

Social media is defined as media that is designed for and centered on social interaction. It is commonly based on web 2.0 technologies and services such as blogs, wikis, media-sharing services, Twitters, Facebooks, collaborative editing tools, and social networking services that enable and empower users to communicate, collaborate, interact, edit, and share information, and achieve common goal in a social environment. According to Bertot et al., social media has four major potential strengths. That are collaboration, participation, empowerment, and time. It allows anyone with access to the Internet the ability to inexpensively publish or broadcast information in near-real time, effectively democratizing media (Bertot et al. 2010). Chinese government blocked most of the popular social media sites from western countries due to political concerns, but similar social media, such as WeChat, QQzone and other local software, are very popular in China.

As the social media are becoming ubiquitous, governments and other organizations use this technology to support their daily functions and emergency responses. For example, extensive use of social media for presidential debates and campaigns in the United States have become a matter of routine (Gibson 2009). In addition, social media was used to "crowdsource" feedback about local policies and services (Cliff et al. 2011). More importantly, social media and other WEB 2.0 technologies are great tools to reduce information asymmetry between public and government and improve citizen capability to engage, rate, and monitor the implementation of public policies, which in turn keep governments more accountable for the results of public policy and the governance more transparent.

Although contribution of social media and other ICTs to enhance transparency and openness of public sector and to promote new forms of accountability has been highlighted by various authors, research on social media for policy development are very few from China. This study is motivated by the needs to meet and gap and rethink how citizen engagement via ICT in China can be deployed to enhance the effectiveness of environmental regulation.

Differential Games

Differential game theory is an effective method in analyzing a rational player's response under the conditions of both conflict and cooperation, and it has been widely used in studies of participants' strategic decisions and dynamic evolution of pollution. In game theory, differential games capture a series of decisions that are dynamically adjusted according to status affected by the earlier decisions made by actors. Early analyses of differential game reflected competitive situations encountered by military interests, which consider two actors—the pursuer and the evader—with diametrically opposed goals. The Stackelberg model is a competition game in which the leading player moves first and then the following player move correspondingly, and both are to maximize their own interests. In recent decades, it has been applied extensively in economic and management literature (Dockner 2000; Sethi and Thompson 2000; Pontryagin 1986), and cooperative games are modeled as well. In game theory, a cooperative game is a game where groups of players may enforce cooperative behaviors, and they play as coalition to maximize their common interests.

Applications of non-cooperative differential games in environmental studies can be traced back to the 1990s, with examples of differential game between a policy maker and business in one country, or game between two neighboring countries for pollution control (Yeung 1992; Dockner and Long 1993). The application of this tool covers the topics of pollution control, CO_2 abatement, and environmental policy instruments (Tahvonen 1994; Feenstra et al. 2001). Cooperative differential games in environmental control are also presented later on (Jørgensen and Zaccour 2001; Breton et al. 2005).

Fernandez developed a differential game to examine the effects of trade liberalization on trans-boundary water pollution. His results show that cooperation and trade liberalization limit emissions from both countries (Fernandez 2002). Yeung published a number of studies concerning trans-boundary environmental control along the border. His research demonstrates the differences of strategic behaviors of local governments and businesses under cooperative and competitive conditions. He concluded that cooperation in environmental control offered the best probability of effective response (Yeung 1992; Yeung 2007; Yeung and Petrosyan 2008).

Various researches show that the tool is effective in environmental domain, however, challenges remain in solving trans-boundary pollution problems. Fernandez compares three game theory scenarios, which include a non-cooperative game, a cooperative game, and a Stackelberg game and shows that pollution stock can grow profusely along with abatement cost and damages if waste water management efforts are delaye (Fernandez 2013). Benchekroun finds that the level of emission would increase when the damage and initial pollution stock are large (Benchekroun and Chaudhuri 2014).

In addition, most game researches focus on benefit distribution between different local governments sharing the same ecological resources, but not on the inherent basis of solving trans-boundary pollution problems. The question of how to maintain the balance between economic and environment interests to ensure sustainable development for human being and wildlife has not been the focus. Moreover, the roles of ICTs in public policy making and implement are yet to be incorporated in game models.

In this paper, we build a stochastic differential game model in which the decisions of upper stream government are constrained by the PES scheme formed with the downstream government. In the upper area, local governments impose pollution tax on the factories and businesses who produce and release pollutions into the river. We add a damage parameter h, which is the cost of environmental degradation (CED), into the local government's payoff function. CED is not only the cost of environment recovery but also includes the loss of social welfare and wellness of the citizens who live there. The cost of environmental recovery could be defined and measured by tangible means. The social welfare and human being wellness part of CED are intangible, and often are underestimated without transparency and closed monitoring. In this study, they can be made apparent when information sharing, citizen engagement and interaction can be assumed.

We assume that the local government in upstream and the business are two independent players first, and we simulate their behaviors using a Stackelberg differential game and cooperation differential game. The simulation results lead to the conclusion that collusion could not be avoided when CED is underestimated. This means local government has inherent reason to collude with business, when the regulatory environment is opaque and there is a lack of transparency and citizen engagement.

Game Formulation

In this section, we explain modeling frame. The local government and businesses in upper stream are two players in these models. The local government cares about the trade-offs between economic development and environmental safety, such as water quality in this area, and it is liable for the transboundary pollution compensation. The businesses are only interested in maximizing their financial gains. Although business and government are two separate entities, the association with each other and common interests may still exist to certain extent.

When the business produces under governments' supervision, Stackelberg stochastic differential game is applied. When the government and businesses decide to collude with each other, cooperative stochastic differential game is applied.

Model Assumption

Under the premise of no significant economic change, the assumptions of models are specified as follows.

1. The observation period is assumed to be the same as the tenure of local officials in this study. Therefore, the feasibility of reasonable tax rate adjustment and policy continuity could be ensured. The objectives of local government include economic growth, social welfare, political stability, and environmental sustainability. Economic benefits in the observation period are the most prominent objective of local governments in China.
2. We assume that the major goal of firms is to maximize profits. The production scale is restricted by the law of diminishing marginal returns of capital, and profits are a strict concave function of production. The firm is assumed to be a monopoly who produces only one kind of goods. Production technology and demand function will not change, and the discount rate of government and firm is a risk-free rate.
3. The Cost of Environmental Degradation (CED) is the measurement of the deterioration of the environment through depletion of resources such as air, water and soil; destruction of ecosystems and extinction of wildlife; and decline in human living conditions. This cost, also called pollution loss, represents environmental damage caused by externalities of industry. In this paper, we assume that the CED includes two elements. One is the financial cost associated with environmental recovery which could be observed. Another part is the potential loss of social welfare and human well-being which could be gauged by feedback from citizens through social media.

State Variable

Industrial production emits pollutants into the environment, and the effluent quantity is determined by production outputs and pollution reduction efforts. The emitted pollutants cause short term local impacts on neighboring areas of the origin of production in the form of passing-by waste in waterways.

Let x(t) denote the level of pollution at time s. The dynamics of pollution stock is governed by the stochastic differential equation:

$$
\begin{aligned}
&\mathrm{d}x(t)=\left\{\alpha q(t)-\beta\left[u(t)+I(t)\right]\times\left[x(t)\right]^{\frac{1}{2}}-\delta x(t)\right\}dt+\sigma x(t)\mathrm{d}z(t),\\
&x(t_0)=x_0 \qquad (1)
\end{aligned}
$$

where σ is a noise parameter and $z(s)$ is the standard Wiener process, αq(t) is the amount added to the pollution stock by one unit of business's output, u (t) and $I(t)$ is the pollution abatement effort made by government and business respectively, $\beta\left[u(t)+I(t)\right]x(t)^{\frac{1}{2}}$ is the amount of pollution reduced by $[u(t)+I(t)]$ unit of abatement efforts, and δ is the natural rate of decay of pollution, nature's capability to replenish the environment. Because the rate of pollution degradation and climate change are subject to certain uncertainty, a stochastic dynamic is applied in pollution stock. The Cost of Environmental Degradation at time t is $hx(t)$.

Firm's Objectives

Considering an economy with only one firm that produces one homogeneous product, the demand function of the output of the firm at time instant t is

$$p(t) = a - bq(t),$$

where $p(t)$ is the price of the product, $q(t)$ is the quantity of the output, a and b are positive constants. The output, $q(t) \in [0, Q]$, is nonnegative and bounded by maximum output constraint Q. The business's profits at time t can be expressed as

$$\pi(t) = \left[a - bq(t)\right]q(t) - c\left[q(t)\right]^2 - v(t)q(t) - \mathrm{d}\left[I(t)\right]^2$$

where $v(t)$ is the tax rate imposed on its output at time t, c is a positive constant which is the product cost coefficient, d is the cost coefficient of pollution abatement, and $q(t)$ and $I(t)$ are two control variables of the business. The business's object is to maximize the total profit in the time zone $t \in [0, T]$.

The business's expected payoff function over the whole planning horizon could be expressed as

$$E_{t_0}\left\{\begin{array}{l}\int_{t_0}^{T}\left\{\left[a - bq(t)\right]q(t) - c\left[q(t)\right]^2 - v(t)q(t) - d\left[I(t)\right]^2\right\}e^{-r(t-t_0)}\mathrm{d}t \\ -k\left[x(T) - A\right]e^{-r(T-t_0)}\end{array}\right\}, \quad (2)$$

where A is the environmental capacity of the river body. At the end of the observe period, pollution stock in the river should be assessed by a third-party organization. The business will be punished if the pollution stock at time T is higher than A; otherwise, it will be rewarded for abatement.

Local Governments' Objectives

The local governments have to grow tax revenue and at the same time handle the financing of the costs associated with pollution cleanup. In particular, government intends to maximize the net gains which is equal to industrial tax minus the sum of expenditures on pollution abatement, cost of environmental degradation and ecological compensation.

The instantaneous objective of government at time t can be expressed as:

$$v(t)q(t)-d\left[u(t)\right]^2-hx(t)$$

where $d[u(t)]^2$ is the cost of employing u amount of pollution abatement effort, and $hx(t)$ is the cost of environmental degradation from $x(t)$ amount of pollution.

The government's planning horizon is $t\in[0,T]$. At time T, pollution stock $x(T)$ would be detected and the final payment is $g[x(T)-A]$, where $g>0$ and $A>0$.

The local governments maximize the integral of its instantaneous objective over the whole planning horizon subject to the state variable pollution stock (1) by controlling the abatement efforts and tax rate. The local government's excepted payoff is,

$$\mathrm{E}_{t_0}\left[\int_{t_0}^{T}\left\{v(t)q(t)-d\left[u(t)\right]^2-hx(t)\right\}e^{-r(t-t_0)}\mathrm{d}t-g\left[x(T)-A\right]e^{-r(T-t_0)}\right], \tag{3}$$

where r is the discount rate.

PES mechanism in china is a direct punishment or reward to the pollutant emission or environmental conservation. Payment amount depends on the difference between the central government mandated standard and water quality for the proposed monitoring section [3].

In this paper, the mandated standard is expressed by A in Eq. (3). According to the PES rule, local government should pay downstream government for pollutant transfer when emission stock level in the river body is higher than the standard. Otherwise, local government will be paid for emission abatement and environmental conversation in the river. Value of PES coefficient, which is not discussed in this paper could be mandated by the central government or could be negotiated between governments in upstream and downstream.

Game Equilibriums

Stackelberg Outcomes

In this section, we discuss the solution to the Stackelberg game which means that the government as the supervisor of the business. In this game, a mandated tax is imposed on the business, therefore, the government is the "leader" and the business is the "follower" in the Stackelberg differential game.

The control variables that the local government could decide are pollution abatement efforts and tax rate. If the government makes tax rate decisions without considering reactions from the business, this is a unilateral optimal control model. If the government considers the reactions from the firm, decision making becomes a continuous process in planning horizon.

In stackelberg game, the "leader" and "follower" make their decisions in different order. We could divide the decision-making process into two steps. First, the government decides tax $v(t)$ at initial time, then the firm chooses the optimal $q^*(t)$ and $I^*(t)$ to maximize its payoff according to this tax rate. The government chooses a new tax rate considering the rational reaction from the firm to maximize its payoff. Both the government and the firm make the optimal dynamic decisions in that order in the whole period.

When government declares a tax rate $v(t)$, the firm decides the optimal strategies which are $q^*(t)$ and $I^*(t)$ based on a given-parameter of $v(t)$. At that moment, differential game becomes a unilateral optimal control problem of the firm whose control variables are $q(t)$ and $I(t)$.

A set of feedback strategies of $q^*(t)$ ad $I^*(t)$ present a Nash equilibrium solution to stochastic control problem (2) if there is a suitably smooth function $V^1(t,x)$ satisfying the following partial differential Eq. (4):

$$
\begin{aligned}
&-V_t^1(t,x)-\frac{1}{2}\sigma^2x^2V_{xx}^1(t,x)\\
&=\max_{q(t),I(t)}\left\{\begin{aligned}&\left[\left[a-bq(t)\right]q(t)-c\left[q(t)\right]^2-v(t)q(t)-d\left[I(t)\right]^2\right]e^{-r(t-t_0)}\\&+V_x^1\left[\alpha q(t)-\beta\left[u(t)+I(t)\right]x^{\frac{1}{2}}-\delta x\right]\end{aligned}\right\}
\end{aligned}
\tag{4}
$$

subjected by $V^1(T,x)=-k(x-A)e^{-r(T-t_0)}$.

The local government could predict that $q^*(t)$ and $I^*(t)$ are rational reactions. So it chooses the next optimal strategies $v^*(t)$, $u^*(t)$, on the given $q^*(t)$, $I^*(t)$. In that case, this game becomes a unilateral optimal control problem of the local government.

A set of feedback strategies $v^*(t)$, $u^*(t)$ provides a Nash equilibrium solution to stochastic control problem (3) if there is a suitably smooth function $V^2(t,x)$ satisfying the following partial differential Eq. (5):

$$
\begin{aligned}
&-V_t^2(t,x)-\frac{1}{2}\sigma^2x^2V_{xx}^2(t,x)\\
&=\max_{u(t),v(t)}\left\{\begin{aligned}&\left[v(t)q(t)-d\left[u(t)\right]^2-hx(t)\right]e^{-r(t-t_0)}\\&+V_x^2\left[\alpha q(t)-\beta\left[u(t)+I(t)\right]x^{\frac{1}{2}}-\delta x\right]\end{aligned}\right\},
\end{aligned}
\tag{5}
$$

subjected by $V^2(T,x)=-g(x-A)e^{-r(T-t_0)}$.

We solve a set of $q^*(t)$, $I^*(t)$, $v^*(t)$, $u^*(t)$, optimal strategies of game (2) (3), by Bellman Equation. Substituting these results into Eqs. (4) and (5), we obtain

Proposition 1
Systems (4)–(5) admit a solution (1 represents government, 2 represents the firm)

$$V^i(t,x)=\left[A_i(t)x(t)+C_i(t)\right]e^{-r(t-t_0)},\ (\mathrm{i}=1{,}2) \tag{6}$$

where A1 and A2 satisfying the following set of constant coefficient quadratic ordinary differential equations:

$$\begin{cases}\dot{A}_1(t)=\dfrac{-\beta^2}{4d}A_1^2(t)+\left(\dfrac{-\beta^2}{2d}A_2(t)+\delta+r\right)A_1(t), A_1(T)=-k\\ \dot{A}_2(t)=\dfrac{-\beta^2}{4d}A_2^2(t)+\left(\dfrac{-\beta^2}{2d}A_1(t)+\delta+r\right)A_2(t)+h, A_2(T)=-g.\end{cases} \tag{7}$$

Nash equilibrium solution to this Stackelberg game are

$$q^*(t)=\frac{a+\alpha\left[A_1(t)+A_2(t)\right]}{4(b+c)}, \tag{8}$$

$$I^*(t)=\frac{-\beta}{2d}\left[x(t)\right]^{\frac{1}{2}}A_1(t), \tag{9}$$

$$v^*(t)=\frac{a+\alpha\left[A_1(t)-A_2(t)\right]}{2}, \tag{10}$$

$$u^*(t)=\frac{-\beta}{2d}\left[x(t)\right]^{\frac{1}{2}}A_2(t), \tag{11}$$

the expected value of pollution stock in river body is

$$\mathrm{E}\left[x(t)\right]=-\frac{\alpha\left[a+\alpha\left(A_1(t)+A_2(t)\right)\right]}{4(b+c)\left[\dfrac{\beta^2}{2d}A_1(t)+\dfrac{\beta^2}{2d}A_2(t)-\delta\right]}+\left(x_0+\frac{\alpha\left[a+\alpha\left(A_1(t)+A_2(t)\right)\right]}{4(b+c)\left[\dfrac{\beta^2}{2d}A_1(t)+\dfrac{\beta^2}{2d}A_2(t)-\delta\right]}\right)e^{\left[\frac{\beta^2}{2d}A_1(t)+\frac{\beta^2}{2d}A_2(t)-\delta\right]t}, \tag{12}$$

According to Eqs. (8)–(11), we could know the following:

(a) There is a positive correlation between output and $[A_1(t)+A_2(t)]$.

(b) The pollution abatement efforts from government and firm are directly impact by $A_1(t), A_2(t)$, abatement efforts decrease when the benefits from pollutant increase.

Cooperative Outcomes

In this section, local government and business agree to cooperate on the whole planning horizon so that group optimum outcomes could be achieved. Because the cooperative scheme should be obeyed throughout the whole game horizon only if both group rationality and individual rationality are satisfied at any point of time. Group rationality is that all potential gains from the cooperation are captured. Individual rationality requires that the participants gain no less than that could have gained from non-cooperative game. Failure to guarantee individual or group rationality leads to failure of cooperative agreement.

In cooperation, local government imposes no tax on the business. but shares the profit with the business at the end of the period. In order to ensure both participants obey agreed-upon scheme, the agreement should be subgame-consistent.

Next, consider the cooperative game as a collusion between the local government and the business. The collusion group maximizes their common expected payoff by choosing their production and pollution abatement efforts. The cooperative stochastic differential games expected payoff is

$$E_{t_0}\left\{\int_{t_0}^{T}\left\{\left[\left[a-bq(t)\right]q(t)-c\left[q(t)\right]^2-d\left[UI(t)\right]^2-hx(t)\right]\right\}e^{-r(t-t_0)}dt \atop -(g+k)\left[x(T)-A\right] \quad e^{-r(T-t_0)}\right\}, \quad (13)$$

subjected by dynamics

$$dx(t)=\left\{\alpha q(t)-\beta UI(t)\times\left[x(t)\right]^{\frac{1}{2}}-\delta x\right\}dt+\sigma\left[x(t)\right]dz(t),\quad x(t_0)=x_0 \quad (14)$$

where $UI(t)$ is the pollution abatement effort from the collusion. A set of feedback strategies $q_0^*(t)$, $UI^*(t)$ provide a Nash equilibrium solution to stochastic control problem Eqs. (13) and (14) if there is a suitably smooth function $W(t,x)$ satisfying the following partial differential equation

$$
\begin{aligned}
&-W_t(t,x)-\frac{1}{2}\sigma^2x^2W_{xx}(t,x)\\
&=\max_{q(t),UI(t)}\left\{\left[a-bq(t)\right]q(t)-c\left[q(t)\right]^2-d\left[UI(t)\right]^2-hx(t)\right\}e^{-r(t-t_0)}\\
&+W_x\left\{\alpha q(t)-\beta\left[UI(t)\right]x(t)^{\frac{1}{2}}-\delta x\right\},
\end{aligned}
\tag{15}
$$

subjected by

$$
W(T,x)=-(g+k)\left[x(T)-A\right]e^{-r(T-t_0)}, \tag{16}
$$

Proposition 2

Systems (15), (16) admit a solution as

$$
W(t,x)=\left[A_0(t)x(t)+C_0(t)\right]e^{-r(t-t_0)}, \tag{17}
$$

where A_0 satisfying the following set of constant coefficient quadratic ordinary differential equations:

$$
\begin{cases}
\dot{A}_0(t)=\dfrac{-\beta^2}{4d}A_0^2(t)+(\delta+r)A_0(t)+h,\\
\dot{C}_0(t)=rC_0(t)-\dfrac{\left[a+\alpha A_0(t)\right]^2}{4(b+c)},
\end{cases}
$$

and

$$
\begin{cases}
A_0(T)=-k-g,\\
C_0(T)=(k+g)A.
\end{cases}
\tag{18}
$$

Nash equilibrium solution to this cooperative game are

$$
q_0^*(t)=\frac{a+\alpha A_0(t)}{2(b+c)}, \tag{19}
$$

$$
UI^*(t)=\frac{-\beta}{2d}\left[x(t)\right]^{\frac{1}{2}}A_0, \tag{20}
$$

the expected value of pollutant in river body is

$$\mathrm{E}\left[x(t)\right] = -\frac{a\alpha+\alpha^2 A_0(t)}{a\alpha+\alpha^2 A_0(t)+2(b+c)\left[\frac{\beta^2}{2d}A_0(t)-\delta\right]} + \left(x_0+\frac{a\alpha+\alpha^2 A_0(t)}{a\alpha+\alpha^2 A_0(t)+2(b+c)\left[\frac{\beta^2}{2d}A_0(t)-\delta\right]}\right)e^{\left[\frac{\beta^2}{2d}A_0(t)-\delta\right]t}.$$

A payment distribution scheme forms over the time in that they share the payment at a certain percentage. $W^1(t_0, x)$ is the payment distributed to the firm, and $W_2(t_0, x)$ is the payment distributed to the government.

$$W_1(t_0,x)=\frac{V^1(t_0,x)}{V^1(t_0,x)+V^2(t_0,x)}W(t_0,x),$$

$$W_2(t_0,x)=\frac{V^2(t_0,x)}{V^1(t_0,x)+V^2(t_0,x)}W(t_0,x).$$

Numerical Example and Analysis

In this section, we try to indicate the characters of the optimal solutions in both game model by BSDE (back-ward stochastic different equation) and numerical simulation. In order to simplify the analytic process without making difference to the result, we set $t_0 = 0$.

Shadow Price

The "Shadow Price" is the marginal benefit caused by pollution which equals to the partial derivative of benefits with respect to the "state variable"—pollution stock. The shadow price reflects the willingness that participants prefer emit pollution. Because it has a close correlation with the optimal solution so that we could uncover the characteristics of the equilibriums by $aA_1(\tau)=0$ nalyzing the shadow price.

The Firm's Shadow Price

$A_1(t)$ is the shadow price of the firm at time t. Let $\tau = T - t$, the BSDE of $A_1(t)$ is

$$\dot{A}_1(\tau) = -\dot{A}_1(t) = \frac{\beta^2}{4d}A_1^2(\tau) + \left(\frac{\beta^2}{2d}A_2(\tau) - \delta - r\right)A_1(\tau)$$

when $\tau=0$, $A_1(\tau)=-k$, $A_2(\tau)=-g$, so $\dot{A}_1(\tau) > 0$.

$A_1(\tau)$ increase during the range of $\tau \in [0, T]$. When it comes to , $\dot{A}_1(\tau) = 0$. As a result, $A_1(\tau)$ does not change till $\tau=T$. That means, $A_1(t)=0$ is the maximum value of $A_1(t)$, it decreases with the increase of t.

The Government's Shadow Price

$A_2(t)$ is the shadow price of the government at time t. It represents the contribution to benefit per unit pollutant at time t.

$A_2(t)$ shows different trends in different case.

1. In the case of $h < h_2 = \frac{\beta^2}{4d}g^2 + g\left(\frac{\beta^2}{2d}k + \delta + r\right)$, $A_2(t)$ is a monotonic decreasing function in the range of $t \in [0, T]$, and the minimum value is $A_2(T) = -g$.
2. In the case of $h \geq h_2 = \frac{\beta^2}{4d}g^2 + g\left(\frac{\beta^2}{2d}k + \delta + r\right)$, $A_2(t)$ is a monotonic increasing function in the range of $t \in [0, T]$, and receives the maximal value when $t = T$.

CED h has a direct effect on government policy implementation. If CED is higher than h_2, what means the government do value the environment and cannot tolerate the sufferings from the environment degradation, pollution has the greater negative effect on government's payoff when $t = 0$, and it goes down along with time goes by. If CED is lower than h_2, what means the government do not value the environment so much and could tolerate the sufferings from the environment degradation, small negative effect was found at the beginning, and it becomes greater with t increases. This trend conforms to the real conditions, in that the government does not estimate the real cost of environmental degradation until the last minute in the period, with the shadow price remaining high in the period and declining sharply to a negative value at the end.

We set parameter values as follows:

Results from numerical analysis are consistent to that from BDE analysis. Keeping other parameters constant, the values of $A_2(t)$ when $h = 0.1$ and $h = 4$ are presented in Fig. 1. As it is shown in Fig. 1, $A_2(t)$ increases over the time interval when $h = 4$, declines over the time interval when $h = 0.1$. However, $A_2(t)$ in the former condition is always less than that in the latter.

The Group's Shadow Price

$A_0(t)$, whose curve is similar to $A_1(t)$, represents the group's shadow price in cooperative. It is important to note that the crucial threshold which determines the direction of curve $A_0(t)$ is very different. We analysis $A_0(t)$ by BDE as the same way as illustrated in the section "The Government's Shadow Price".

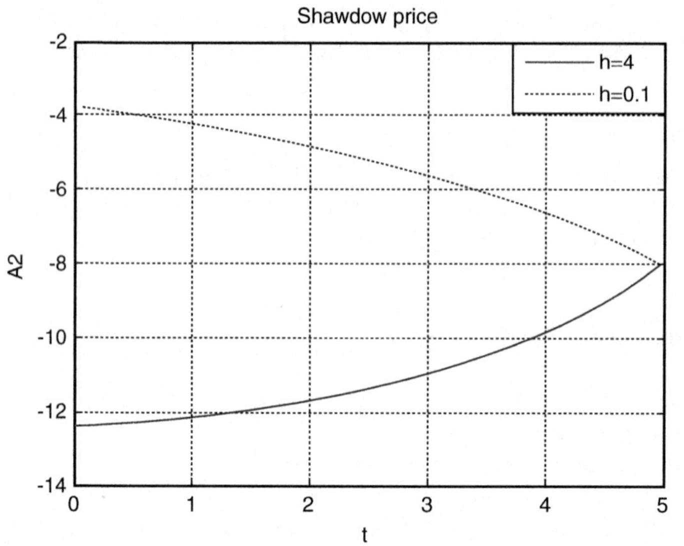

Fig. 1 The government's shadow price in different *h*

Table 1 Table parameter values in numerical simulation

k	g	r	a	b	c	d	h	α	β	δ
2	8	0.05	50	1	1	0.5	4	0.5	0.2	0.01

1. In the case of $h < h_0 = \frac{\beta^2}{4d}(g+k)^2 + (g+k)(\delta+r)$, $A_0(t)$ is a monotonic decreasing function in the range of $t \in [0, T]$.
2. In the case of $h \geq h_0 = \frac{\beta^2}{4d}(g+k)^2 + (g+k)(\delta+r)$, $A_0(t)$ is a monotonic increasing function in the range of $t \in [0, T]$.

   $h > h_0 = \frac{\beta^2}{4d}(g+k)^2 + (g+k)(\delta+r)$, and vice versa.

Because the $h_0(T)$ in both situations achieve the same value, $h_0(t)$ remains higher if h is lower in the whole period. But more remarkably, threshold h_0 which may change curve-direction of shadow price in cooperative game is much higher than h_2 in a Stackelberg game. This implies that it is difficult for the group to give up emission even though the CED is not underestimated. So they would rather make less effort for emission abatement than that would have been done by local government in a Stackelberg game.

Under the given parameter in Table 1, the shadow price is higher than cooperative too. It is presented by Fig. 2.

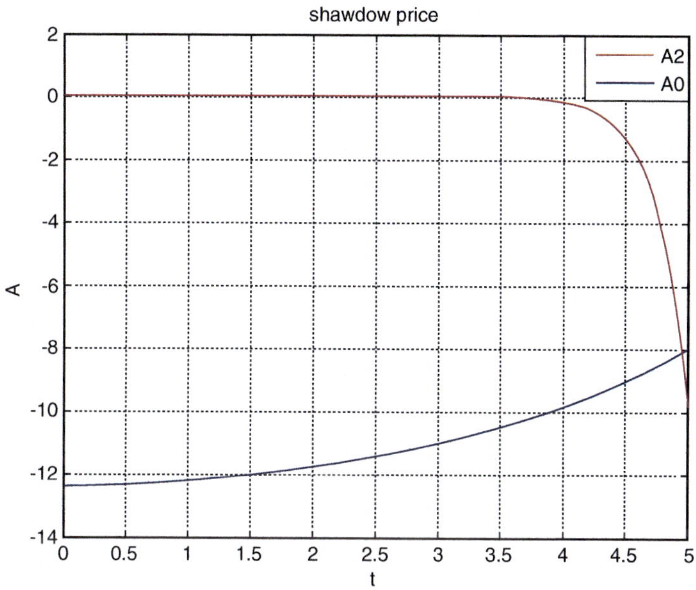

Fig. 2 The government's and group's shadow price

Fig. 3 Comparison in payoff

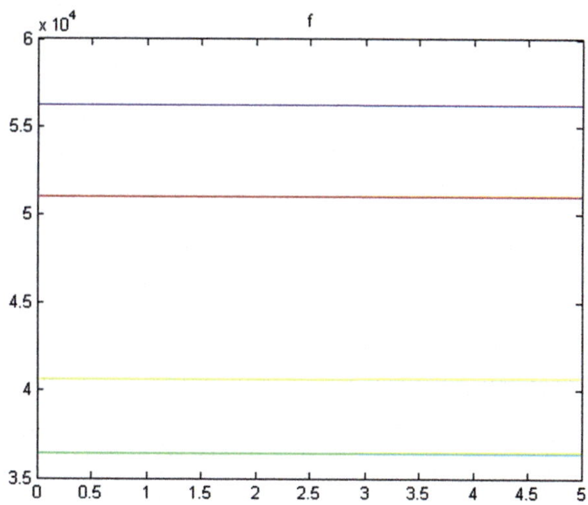

Comparison in Payoff

In Fig. 3, four straight lines represent total payoff of the business in cooperative game and in stackelberg game, with the total payoffs for government in cooperative game and in stackelberg game laid out separately. Both the government and

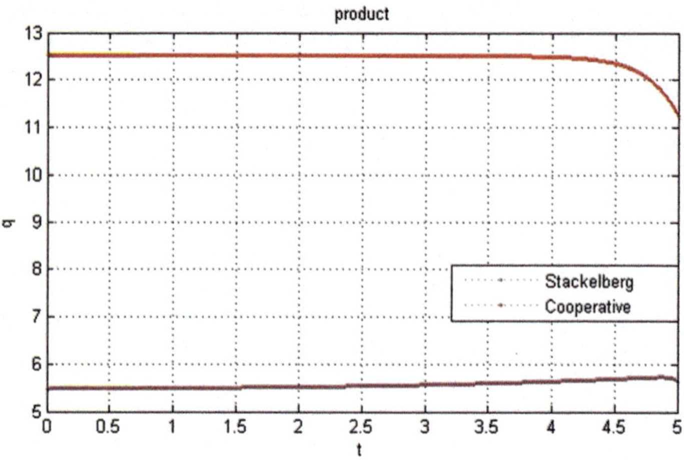

Fig. 4 comparison in outputs

business gain more in the cooperative game than they do in the Stackelberg game. This explains why they have the motivations to collude with each other.

Comparison in Outputs and Emission

In cooperative game, both outputs level and emission level are much higher that in the Stackelberg. This lead to the conclusion that government engages into production decision making and profit sharing with the business in cooperative game, and fail to regulate. This leads to the damage to the environment and reduce the health and social welfare of human being. Higher pollution cannot be avoided in the cooperative game, although there is PES constraint. The simulation results are presented in Figs. 4 and 5.

Conclusion

By analyzing the equilibrium of the two game scenarios, we could make two propositions. First, the cost of environmental degradation (CED) has a direct impact on the government's decisions. When citizens were able to be involved in monitoring the environmental performance of public policies, government are more accountable for not grossly underestimating the CED. As can be seen in the simulations, government would make stricter policy when CED is estimated to be higher in a Stackelberg game. In response, the business would choose to decrease emission and the local government would invest in more efforts in pollution reduction. This means more citizen engagement and transparency can be critical to keep public sector accountable for environmental results.

Fig. 5 Comparison in emission

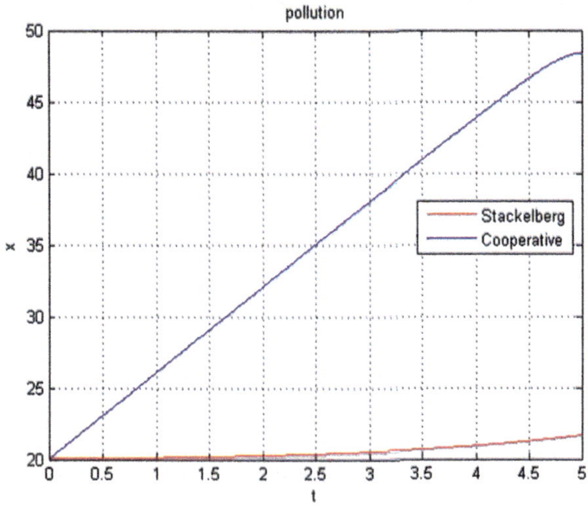

It should be highlighted that only if CED is correctly estimated, especially taking account for the intangible cost of environmental deterioration, human well-being and environmental sustainability could not be achieved. In China, the Central Government and the public have the interest to keep environmental goal in check again the economic interests of local governments. However, because of the lack of formal disclosure regulations and enforcement power to keep environmental information public in a timely manner in China, social media might step up to serve as an useful tool to monitor the performance of local governments. It helps eliminate information asymmetry among key stakeholders and build more effective environmental assessment standards.

Second, it is important to be realistic about the tendency of collusion between local governments and businesses in developing CED scheme and policies. Local governments and businesses prefer collusion disregard of the levels of CED, according to the analyses. This implies that inherent interests exist that lead them to collude and share the benefits from collusion. Public, central government, local government are different stakeholder groups and have different interests. Without adequate level of supervision, local governments may choose collusion to pursue more economic interests. This especially highlight the importance of supervision and monitoring from a third-party or the public to check the tendency of collusion.

Policy Suggestions

According to the analyses made in this paper, we developed two policy suggestions. First, greater level of transparency in governance is required to ensure the success of PES scheme. Many problems incurred in enforcing environmental regulations is cause by the lack of transparency in government. The evaluation and promotion decision of local officials are not based on the long run effect of their decisions and actions. Rather, local officials are only rewarded and promoted when they can demonstrate visible growth to the local economy in the short term. In the information era, social media is a cost-effective way to engage the citizens in actively participating in providing feedback, voicing environmental concerns, and monitoring actions of governments. It is possible to reform performance appraisal system by extending the period of assessment, aligning environmental quality feedback from public with official performance assessment, and practicing One-Vote Veto based on their contribution to the environmental performance. These would enhance the stake for local governments to estimate the true cost of environmental degeneration more judiciously.

The second suggestion is made to the Central Government. It could strengthen the supervision of local environment by empowering citizens' active participation in self-governance of environment. There should be credible punishments and extra penalty to local government once exposed by social media. Making local officials accountable to the feedback received from the public could be a way to reduce and eliminate collusions and corruptions.

This paper is a preliminary theoretical model development in the PES and e-government environmental information transparency domain. The limitation of this paper is that only mathematical analysis and numerical simulations are applied in understanding the government decision making process and tendencies. Empirical efforts using data from multiple industries are recommended as they could play a significant role in the future research.

Acknowledgement We would like to express our gratitude to the reviewers and editors who helped us polishing this chapter. This research is supported by the Chinese National Social Science Foundation, Grant# 15CJY014.

References

Benchekroun H, Chaudhuri AR (2014) Transboundary pollution and clean technologies. Resour Energy Econ 36(2):601–619. doi:10.1016/j.reseneeco.2013.09.004

Bertot JC, Jaeger PT, Grimes JM (2010) Using ICTs to create a culture of transparency: E-government and social media as openness and anti-corruption tools for societies. Gov Inf Q 27(3):264–271. doi:10.1016/j.giq.2010.03.001

Bertot JC, Jaeger PT, Grimes JM (2012) Promoting transparency and accountability through ICTs, social media, and collaborative e-government. Transforming Gov People Process Policy 6(1):78–91. doi:10.1108/17506161211214831

Breton M, Zaccour G, Zahaf M (2005) A differential game of joint implementation of environmental projects. Automatica 41(10):1737–1749. doi:10.1016/j.automatica.2005.05.004

Cliff L, LaRose R, Steinfield C, DeMaagd K (2011) Inherent barriers to the use of social media for public policy informatics. Innovation J Public Sect Innovation J 16(1):17

Dockner E (2000) Differential games in economics and management science. Cambridge University Press, Cambridge

Dockner EJ, Long NV (1993) International pollution control: cooperative versus noncooperative strategies. J Environ Econ Manag 25(1):13–29. doi:10.1006/jeem.1993.1023

Ecosystem Assessment Millennium (2005) Millennium ecosystem assessment: ecosystems and human well-being (synthesis report). 2005. Washington
Engel S, Pagiola S, Wunder S (2008) Designing payments for environmental services in theory and practice: an overview of the issues. Ecol Econ 65(4):663–674. doi:10.1016/j.ecolecon.2008.03.011
Feenstra T, Kort PM, de Zeeuw A (2001) Environmental policy instruments in an international duopoly with feedback investment strategies. J Econ Dyn Control 25(10):1665–1687. doi:10.1016/S0165-1889(99)00076-7
Fernandez L (2002) Trade's dynamic solutions to transboundary pollution. J Environ Econ Manag 43(3):386–411. doi:10.1006/jeem.2001.1187
Fernandez L (2013) Transboundary water institutions in action. Water Resour Econ 1:20–35. doi:10.1016/j.wre.2013.03.005.
Gibson RK (2009) New media and the revitalisation of politics. Representations 45(3):289–299. doi:10.1080/00344890903129566
Jørgensen S, Zaccour G (2001) Time consistent side payments in a dynamic game of downstream pollution. J Econ Dyn Control 25(12):1973–1987. doi:10.1016/S0165-1889(00)00013-0
Jaffe AB, Newell RG, Stavins RN (2002) Environmental policy and technological change. Environ Resour Econ 22(1):41–70. doi:10.1023/a:1015519401088
Liu G, Wan J, Zhang H, Cai L (2008) Eco-compensation policies and mechanisms in China. Rev Eur Community Int Environ Law 17(2):234–242. doi:10.1111/j.1467-9388.2008.00600.x
Liu S (2011) Reviews on watershed Eco-compensation of China. Truth Seeking 3:13
Pontryagin LS (1986) Mathematical theory of optimal processes. Gordon and Breach Science Publishers, New York
SP Sethi, GL Thompson (2000) What is optimal control theory? In: Optimal control theory. Springer US, pp 1–22
Stavins RN (2007) Environmental Economics. SSRN Electron J 10(2–4):109
Tahvonen O (1994) Carbon dioxide abatement as a differential game. Eur J Polit Econ 10(4):685–705. doi:10.1016/0176-2680(94)90033-7
Travis B, Bruk B, Kim D, Lee KT (2010) Enhancing public sector innovation: examining the network-innovation relationship. Innovation J Public Sect Innovation J 15(3):18
Van Hecken G, Bastiaensen J (2010) Payments for ecosystem services: justified or not? A political view. Environ Sci Pol 13(8):785–792. doi:10.1016/j.envsci.2010.09.006
Xiong Y, Wang K (2010) Eco-compensation effects of the wetland recovery in Dongting Lake area. J Geogr Sci 20(3):389–405. doi:10.1007/s11442-010-0389-1
Yeung DWK (2007) Dynamically consistent cooperative solution in a differential game of transboundary industrial pollution. J Optim Theory Appl 134(1):143–160. doi:10.1007/s10957-007-9240-y
Yeung D (1992) A differential game of industrial pollution management. Ann Oper Res 37(1):297–311
Yeung DWK, Petrosyan LA (2008) A cooperative stochastic differential game of transboundary industrial pollution. Automatica 44(6):1532–1544. doi:10.1016/j.automatica.2008.03.005

Danlu Jiang is a lecturer at the School of Business Administration and Management at Chongqing University of Science and Technology. She also is a scholar at the Center of Chongqing Energy Economics Research. She received her Ph.D Degree of Management at Chongqing University in 2012. Her research focuses on environmental and energy economics management, pollution management and E-government.

Jing Zhang is a Professor of Management at Clark University. Her research focuses on inter-organizational information and knowledge sharing, organizational impact of technology and innovation, and the information technology support for sustainable consumption and sustainable supply chain. She served as the program co-chair for dg.o conference from 2012–2015, and she is an elected board member of the Digital Government Society. Her publications appear in European Journal of Information Systems, Government Information Quarterly, Public Performance and Management Review, Information Technology and Management, among others.

Editors' Biography

J. Ramon Gil-Garcia is an Associate Professor of Public Administration and Policy and the Research Director of the Center for Technology in Government, University at Albany, State University of New York (SUNY). Dr. Gil-Garcia is a member of the Mexican Academy of Sciences and of the Mexican National System of Researchers as Researcher Level III, which is the highest distinction a researcher can obtain before becoming Researcher Emeritus as a result of a life-long career of research contributions. In 2009, he was considered the most prolific author in the field of digital government research worldwide and in 2013 he was selected for the Research Award, which is "the highest distinction given annually by the Mexican Academy of Sciences to outstanding young researchers." Currently, he is a Faculty Affiliate at the National Center for Digital Government, University of Massachusetts Amherst and also an Affiliated Faculty member of the Informatics Doctorate Program at the College of Engineering and Applied Sciences, University at Albany. He is also a Professor (on leave) in the Department of Public Administration at Centro de Investigación y Docencia Económicas (CIDE) in Mexico City. Dr. Gil-Garcia is the author or co-author of articles in prestigious international journals in Public Administration, Information Systems, and Digital Government and some of his publications are among the most cited in the field of digital government research worldwide. His research interests include collaborative electronic government, inter-organizational information integration, smart cities and smart governments, adoption and implementation of emergent technologies, information technologies and organizations, information technologies and education, digital divide policies, education administration and policy, new public management, public policy evaluation, and multi-method research approaches. Dr. Gil-Garcia has extensive teaching experience and has collaborated with 11 universities, including departments of Public Administration, Political Science, Social Science, Information Studies, and Management Information Systems. Dr. Gil-Garcia also has many years of experience as a consultant for federal, state, and local government agencies.

J.R. Gil-Garcia et al. (eds.), *Policy Analytics, Modelling, and Informatics*, Public Administration and Information Technology 24,
DOI 10.1007/978-3-319-61762-6

Theresa A. Pardo, Ph.D. serves as Director of the Center for Technology in Government at the University at Albany, State University of New York, where she is also a full research professor in Public Administration and Policy. Under her leadership, the Center works closely with multi-sector, multi-disciplinary teams from the U.S. and around the world to carry out applied research and strategic consulting projects focused on the intersections of policy, management, and technology in the public sector.

Dr. Pardo serves as OpenNY Adviser to New York State's Governor Andrew Cuomo and as Chair of the U.S. Environmental Protection Agency's National Advisory Committee. Dr. Pardo is a member of the steering committee of the Northeast Big Data Hub and the U.S. Government Accountability Office Executive Council on Information, Management, and Technology. She is also a member of the User Working Group of the NASA Socioeconomic Data and Applications Center (SEDAC), operated by the Center for International Earth Science Information Network at Columbia University and a member of the City of Schenectady's Smart Cities Advisory Commission. Dr. Pardo is the Founder of the Global Smart Cities Smart Government Research Practice Consortium and a co-developer of the top ranked academic program in Government Information Strategy and Management Program offered by Rockefeller College. She has served as first female Jury Chair in 2016 for Oman's Sultan Qaboos E-government Award.

Dr. Pardo is ranked among the top five scholars in her field in terms of citations to her published work. Her work has been recognized through numerous journal and conference best paper awards. She has also received various awards for the following: University at Albany's Excellence in Teaching Award, in 2016 Distinguished Alumni Award, the Rockefeller College's Distinguished Service Award, and *Government Technology Magazine*'s Top 25 Doers, Drivers, and Dreamers Award.

Luis F. Luna-Reyes is an Associate Professor in the Department of Public Administration and Policy and Affiliated Faculty in the Department of Information Science at the University at Albany. He has been a Fulbright Scholar and he is currently Faculty Fellow at the Center for Technology in Government and also a member of the Mexican National Research System. His research is at the intersection of Public Administration, Information Systems, and Systems Sciences. More specifically, Luna-Reyes uses multi-method approaches to contribute to a better understanding of collaboration and governance processes in the development of information technologies across functional and organizational boundaries in government. His research interests are related to areas such as inter-organizational collaboration, information sharing, success of government-wide Web sites, modeling and simulation, and information policy. He is the author or co-author of more than 100 articles published in leading journals and academic conferences such as Government Information Quarterly, European Journal of Information Systems, System Dynamics Review, Information Polity, and Information Technology and Management among others, working with a network of more than 60 co-authors around the world. His research has been funded by the Consejo Nacional de Ciencia y Tecnología in México, the National Science Foundation, the Fondo de Información y Documentación para la Industria (Infotec) in México, and the Economic Commission for Latin America and the Caribbean (ECLAC).

Index

J.R. Gil-Garcia et al. (eds.), *Policy Analytics, Modelling, and Informatics*, Public Administration and Information Technology 24,
DOI 10.1007/978-3-319-61762-6

Printed by Printforce, the Netherlands